Electrocatalytic Materials for Renewable Energy

Scrivener Publishing
100 Cummings Center, Suite 541J
Beverly, MA 01915-6106

Publishers at Scrivener
Martin Scrivener (martin@scrivenerpublishing.com)
Phillip Carmical (pcarmical@scrivenerpublishing.com)

Electrocatalytic Materials for Renewable Energy

Edited by

Sudheesh K. Shukla
School of Chemical Engineering and Physical Sciences, Lovely Professional University, Jalandhar, Punjab, India
Department of Chemical Science, (Doornfontein Campus), University of Johannesburg, Johannesburg, South Africa

Chaudhery Mustansar Hussain
Department of Chemistry and Environmental Sciences, New Jersey Institute of Technology, Newark, New Jersey, United States

Santanu Patra
IDUN, Department of Health Technology, Technical University of Denmark, Kgs. Lyngby, Denmark

Meenakshi Choudhary
School of Chemical Engineering and Physical Sciences, Lovely Professional University, Jalandhar, Punjab, India

Scrivener Publishing

WILEY

This edition first published 2024 by John Wiley & Sons, Inc., 111 River Street, Hoboken, NJ 07030, USA and Scrivener Publishing LLC, 100 Cummings Center, Suite 541J, Beverly, MA 01915, USA

For more information about Scrivener publications please visit www.scrivenerpublishing.com.

Wiley Global Headquarters
111 River Street, Hoboken, NJ 07030, USA

For details of our global editorial offices, customer services, and more information about Wiley products visit us at www.wiley.com.

Library of Congress Cataloging-in-Publication Data

ISBN 9781119901051

Cover image: Pixabay.Com
Cover design by Russell Richardson

Set in size of 11pt and Minion Pro by Manila Typesetting Company, Makati, Philippines

Printed in the USA

10 9 8 7 6 5 4 3 2 1

Contents

Preface

In the current era of climate change and environmental degradation, there is an urgent need to shift toward sustainable and clean energy sources. The Sustainable Development Goals (SDGs) adopted by the United Nations provide a roadmap for achieving sustainable development by 2030. SDG 7 aims to ensure access to affordable, reliable, sustainable, and modern energy for all. Moreover, to limit global warming to 1.5 degrees Celsius, the European Union has set a target of achieving climate neutrality by 2050. Renewable energy sources such as solar, wind, and hydroelectric power are crucial in achieving these goals. It is an essential tool for reducing greenhouse gas emissions, improving energy security, and promoting a sustainable and clean energy future for all. However, the intermittent and variable nature of these sources poses a significant challenge to their efficient and reliable utilization. Electrocatalytic materials have emerged as a promising solution to this challenge.

Electrocatalytic materials are used to accelerate electrochemical reactions, which are essential in various renewable energy applications, such as fuel cells, water splitting, and carbon dioxide reduction. These materials play a crucial role in enhancing the efficiency, stability, and selectivity of these reactions. The field of electrocatalysis has undergone tremendous advancement in the past few decades, in part owing to improvements in catalyst design at the nanoscale. These developments have been crucial for the realization of an improvement in alternative energy technologies based on electrochemical reactions. In this view, electrocatalytic materials play a crucial role, since they enable the lowering of the over-potentials needed for the operation of electrochemical devices.

As an important branch of catalysts, electrocatalytic materials exhibited important catalytic reactions that are able to convert and store energy through reactions involving electron transfer. However, the study of electrocatalytic materials presents a huge challenge due to the highly complicated reaction network, the variety of reaction selectivity, and the puzzling reaction mechanisms. Tremendous research efforts have been made toward the fabrication of efficient electrocatalytic materials that can be used in the energy sectors. For example, in the field of fuel cells, electrocatalysts such as platinum-based materials have been widely studied due to their high catalytic activity and stability. However, the high cost and limited availability of these materials have hindered their widespread commercialization. As such, efforts have been made to develop alternative electrocatalytic materials such as transition metal oxides, nitrides, and sulfides, which exhibit comparable or even superior catalytic activity to traditional platinum-based materials. Moreover, in the field of rechargeable batteries, electrocatalytic materials have been utilized as active materials in the cathodes and anodes. Overall, the development of efficient electrocatalytic materials has played a critical role in the advancement of various energy technologies, and their continued optimization and development will be essential for the future of sustainable and clean energy.

This book provides a comprehensive overview of various electrocatalytic materials and their applications in renewable energy. The book covers a wide range of topics, including the synthesis, characterization, and performance evaluation of electrocatalytic materials for different renewable energy applications. Furthermore, the book discusses the challenges and opportunities associated with the development and utilization of electrocatalytic materials for renewable energy. The future utility of different electrocatalytic materials is also well-defined in the context of the renewable energy approach.

The contributors to this book are leading experts in the field of electrocatalytic materials for renewable energy, including scientists and engineers from academia, industry, and national laboratories. Their collective expertise and knowledge provide valuable insights into the latest advances in electrocatalysis for renewable energy applications.

This book is intended for researchers, students, and professionals in the fields of materials science, chemistry, physics, and engineering who are interested in the development and utilization of electrocatalytic materials for renewable energy. It is hoped that this book will contribute to the advancement of electrocatalytic materials and their applications in renewable energy, thereby promoting a sustainable and clean energy future for all.

We wish to express our gratitude to everyone involved in this project for their efforts, as well as Wiley and Scrivener Publishing for their cooperation and assistance in the publication of this book.

Sudheesh K. Shukla, Ph.D.
Chaudhery Mustansar Hussain, Ph.D.
Santanu Patra, Ph.D.
Meenakshi Choudhary, Ph.D.

1

An Introduction to the Exploration of the Electronic Structure Properties of Biologically Active Natural Compounds Using Quantum Chemical Methods

Ashok Kumar Mishra[1]*, Satya Prakash Tewari[1] and Aniket Kumar[2]

[1]Department of Physics, Dr. Shakuntala Misra National Rehabilitation University, Lucknow, India
[2]Shobhit Institute of Engineering & Technology, Meerut, India

Abstract
Various naturally occurring bioactive compounds have already been isolated from plants of different species and characterized for the identification of their molecular structure, but the phenomena correlated with their activity need to be explored for the prediction of their derived structure and related noble applications. The viability of quantum chemical calculation for exploring the electronic properties of such macromolecules has been substantiated through this recent study. The present chapter underlies a brief history of some of such bioactive natural compounds and discusses precisely the quantum chemical approaches feasible for the study of these natural compounds. The *ab-initio* methods and the most recent density functional theory are systematically presented along with the molecular docking approach.

***Keywords*:** Natural compound, molecular mechanics, DFT, molecular docking, QSAR, QSPR

1.1 Natural Compounds: Past, Present, and Future

The study of naturally occurring compounds has been a course for scientists since the ancient period, and its records are available in the *Vedic system*

**Corresponding author*: akmishra2k5@gmail.com

Sudheesh K. Shukla, Chaudhery Mustansar Hussain, Santanu Patra, Meenakshi Choudhary (eds.) Electrocatalytic Materials for Renewable Energy, (1–26) © 2024 Scrivener Publishing LLC

of medicine, i.e., Ayurveda 1,000 BC. The Ayurvedacharya (medical scientist) like *Charaka* and *Sushruta* described the various types of mineral and natural products having medicinal properties and their applications [1]. In *Sushruta Sanhita* written by *Acharya Sushruta*, 700 types of medicinal plants, 64 types of medicines extracted from minerals, and 57 types of medicines extracted from animal sources have been described wherein the application of herbs like turmeric was mentioned to have begun in 1,900 BC. In the sequence of ancient records, we found the books, namely, De Materia Medica and Theophrastus Historia Plantarum authored in Greece (Europe) during 400 BC. Shanog Benckawo investigated the Maa-Huwang plant supposed to be the source of the modern ephedrine drug. The Canon of Medicine authored by Awsena in 1025 AD described 800 types of natural products and modern herbal medicines. Although the modern system of medicine (allopathy) based on synthetic chemicals and biomolecules has a rapid impact on health but is associated with toxicity issues, i.e., side effects, natural product-based drugs would be more user-friendly with either no or fewer harmful effects.

Commencing from the Vedic period (ancient period) to the recent time, the investigation of naturally occurring biomolecules has succeeded to find about 2,000 natural drugs derived from India's different traditional medicine systems and folklore practices where plants played a major role, included in the Indian Materia Medica. Reports pertaining to the sources of new drugs exhibit that about 50% of the approved drugs are derived from natural products, and the World Health Organization (WHO) has also been engaged in formulating guidelines for herbal medicines. The Council of Scientific and Industrial Research (CSIR), Govt. of India, took an initiative in 1996 to carry out research focusing on new bioactive molecules that are occurring in natural sources like plants, fungi, microbes, insects, etc., and the Central Drugs Research Institute (CDRI), Lucknow, is one of those CSIR laboratories which are part of this initiative, with an objective to standardize the Ayurvedic, Unani, Siddha, and Homeopathy (AUSH) system of medicine along with developing new natural drugs of national and global importance. Plants or plant-derived phytochemicals have been used for the treatment of various chronic diseases like cancer for a long time. Hartwell listed more than 3,000 of such plant species and 60% of the currently used anticancer agents are derived directly or indirectly from plants, marine organisms, and microorganisms, and plant-derived bioactive natural molecules are beneficial to combat cancer [2] over synthetic drugs associated with harmful side effects. The phytochemicals contained in fruits and vegetables are found to be active as antioxidants scavenging free radicals, modulating enzymes, regulating gene expression, etc. The various bioactive natural biomolecules exhibiting anticancer activity have already been marketized, and semisynthetic derivatives of epipodophyllotoxin isolated from the roots of *Podophyllum* species belonging to the Berberidaceae family are used as an anticancer agent. Many plant-derived bioactive natural molecules have also been reported, possessing anti-inflammatory, antiviral, antimicrobial, and antidepressant activities. Recently, research has been carried out to report the biological activity and characterization of the various naturally occurring bioactive molecules isolated from *Castanopsis indica* and *Anthocephalus cadamba* plants possessing antidiabetic, anticancer, and anti-osteoporosis activities [3, 4].

The future scope for the study on naturally occurring biomolecules is full of probability of developing a low-cost, user-friendly, and multifunctional drug agent with the lowest toxicity for various fatal diseases for which the theoretical study on such biomolecules is needed to be carried out for tracking the pathway of bioactivity at the molecular level.

The theoretical research for establishing the scientific basis of the inherent bioactivity of such naturally occurring biomolecules using existing physical principles at the molecular level can now be carried out in the present century as high-level quantum computational techniques are available [5]. Moreover, the references for exploiting these computational techniques for carrying out such research on biologically active molecules are also available, and the theoretical investigations pertaining to the electronic structure and properties of some of such biomolecules contained in the leaves and fruits of *Anthocephalus chinensis* (Lamk) syn. *Anthocephalus cadamba* (Roxb) Miq. (family Rubiaceae) widely found in India, Nepal, Sri Lanka, Bangladesh, Burma, Thailand, China, and eastward through Malaysia to New Guinea have already been carried out using recent computational techniques [6–8]. These computational approaches are based on electronic structure methods to explore the chemistry of these biologically active molecules by using quantum mechanical theories. The theoretical framework useful for setting up quantum chemical calculations is discussed below.

1.2 Theoretical Framework for Quantum Chemical Calculations

While going through the available literature pertaining to the quantum mechanics calculations on molecules, we are inspired by the opinion of Dirac, which we quote as follows:

> *"The underlying physical law necessary for the mathematical theory of a large part of physical and the whole of chemistry are thus completely known, and the difficulty is only that the exact application of these laws leads to equations much to complicated to be soluble"* (P A M Dirac 1929).

Presently, there are mainly five approaches for carrying out the quantum chemical calculations for the theoretical study of any molecular system at the atomic scale, namely:

1. ***Ab-Initio* Methods**
2. **Semiempirical Methods**
3. **Molecular Mechanics**
4. **Molecular Dynamics**
5. **Density Functional Theory**

1.2.1 *Ab-Initio* Methods

The computation derived directly from the theoretical principles excluding the experimental data is categorized into *ab-initio* methods, and this includes approximate quantum mechanical calculations, namely, the Moller–Plesset (MP) perturbation theory (PT), the Hartree–Fock theory, configuration interaction (CI), coupled cluster method, multiconfigurational self-consistent field theory, and the quantum Monte Carlo method.

1.2.1.1 Hartree–Fock (HF) Theory

This theory includes two approximations [9] wherein the first approximation accounts for the estimation of the central field, which takes into account the coulombic electron–electron repulsion and provides the average effect of repulsion but not the explicit repulsion interaction. The approximate energies calculated are all equal to or greater than the exact energy in a unit called Hartree (1 Hartree = 27.2116 eV) and tend to a limiting value called the Hartree–Fock limit as the basis set is improved. The many-electron Schrodinger equation is transformed into simpler one-electron equations. Single-electron wave functions, called orbitals, and eigenvalues, called orbital energy, are yielded by solving each one-electron equation. The behavior of an electron in the net field of all the other electrons is described by the orbital. A second approximation is related to the wave function, which is known exactly for only a few one-electron systems. Linear combinations of Gaussian-type orbitals e^{-ar^2} are used most often as functions. The wave function is formed from linear combinations of atomic orbitals or mathematically linear combinations of basis functions.

The wave function of a particle is described by the well-known Schrödinger equation [10], which is the fundamental equation of motion in quantum mechanics.

$$\left\{\frac{-\hbar^2}{2m}\nabla^2 + V\right\}\psi = i\hbar\frac{\partial\Psi}{\partial t} \tag{1.1}$$

The Schrödinger equation for Ψ, subject to the appropriate boundary conditions, can yield the energy and many other properties of the particle. Different stationary states of the system correspond to many different wave functions that could be the solutions.

If the wavefunction is written as the product of a spatial function and a time function:

$$\Psi\ (r,t) = \Psi r\ (r)\,\Psi t(t) \tag{1.2}$$

The two equations will be obtained by Equation (1.1) if these new functions are substituted, one of which depends on the position of the particle independent of time and the other of which is a function of time alone. For the problems in which we are interested, this separation is valid, and the familiar time-independent Schrödinger equation is focused entirely on.

$$H\Psi\ (r) = E\Psi\ (r) \tag{1.3}$$

Where H is the Hamiltonian operator and E is the energy of the particle, given by:

$$H = \frac{-\hbar^2}{2m}\nabla^2 + V \tag{1.4}$$

A multiple of the function itself is produced as the result when a function is acted upon by the operator, having the general form:

$$f\Psi = \lambda\Psi \tag{1.5}$$

In the Schrödinger approach, the energies corresponding to the various stationary states of the molecular structure are the eigenvalues of the Schrödinger equation. The Hamiltonian is composed of kinetic and potential energy terms, where each nucleus is studied as an accumulated entity and each electron is investigated independently:

$$H = T + V \tag{1.6}$$

The K.E. is a summation of ∇^2 over all the particles in the molecule

$$T = -\frac{\hbar^2}{2}\sum_{k}^{\beta}\frac{1}{m_k}\left(\frac{\partial^2}{\partial x_k^2} + \frac{\partial^2}{\partial y_k^2} + \frac{\partial^2}{\partial z_k^2}\right) \tag{1.7}$$

The Coulomb repulsion between each pair of charged entities (treating each atomic nucleus as a single charged mass) is the potential energy component:

$$V = \frac{1}{4\pi\varepsilon_0}\sum_j\sum_{k<j}\frac{e_j e_k}{\Delta r_{jk}} \tag{1.8}$$

Where Δr_{jk} is the distance, and e_j and e_k are the charges on electrons j and k, respectively. The potential energy compound for a molecular structure is as follows:

$$V = \frac{1}{4\pi\varepsilon_0}\left(-\Sigma_i^{electrons\ nuclei}\Sigma_l\left(\frac{Z_l e^2}{\Delta r_{il}}\right) + \Sigma_i^{electrons}\Sigma_{j<i}\left(\frac{e^2}{\Delta r_{ij}}\right) + \Sigma_l^{nuclei}\Sigma_{j<l}\left(\frac{Z_l Z_j e^2}{\Delta R_{ij}}\right)\right) \tag{1.9}$$

Electron–nuclear attraction corresponds to the first term, electron–electron repulsion to the second term, and nuclear–nuclear repulsion to the third term.

The Born–Oppenheimer Approximation

The first approximation used to simplify the solution of the Schrödinger equation is related to the general molecular problem by separating nuclear and electronic motions because the mass of the typical nucleus is a thousand times greater than that of an electron. It assumes that electron distribution within the molecule depends on the positions of the nuclei and not on their velocities, and electronic motion can be supposed to take place in a field of fixed nuclei.

The Hamiltonian can be written as for the molecular system:

$$H = T^{electron}(\vec{r}) + T^{nuclei}(\vec{R}) + V^{nuclei-electron}(\vec{R}\vec{r}) + V^{electron}(\vec{r}) + V^{nuclei}(\vec{R}) \tag{1.10}$$

The K.E. term for the nuclei is neglected by the electronic Hamiltonian:

$$H^{electron} = -\frac{1}{2}\sum_{i}^{electrons}\left(\frac{\partial^2}{\partial x_i^2} + \frac{\partial^2}{\partial y_i^2} + \frac{\partial^2}{\partial z_i^2}\right) - \sum_{i}^{electrons\ nuclei} \Sigma_I\left(\frac{Z_I}{\left|\overrightarrow{R_I} - \vec{r_i}\right|}\right)$$
$$+ \Sigma_i^{electrons}\, \Sigma_{j<i}\left(\frac{1}{\left|\vec{r_i} - \vec{r_i}\right|}\right) + \Sigma_I^{nuclei}\, \Sigma_{J<I}\left(\frac{Z_I Z_J}{\left|\overrightarrow{R_I} - \overrightarrow{R_J}\right|}\right) \tag{1.11}$$

The motion for Hamiltonian of electrons in the field of fixed nuclei is described by the Schrödinger equation:

$$H^{electron}\Psi^{electron}(\vec{r}\vec{R}) = E^{eff}(\overrightarrow{R_I})\Psi^{electron}(\vec{r}\vec{R}) \tag{1.12}$$

Condition on the Electronic Wave Function

There are two important restrictions to be imposed on the wave function Ψ: first is that it is normalized to represent the probability which should be the number of particles, i.e.,

$$\int_{-\infty}^{\infty} |c\Psi|^2\, dv = n_{particles} \tag{1.13}$$

Secondly, for a simple function, the following relation holds: ψ must be antisymmetric.

$$f(i,j) = -f(j,i) \tag{1.14}$$

The following condition must be satisfied by any valid wavefunction: It is essentially a requirement that the results of experimental physics must agree with by ψ.

$$\Psi(\vec{r_1},\vec{r_2},\ldots,\overrightarrow{r_{n-1}},\vec{r_n}) = -\Psi(\vec{r_2},\vec{r_1},\ldots,\overrightarrow{r_{n-1}},\vec{r_n}) \tag{1.15}$$

The molecular orbital theory decomposes ψ into a combination of molecular orbitals: $\phi_1,\phi_2,\phi_3,\ldots\ldots\ldots\ldots$ and we choose the molecular orbitals of a normalized, orthogonal set:

$$\begin{gathered}\iiint \phi_i^* \phi_i dxdydz = 1 \\ \iiint \phi_i^* \phi_j dxdydz = 0;\, i \neq j\end{gathered} \tag{1.16}$$

The Hartree product is formed by the molecular orbitals.

$$\Psi\vec{r} = \phi_1(\vec{r_1})\phi_2(\vec{r_2})\ldots\ldots\ldots\phi_n(\vec{r_n}) \tag{1.17}$$

This Hartree product is an inadequate wave function which improved to an antisymmetric function by introducing the concept of determinant. It is a combination of molecular orbitals in which electrons can have spin up $\left(+\frac{1}{2}\right)$ or down $\left(-\frac{1}{2}\right)$.

We discussed two spin relations, α and β, as follows:

$$\begin{aligned}\alpha(\uparrow) &= 1 \qquad & \alpha(\downarrow) &= 0 \\ \beta(\uparrow) &= 0 \qquad & \beta(\downarrow) &= 1\end{aligned} \tag{1.18}$$

The value of 1 is assigned to the α function for a spin-up electron, and the value of 1 is assigned to the β function for a spin-down electron. The values of α and β for electron i will be designated as $\alpha(i)$ and $\beta(i)$, respectively.

The electrons to these orbitals in pairs of opposite spin:

$$\Psi(\vec{r}) = \frac{1}{\sqrt{n!}} \begin{vmatrix} \phi_1(\vec{r_1})\alpha(1)\phi_1(\vec{r_1})\beta(1)\phi_2(\vec{r_1})\alpha(1)\phi_2(\vec{r_1})\beta(1)\ldots\phi_{\frac{n}{2}}(\vec{r_1})\alpha(1)\phi_{\frac{n}{2}}(\vec{r_1})\beta(1) \\ \phi_1(\vec{r_2})\alpha(2)\phi_1(\vec{r_2})\beta(2)\phi_2(\vec{r_2})\alpha(2)\phi_2(\vec{r_2})\beta(2)\ldots\phi_{\frac{n}{2}}(\vec{r_2})\alpha(2)\phi_{\frac{n}{2}}(\vec{r_2})\beta(2) \\ \cdot \\ \cdot \\ \cdot \\ \phi_1(\vec{r_i})\alpha(i)\phi_1(\vec{r_i})\beta(i)\phi_2(\vec{r_i})\alpha(i)\phi_2(\vec{r_i})\beta(i)\ldots\phi_{\frac{n}{2}}(\vec{r_i})\alpha(i)\phi_{\frac{n}{2}}(\vec{r_i})\beta(i) \\ \phi_1(\vec{r_j})\alpha(j)\phi_1(\vec{r_j})\beta(j)\phi_2(\vec{r_j})\alpha(j)\phi_2(\vec{r_j})\beta(j)\ldots\phi_{\frac{n}{2}}(\vec{r_j})\alpha(j)\phi_{\frac{n}{2}}(\vec{r_j})\beta(j) \\ \cdot \\ \cdot \\ \cdot \\ \phi_1(\vec{r_n})\alpha(n)\phi_1(\vec{r_n})\beta(n)\phi_2(\vec{r_n})\alpha(n)\phi_2(\vec{r_n})\beta(n)\ldots\phi_{\frac{n}{2}}(\vec{r_n})\alpha(n)\phi_{\frac{n}{2}}(\vec{r_n})\beta(n) \end{vmatrix} \quad (1.19)$$

The wave function is formed by mixing all of the possible orbitals of all of the electrons in the molecular system using this determinant.

Basis Sets

The molecular orbitals are expressed as a linear combination of a predefined set of one-electron functions known as basis functions in the next approximation. Usually, these basis functions are centered on the atomic nuclei and bear some resemblance to atomic orbitals. However, any set of appropriately defined functions may be used in the actual mathematical treatment, which is more general than this.

An individual molecular orbital is defined as:

$$\phi_l = \sum_{l=1}^{N} c_{nl}\chi_n \quad (1.20)$$

Where the molecular orbital expansion coefficients are known as the coefficients $c_{\mu i}$

$$g(\alpha,\vec{r}) = cx^n y^m z^l e^{-\alpha r^2} \quad (1.21)$$

A constant for normalization, so that:

$$\int g^2 = 1 \tag{1.22}$$

Thus, c depends on $\alpha, l, m,$ and n.

The Gaussian functions are generally known to be represented as suffix s, p_y, and d_{xy} types as written down below

$$g_s(\alpha,\vec{r}) = \left(\frac{2\alpha}{\pi}\right)^{\frac{3}{4}} e^{-ar^2}$$

$$g_y(\alpha,\vec{r}) = \left(\frac{128\alpha^5}{\pi^3}\right)^{\frac{1}{4}} ye^{-ar^2}$$

$$g_{xy}(\alpha,\vec{r}) = \left(\frac{2048\alpha^7}{\pi^3}\right)^{\frac{1}{4}} xye^{-ar^2} \tag{1.23}$$

1.2.1.2 Moller–Plesset (MP) Perturbation Theory

The Moller–Plesset perturbation theory is called an approach to add the electron correlation and extend it with the size of the molecular system under investigation. The HF wave function is mapped onto a perturbation theory formulation in which HF becomes a first-order perturbation. The second-order MP2 method adds a minimal amount of correlation. The third-order (MP3) and fourth-order (MP4) calculations are also common, while MP5 and higher calculations are seldom done due to high computational requirements [11]. The Moller–Plesset calculations are not variational.

Dividing the Hamiltonian into two parts forms the basis of the perturbation theory:

$$H = H_0 + \lambda H' \tag{1.24}$$

Such that H_0 is soluble exactly, and $\lambda H'$ is a perturbation applied to H_0, a correction which is assumed to be small in comparison to it.

The assumption that H' is a small perturbation to H_0 suggests that the perturbed wavefunction and energy can be expressed as a power series in terms of the parameter λ as follows:

$$\begin{aligned} \Psi &= \Psi^{(0)} + \lambda\Psi^{(1)} + \lambda^2\Psi^{(2)} + \lambda^3\Psi^{(3)} + \ldots \\ E &= E^{(0)} + \lambda E^{(1)} + \lambda^2 E^{(2)} + \lambda^3 E^{(3)} + \ldots \end{aligned} \tag{1.25}$$

The Schrödinger equation is where the perturbed wavefunction and energy are substituted back into.

$$(H_0 + \lambda H')(\Psi^{(0)} + \lambda\Psi^{(1)} + \cdots) = E^{(0)} + \lambda E^{(1)} + \cdots)(\Psi^{(0)} + \lambda\Psi^{(1)} + \cdots) \tag{1.26}$$

We equate the coefficients of each power of λ, which gives us a series of relations representing successively higher orders of perturbation. The first three such equations corresponding to powers of 0, 1, and 2 of λ are given by:

$$\begin{aligned} (H_0 - E^0)\Psi^{(0)} &= 0 \\ (H_0 - E^0)\Psi^{(1)} &= (E^{(1)} - H')\Psi^{(0)} \\ (H_0 - E^0)\Psi^{(2)} &= (E^{(1)} - H')\Psi^{(1)} + E^{(2)}\Psi^{(0)} \end{aligned} \tag{1.27}$$

The MP perturbation theory defines H_0 as the sum of the one-electron Fock operation:

$$H_0 = \Sigma_i F^i \tag{1.28}$$

Where F^i is the Fock operator action on the ith electron.
Thus,

$$H_0\Psi_s = E_s\Psi_s \tag{1.29}$$

for all substituted determinant wavefunctions.

$$E^0 = \langle\Psi^{(0)}|H_0|\Psi^{(0)}\rangle = \Sigma_i\, \varepsilon_i \tag{1.30}$$

To obtain the expression for $E^{(1)}$, we can use simple linear algebra. We begin by forming the inner product of both sides of the second relation in Equation (1.28) with a suitable wavefunction:

$<\Psi^{(0)}$:

$$\langle\Psi^{(0)}|H_0 - E^{(0)}|\Psi^{(1)} = \langle\Psi^{(0)}|E^{(1)} - H'|\Psi^{(0)}\rangle \Rightarrow$$
$$\langle\Psi^{(0)}|H_0|\Psi^{(1)}\rangle - E^{(0)}\langle\Psi^{(0)}|\Psi^{(1)}\rangle = E^{(1)}\langle\Psi^{(0)}|\Psi^{(1)}\rangle - \langle\Psi^{(0)}|H'|\Psi^{(0)}\rangle \tag{1.31}$$

Leaving this expression for $E^{(1)}$:

$$E^1 = \Psi^{(0)}|H'|\Psi^{(0)} \tag{1.32}$$

Summation $E^{(0)}$ and $E^{(1)}$ yields the HF energy (since H_0 + V is the full Hamiltonian):

$$E^0 + E^1 = \langle\Psi^{(0)}|H_0|\Psi^{(0)}\rangle + \langle\Psi^{(0)}|H'|\Psi^{(0)}\rangle$$
$$= \langle\Psi^{(0)}|H_0 + H'|\Psi^{(1)}\rangle = \langle\Psi^{(0)}|H_0|\Psi^{(0)}\rangle = E^{HF} \tag{1.33}$$

We will begin to examine the third relation in Equation (1.28) in the same way:

$$\langle\Psi^{(0)}|H_0 - E^{(0)}|\Psi^{(2)}\rangle = \langle\Psi^{(0)}|E^{(1)} - H'|\Psi^{(1)}\rangle + E^{(2)}\langle\Psi^{(0)}|\Psi^{(1)}\rangle \Rightarrow$$
$$E^{(2)} = \langle\Psi^{(0)}|V - E^{(1)}|\Psi^{(1)} = \langle\Psi^{(0)}|V|\Psi^{(1)}\rangle \tag{1.34}$$

Before we can calculate $E^{(2)}$, we need to find $\Psi^{(1)}$, and we will form it as a linear combination of substituted wavefunctions and solve for the coefficients.

$$\Psi^{(1)} = \sum_x a_s\Psi_s \ni H_0\Psi_s = E_s\Psi_s \tag{1.35}$$

This time, we will use the second relation in Equation (1.27) to find the coefficients for $\Psi^{(1)}$:

$$(H_0 - E^0)\sum_s a_s \Psi_s = \left(E^{(1)} - H'\right)\Psi^{(0)} \tag{1.36}$$

The inner product of both sides of Equation (1.36) with an arbitrary substituted wavefunction Ψ_t will be formed, and then a_t will be solved for:

$$\langle \Psi_t | (H_0 - E^0) | \sum_s a_s \Psi_s \rangle = \langle ¤_t | E^1 - H' | \Psi^{(0)} \rangle \Rightarrow$$

$$\sum_s a_s \langle \Psi_t | (H_0 - E^0) | \Psi_s \rangle = E^{(1)} \langle \Psi_t | \Psi^{(0)} \rangle - \langle \Psi_t | H' | \Psi^{(0)} \rangle \Rightarrow$$

$$\sum_s a_s \langle \Psi_t | (H_0) | \Psi_s \rangle \, \langle \Psi_t | (E^0) | \Psi_s \rangle = E^{(1)} \langle \Psi_t | \Psi^{(0)} \rangle - \langle \Psi_t | H' | \Psi^{(0)} \rangle \tag{1.37}$$

The left side of the final Equation (1.37) yields a non-zero value only when $s = t$.

$$a_t \left(E_t - E^{(0)}\right) = - \langle \Psi_t | H' | \Psi^{(0)} \rangle \Rightarrow$$

$$a_t = \frac{\langle \Psi_t | H' | \Psi^{(0)} \rangle}{E^{(0)} E_t} \tag{1.38}$$

It is indicated by the result in Equation (1.38) that larger contributions to the perturbation are made by substitutions close in energy to the ground state.

The following expression for Ψ^1 is obtained from these coefficients:

$$\Psi^{(1)} = \sum_t \left(\frac{\langle \Psi_t | H' | \Psi^{(0)} \rangle}{E^{(0)} E_t} \right) \Psi_t \tag{1.39}$$

The expression for E^2:

$$E^{(2)} = \langle \Psi^{(0)} | H' | \Psi^{(1)} \rangle = \langle \Psi^{(0)} | H' | \sum_t a_t \Psi_l \rangle = \sum_t a_t \langle \Psi^0 | H' | \Psi_t \rangle$$
$$= \sum_t \frac{\langle \Psi^0 | H' | \Psi_t \rangle \langle \Psi_t | H' | \Psi^0 \rangle}{E^0 - E_t}$$
$$= -\sum_t \frac{|\langle \Psi^0 | H' | \Psi_t \rangle|^2}{E_t - E^0} \tag{1.40}$$

It is known that both the numerator and denominator in the final expression are always positive expressions, with the denominator being always positive because E^0 is the lowest energy eigenvalue of the unperturbed system.

1.2.1.3 Configuration Interaction (CI) Method

A multiple-determinant wave function, known as a configuration interaction wave function, is constructed by starting with the HF wave function and making new determinants by promoting electrons from the occupied to unoccupied orbitals. Configuration interaction calculations, which can be very accurate, are classified by the number of excitations used to make each determinant. A configuration interaction single-excitation (CIS) calculation is performed when only one electron has been moved for each determinant. CIS calculations give an approximation to the excited states of the molecule but do not change the ground-state energy. A ground-state energy that has been corrected for correlation can be obtained by performing configuration interaction single- and double-excitation (CISD) calculations. Configuration interaction single-double and triple excitation (CISDT) and configuration interaction single-double-triple-quadruple-excitation (CISDTQ) calculations are done only when very-high-accuracy results are desired. However, the cost in CPU time for these calculations is very high.

The term used for a configuration interaction calculation that involves all possible excitations is full CI. Performing a full CI calculation with an infinitely large basis set yields an exact quantum mechanical result. Nevertheless, due to the immense amount of computer power required, full CI calculations are rarely carried out. The full CI method involves forming the wavefunction ψ as a linear combination of the Hartree–Fock determinant and all possible substituted determinants:

$$\Psi = a_0\Psi_0 + \sum_{l>0} a_l\Psi_l \tag{1.41}$$

The Hartree–Fock level is represented by the 0-indexed term.

1.2.1.4 Coupled Cluster Method (CCM)

Coupled cluster calculations involve constructing a wave function as a linear combination of many determinants, and they share this aspect with configuration interaction calculations. However, the method used to choose the determinants in a coupled cluster calculation is more complex than in a CI calculation. The orders of the CC expansion, such as CCSD, CCSDT, etc., are similar to those in CI. In a CCSD (T) calculation, the triple excitations are included perturbatively rather than exactly.

1.2.2 Semiempirical Methods

Semiempirical methods do not include the core electrons in the calculation and use only a minimal basis set. The omitted values are obtained by fitting the results to experimental data or *ab-initio* calculations, and often, these parameters replace some of the excluded integrals. Semiempirical calculations have the advantage of being much faster than *ab-initio* calculations. However, the disadvantage is that the results can be erratic and fewer properties can be predicted reliably. If the molecule being computed is similar to the molecules in the database used to parameterize the method, the results may be very good. Conversely, if the molecule being computed is significantly different from anything in the parameterization set, the results may be very poor.

The semiempirical method is a computational approach that starts from the Hückel theory and has been extended to various other methods, including PPP, CNDO, INDO, MINDO, MNDO, SINDO1, PRDO, AM1, PM3, PHM, TNDO, SAM1, and the Gaussian theory. These methods use a minimal basis set and do not include core electrons in the calculations. Instead, they use parameters that are fitted to experimental data or *ab-initio* calculations to replace excluded integrals. Semiempirical calculations are much faster than *ab-initio* calculations but can produce erratic results and predict fewer properties reliably [12].

The Hückel calculation, which is the basis for the semiempirical methods, models only the p valence electrons in a planar conjugated hydrocarbon and uses a parameter to describe the interaction between bonded atoms. However, there are no second atom effects. While Hückel calculations can provide qualitative insights and reflect orbital symmetry and coefficients, they are limited in their ability to deal with large molecular systems and give only crude quantitative information. The PPP method is an extension of the Hückel method that can handle heteroatoms other than hydrogen, and it is still occasionally used when only minimal amounts of electronic effects are required. It is also useful for developing simple parameterized analytic expressions for molecular properties.

The MINDO method has three variations: MINDO/1, MINDO/2, and MINDO/3. However, these have been replaced by more accurate methods such as AM1 and PM3. MINDO/3 is still sometimes used as a starting point for *ab-initio* calculations. The MNDO method is popular for its ability to give reasonable qualitative results for many organic systems and has been incorporated into several semiempirical programs.

An extrapolation method called the G1 method, which uses *ab-initio* calculations, is seldom used due to its lower accuracy compared with the G2 method. G2 is an accurate way to model small organic molecules, but its accuracy decreases when applied to chlorofluorocarbons [13]. The G3 method shows some improvement in accuracy, particularly for chlorofluorocarbons. These extrapolation methods use empirically defined equations parameterized to accurately reproduce results from a test set of molecules [14].

1.2.3 Molecular Mechanics

It has been found possible to model the behavior of a large molecular system avoiding the quantum mechanics treatment pertaining to the *ab-initio* and semiempirical calculations totally by using molecular mechanics. Bypassing these quantum mechanical treatments becomes significant in view of the limitations imposed with *ab-initio* and semiempirical calculations on the large molecular and biomolecular system.

The equations describe various aspects of the molecule, such as bond stretching, bond bending, torsions, electrostatic interactions, van der Waals forces, and hydrogen bonding. We call a set of equations with their associated constants a "force field." Since this method does not explicitly include electrons, it cannot model electronic processes.

The table below presents the mathematical form of energy terms that are frequently used in popular force fields:

Name of force field	Parameter used	Energy term
Harmonic	Bond stretching	$k\,(l - l_0)^2$
Harmonic	Angle bending	$k\,(\theta - \theta_0)^2$
Cosine	Torsion	$k\,[1 + \cos(n\,\theta)]$
Leonard-Jones 6–12	van der Waals	$4k\,(A/r)^{12} - (B/r)^6$
Leonard-Jones 10–12	van der Waals	$4k\,(A/r)^{12} - (B/r)^{10}$
Coulomb	Electrostatic interaction	$\frac{q_1 q_2}{4\pi\varepsilon_0 r}$
Taylor	Stretching-bending	$k\,(\theta - \theta_0)\,[(l_1 - l_{10})\,(l_2 - l_{20})$
Morse	Bond stretching	$D_e\,[1 - e^{-a(l-l0)}]^2$

Abbreviations: *k*, *α*, *A*, *B*: constants particular to the elements in a certain hybridization state; *l*: bond length; *θ*: bond angle; *n*: an integer; *q*: charge; D_e: dissociation energy.
*Source: Ref. [5].

The constants in these equations must be obtained from experimental data or an *ab-initio* calculation. The success of the method is dependent on the database of compounds used for parameterization. Most force fields simplify the calculations by omitting most of the hydrogen atoms and modifying the parameters describing each backbone atom to account for the behavior of the attached hydrogens. This approach is known as united atom force fields or intrinsic hydrogen methods and is commonly used to describe large biomolecules. Originally developed for proteins and nucleic acids, it is now applied to a range of studies such as molecular dynamics, salvation, crystal packing, vibrational analysis, and QM/MM studies. Five valence terms, including an electrostatic term, are used in CHARMM [15]. DREIDING is another important force field program used for simulating biomolecules.

1.2.4 Molecular Dynamics

The time-dependent behavior of a molecular system, such as vibrational motion or Brownian motion, is simulated in molecular dynamics, which requires a way to compute the energy of the system, most often using a molecular mechanics calculation. The forces on the atoms for any given geometry are computed using this energy expression. An equilibrium system can be simulated using a well-defined procedure for executing molecular dynamics. If the purpose of the simulation is to search conformation space for geometry optimization, the force field is designed to describe intermolecular forces and vibrations away from equilibrium. Computer programs embedded with force fields like GROMOS or OPLS are available for simulating bulk systems. Several equations of motion such as the Verlet algorithm, velocity, Beeman's algorithm, and Gear predictor-corrector algorithm, to name a few, are used for molecular dynamics simulations. Predictor corrector algorithms give accurate integration but are seldom used due to their large computational needs.

To verify the simulation results obtained for a synthetic system, it is important to execute the aforesaid algorithm for a system with a constant volume, number of particles, and temperature.

1.2.5 Density Functional Theory (DFT)

The popularity of density functional theory (DFT) in recent years is justified by the pragmatic observation that it is less computationally intensive than other methods with similar accuracy. The electronic energy in current DFT methods is partitioned into several terms using an approximate functional:

$$E = E^T + E^V + E^J + E^{XC} \tag{1.42}$$

The electronic energy in current DFT methods is partitioned into several terms as follows: The kinetic energy term, arising from the motion of the electrons, is denoted by E^T. The potential energy of the nuclear–electron attraction and the repulsion between pairs of nuclei are included in E^V. The electron–electron density repulsion term, also known as the Coulomb self-interaction of the electron density, is denoted by E^J, and the exchange-correlation term, which includes the remaining part of the electron–electron interactions, is denoted by E^{XC}.

E^J is given by the following expression.

$$E^J = \frac{1}{2}\iint \rho(\vec{r_1})(\Delta r_{12})^{-1}\rho(\vec{r_2})d\vec{r_1}\vec{r_2} \tag{1.43}$$

An integral involving only the spin densities and possibly their gradients is usually used to approximate E^{XC} in practice:

$$E^{XC}(\rho) = \int f(\rho_\alpha(\vec{r}), \rho_\beta(\vec{r}).\nabla\rho_\alpha(\vec{r})\nabla\rho_\beta(\vec{r}))d^3\vec{r} \tag{1.44}$$

The α spin density is denoted by ρ_α, the β spin density is denoted by ρ_β, and the total electron density $(\rho_\alpha + \rho_\beta)$ is denoted by ρ. E^{XC} is divided into separate parts, corresponding to same-spin and mixed-spin interactions and referred to as the exchange and correlation parts, respectively:

$$E^{XC}(\rho) = E^X(\rho) + E^C(\rho) \tag{1.45}$$

The electron density determines the nature of all three terms, and the functionals that define the two components on the right side of Equation (1.45) are known as exchange functionals and correlation functionals, respectively. Both components can be of two types: local functionals, which depend only on the electron density, and gradient-corrected functionals, which depend on both ρ and its gradient, ∇_ρ.

In the sample functional, the local exchange functional is virtually always defined as follows:

$$E^X_{\text{LDA}} = -\frac{3}{2}\left(\frac{3}{4\pi}\right)^{\frac{1}{3}} \int \rho^{\frac{4}{3}} d^3\vec{r} \tag{1.46}$$

Of course, ρ is a function of r. The local exchange functional in the form mentioned above was developed to reproduce the exchange energy of a uniform electron gas. However, this functional has weaknesses when it comes to describing molecular systems. Becke formulated the following gradient-corrected exchange functional, which is based on the local density approximation (LDA) exchange functional and is now widely used:

$$E^X_{\text{Becke88}} = E^X_{\text{LDA}} - \gamma \int \frac{\rho^{\frac{4}{3}} x^2}{1 + 6\gamma \sinh^{-1} x} d^3\vec{r} \tag{1.47}$$

Here, $x = \rho^{-\frac{4}{3}} |\nabla_\rho|$. The parameter γ is chosen to fit the known exchange energies of the inert gas atoms, and Becke defines its value as

0.0042 Hartrees. As Equation (1.47) shows, Becke's functional is a correction to the local density exchange functional and successfully remedies many of the deficiencies of the LDA functional.

There are also local and gradient-corrected correlation functionals. For instance, Perdew and Wang formulated the following local part of their 1992 correlation functional [16]:

$$E^C = \int \rho \varepsilon_C \left(r_s \left(\rho(\vec{r}) \right), \zeta \right) d^3\vec{r}$$

$$r_s = \left[\frac{3}{4\pi\rho} \right]^{\frac{1}{3}}$$

$$\zeta = \frac{\rho_\alpha - \rho_\beta}{\rho_\alpha + \rho_\beta}$$

$$\varepsilon_C(r_s,\zeta) = \varepsilon_C(\rho,0) + a_C(r_s)\frac{f(\zeta)}{f''(0)}(1-\zeta^4) + [\varepsilon_C(\rho,1) - \varepsilon_C(\rho,0)]f(\zeta)\zeta^4$$

$$f(\zeta) = \frac{[(1+\zeta)^{4/3} + (1-\zeta)^{4/3} - 2]}{(2^{\frac{4}{3}} - 2)} \tag{1.48}$$

r_s is termed the density parameter. ζ is the relative spin polarization. $\zeta = 0$ corresponds to equal α and β densities, $\zeta = 1$ corresponds to all α density, and $\zeta = -1$ corresponds to all β density. It is noted that $f(0) = 0$ and $f(\pm 1) = 1$.

The general expression for ε_C involves both r_s and ζ its final term performs an interpolation for mixed spin cases. The following function G is used to compute the values of $\varepsilon_C(r_s,0), \varepsilon_C(r_s,1)$, and $-a_C(r_s)$ as:

$$G(r_s, A, \alpha_1, \beta_1, \beta_2, \beta_3, \beta_4, P) = -2A(1+\alpha_1 r_s) in \left(1 + \frac{1}{2A(\beta_1 r_s^{\frac{1}{2}} + \beta_2 r_s + \beta_3 r_s^{\frac{3}{2}} + \beta_4 r_s^{P+1}} \right) \tag{1.49}$$

In Equation (1.47), the parameters for G, except for r_s, were chosen by Perdew and Wang to reproduce accurate calculations on uniform electron gases. When G is used to evaluate each of $\varepsilon_C(r_s,0), \varepsilon_C(r_s,1)$, and $-a_C(r_s)$, the parameter sets differ. Similar to the exchange functional examined earlier, a gradient correction can improve a local correlation functional.

Pairing an exchange functional with a correlation functional defines pure DFT methods. For instance, the well-known BLYP functional pairs Becke's gradient-corrected exchange functional with the gradient-corrected correlation functional of Lee, Yang, and Parr [17, 18].

1.3 Theoretical Framework for Biological Activity

Great commercial interest exists in predicting the activity of a compound in a biological system, encompassing both desired properties like drug activity and undesired properties like toxicity. The complexity of biological systems makes such predictions very challenging. Molecular simulation techniques can be utilized to predict how a compound will interact with a specific active site of a biological molecule. These techniques are categorized as QSAR, QSPR, and molecular docking, which are discussed below.

1.3.1 Quantitative Structure–Activity Relationship (QSAR)

QSAR is widely used to study the biological activity of a molecular system. It involves creating an equation that predicts biological activity from the properties of the individual molecule only, using a curve fitting technique. To parameterize a QSAR equation, a quantified activity for a set of compounds must be known, typically referred to as lead compounds in the pharmaceutical industry. However, test results are usually available for only a small number of compounds, making it challenging to select a number of descriptors that will provide useful results without fitting to anomalies in the test set. Normally, three to five lead compounds per descriptor in the QSAR equation are considered adequate.

It may be desirable to use parabolic functions instead of linear functions in drug design, as the descriptor for an ideal drug candidate often has an optimum value. In such cases, drug activity will decrease when the value is either larger or smaller than the optimum value, and this functional form can be described by a parabola instead of a linear relationship. QSAR is useful for determining general criteria for activity, but it does not readily yield detailed structural predictions.

1.3.2 Quantitative Structure–Property Relationship (QSPR)

To develop a QSPR equation, one should begin by compiling a list of compounds with known experimentally determined properties. It is ideal to have a large list, often including thousands of compounds. If there are fewer compounds than parameters to be fitted in the equation, the curve fitting will fail. An exact fit obtained by fitting the equation to all the anomalies in the data can be misleading and may not reflect the correct trends needed for a predictive method. To ensure that the method will be predictive, there should be ideally 10 times as many test compounds as the fitted parameters. The choice of compounds is crucial, as fitting the equation with only hydrocarbon data will only be reliable for predicting hydrocarbon properties.

The next step is to obtain geometries for the molecules. While crystal structure geometries can be used, it is better to use theoretically optimized geometries. Theoretical geometries cancel out any systematic errors in computation and enable the method to predict compounds that have not been synthesized yet. Some of the simpler methods require only connectivity.

To compute molecular descriptors, researchers must use any numerical value that describes the molecule. They can obtain many descriptors from molecular mechanics or semiempirical calculations, such as energies, population analysis, and vibrational frequency analysis with associated thermodynamic quantities. After computing the descriptors, researchers must decide which ones to use by computing correlation coefficients. Correlation coefficients measure the linear relationship between the descriptor and the property. A correlation coefficient of 1 indicates an exact description of the property, while a coefficient of zero means the descriptor is irrelevant. Researchers should choose descriptors with the largest correlation coefficients to create a property prediction equation through curve fitting.

Such as

$$\text{Property} = c_0 + c_1 d_1 + c_2 d_2 \ldots\ldots\ldots \tag{1.50}$$

Where c_i are the fitted parameters and d_i the descriptors. The process described in the preceding paragraphs has been automated very well in the more sophisticated QSPR programs.

1.3.3 Molecular Docking

Computational docking methods like DOCK, FLEXL, and GOLD in combination with empirical screening functions are used to predict the orientation of ligands on protein binding sites and their binding affinity. These methods are used to predict the protein–ligand interaction study, including the binding energy of the complex and the types of signals produced, which have significant applications in medicine.

The docking methods are categorized as lock and key/rigid docking, in which both the internal geometry of the receptor and ligand is kept fixed, and induced fit/flexible docking, in which enumeration on the rotations of one of the molecules (usually the smaller one) is performed, and energy is calculated. Later, the most optimum pose is selected. Several program packages are available for molecule docking, including Sanjeevini, Schrodinger, Dock Autolock Tools [19], Discovery Studio, and iGemDock.

1.4 Future Scope

The future study may be focused to investigate the electronic properties and conformation by using quantum mechanical computation which has not been ever carried out on the biomolecules naturally contained in the leaves and fruits of various natural sources already isolated and characterized in order to explore their novel applications and their derivatives.

Due to the naturally occurring biomolecules' large size, DFT-based quantum chemical calculations have been found to be more dominant than *ab-initio* and semiempirical methods due to their consistency with experimental measurements. Recent reports have shown that the DFT approach is best suited for quantum chemical calculations on biomolecules [20, 21]. Biophysical studies using docking approaches have also been carried out to investigate the molecular forces controlling the biological interactions [22, 23] of these biomolecules with protein receptors and to calculate binding energy and subsequent bioactivity, as this approach has been found to be most appropriate for this purpose [24]. Although these naturally occurring biomolecules have been reported to be biologically active upon isolation from the aforementioned plant, the physical principles behind such bioactivity are yet to be elaborated. Therefore, quantum chemical calculations and molecular docking are carried out to explore the physical principles behind the bioactivity and prediction of novel applications. The development of computational techniques has opened up avenues for such investigations in the light of existing theories of physics.

We intend to utilize DFT-based quantum chemical calculations to determine the optimized geometry, Mulliken atomic charge distribution, HOMO–LUMO energy band gap, global reactivity descriptors, molecular electrostatic potential (MESP) surface, non-linear optical (NLO) properties, and spectroscopic characteristics (IR, Raman, NMR, and UV-visible spectra) [25, 26]. Additionally, we propose conducting a molecular docking study to investigate protein–ligand interaction and compute the free energy of binding. Furthermore, the natural bond orbital (NBO) analysis and local reactivity descriptors, such as Fukui function ($f_k^{\pm}$), local softness ($s_k^{\pm}$), and local electrophilicity ($\omega_k^{\pm}$), will be computed using DFT-based quantum chemical calculations on biologically active natural compounds. This article offers several research ideas that could help to elucidate the physical principles behind the bioactivity of these natural compounds and to identify potential novel applications.

References

1. Dwivedi, G. and Dwivedi, S., History of medicine: Sushruta – the clinician – teacher par excellence. *Indian J. Chest Diseases Allied Sci.*, 49, 243–244, 2007.
2. Prakash, O., Kumar, A., Kumar, P., Ajeet, Anticancer potential of plants and natural products: A review. *Am. J. Pharmacol. Sci.*, 1, 6, 104–115, 2013.
3. Mishra, D.P. and Maurya, R., Isolation and Characterization of Bioactive Natural Products from Indian Medicinal Plants PhD Thesis, CDRI, India, 2014.
4. Dolai, N., Haldar, P.K., Islam, A., Anti Proliferative Activity of Natural Products Against Different Cancer Cell Lines Through Induction of Apoptosis, PhD Thesis, Jadavpur University, India, 2015.
5. Young, D.C., Jhon Wiley and sons, New York, 2001.
6. Tewari, S.P. and Mishra, A.K., Computational Study of Some Naturally Occurring Biomolecules, PhD Thesis, Dr. Shakuntla Misra National Rehabilitation University, Lucknow, India, 2020, https://shodhganga.inflibnet. ac.in/simple-search?query=Satya+Prakash+Tewari.
7. Mishra, A.K. and Tewari, S.P., Structure, spectra and bioactivity of Pentyl Ester of Chlorogenic acid: DFT study. *Emerging Mater. Res.*, 8, 651–662, 2019.
8. Mishra, A.K. and Tewari, S.P., Density functional theory calculations of spectral, NLO, reactivity, NBO properties and docking study of Vincosamide N Oxide active against lung cancer cell lines H1299. *SN Appl. Sci.*, 2, 1021, 2020.
9. Hartree, D.R., The wave mechanics of an atom with a non-coulomb central field.part III. term values and intensities in series in optical spectra. *Proc.-Cambridge, Phil. Soc.*, 24, 3, 426–437, 1928.
10. Schrodinger, E., Quantisierung als eigenwertproblem annalen der physik. *Annalen Der Physik*, 79, 361–376, 1926.
11. Raghavachari, K. and Anderson, J.B., Electron correlation effects in molecules. *J. Phys. Chem.*, 100, 12960–12973, 1996.

12. Leach, A.R., Molecular modelling principles and applications longman. *Essex*, 595, 1996.
13. https://shodhganga.inflibnet.ac.in/bitstream/10603/124816/6/06_chapter%201.pdf
14. https://chempedia.info/info/chlorofiuorocarbon/
15. Brooks, B.R., Brooks 3rd, C.L., Mackerell Jr,.A.D., Nilsson, L., Petrella, R.J., Won, Y., Archontis, G., Bartels, C., Boresch, S., Caflisch, A., Caves, L., Cyi, Q., Dinner, A.R., Feig, M., Fischer, S., Gao, J., Hodoscek, M., Im, W., Kuczera, K., Lazaridis, T., Ma, J., Ovchinnikov, V., Paci, E., Pastor, R.W., Post, C.B., Pu, J.Z., Schaefer, M., Tidor, B., Venable, R.M., Woodcock, H.L., Yang, W., York, D.M., Karplus, M., CHARMM: The biomolecular simulation program. *J. Compu. Chem.*, 30, 10, 1545–1614, 2009.
16. Perdew, J.P. and Wang, Accurate and simple analytic representation of the electron gas correlation energy. *Phys. Rev. B*, 45, 23, 13244, 1992.
17. Becke, A.D., Density-functional thermochemistry: iii the role of exact exchange. *J. Chem. Phys.*, 98, 5648, 1993.
18. Lee, C., Yang, W., Parr, R.G., Development of the colle-salvetti correlation-energy formula into a functional of the electron density. *Phys. Rev. B*, 37, 785, 1988.
19. http://autodock.scripps.edu/
20. Mishra, A.K., Gupta, V., Tewari, S.P., *In Silico* screening of some naturally occurring bioactive compounds predicts potential inhibitors against SARS-COV-2 (COVID19) protease. *Indian J. Biochem. Biophysics*, 57, 416–425, 2021.
21. Mishra, A.K. and Tewari, S.P., Theoretical study on structure, spectroscopic and bioactivity characteristics of isodihydrocadambine. *Sensors Lett.*, 17, 10, 822–825, 2019.
22. Mishra, A.K. and Tewari, S.P., Density functional theory approach towards bioactivity analysis of Isovallesiachotamine natural bio molecule. *Int. J. Sci. Res. Phy. And Appl. Sci.*, 7, 118–131, 2019.
23. Mishra, A.K. and Tewari, S.P., 7-Hydroxy-6-Methoxy- Coumarian to be a multifunctional bioactive natural molecule: A theoretical study. *Mater. Today: Proc.*, 15, 400–408, 2019.
24. Mishra, A.K. and Tewari, S.P., Theoretical evaluation of the bioactivity of a plant-derived natural molecule- D-Pinitol and other derived structure. *AIP Conf. Proc.*, 2142, 150019, 2019.
25. Srivastava, A.K., Pandey, A.K., Jain, S., Misra, N., FT-IR spectroscopy, intra-molecularC-H-Ointeractions, HOMO, LUMO, MESP analysis and biological activity of two natural products, triclisine and rufescine: FT and QTAIM approaches. *Spectrochim Acta A*, 136, 682–689, 2015.
26. Zhongqiang, L., Zhao-Xu, C., Biaobing, J., Density functional theory studies on the structures and vibrational spectroscopic characteristics of nickel, copper and zinc naphthalocyanines. *Spectrochimica Acta Part A*, 217, 8–17, 2019.

2

Facile Synthesis of Hybrid Fe_3O_4/ZnO Nanosphere Composites and Their Potential Applications in Dye-Sensitized Solar Cells

Y. Prapawasit[1], P. Hemnil[1], V. Karthikeyan[2,3], T. Wongwuttanasatian[1,4,5], Müslüm Arici[6] and V. Seithtanabutara[5,7]*

[1]*Energy Engineering Program, Faculty of Engineering, Khon Kaen University, Khon Kaen, Thailand*
[2]*Department of Physics, Faculty of Science, Naresuan University, Phitsanulok, Thailand*
[3]*Smart Energy System Integration Research Unit, Department of Physics, Faculty of Science, Naresuan University, Phitsanulok, Thailand*
[4]*Mechanical Engineering Division, Khon Kaen University, Khon Kaen, Thailand*
[5]*Center for Alternative Energy Research and Development, Khon Kaen University, Khon Kaen, Thailand*
[6]*Mechanical Engineering Department, Engineering Faculty, Kocaeli University, Kocaeli, Turkey*
[7]*Department of Chemical Engineering, Division, Khon Kaen University, Khon Kaen, Thailand*

Abstract
Electronic band structure and band gap energy play a vital role in the performance of photocatalysts. At present, magnetite/zinc oxide (Fe_3O_4/ZnO) composite photocatalysts are widely examined for dye-sensitized solar cell (DSSC) application as they are thermally and chemically stable. In this study, iron oxide (Fe_3O_4) and zinc oxide (ZnO) composites were synthesized using the precipitation–calcination (PC) and hydrothermal (HT) methods. Fe_3O_4 was synthesized by performing several chemical processing treatments and mixed with ZnO. To obtain Fe_3O_4/ZnO-PC from the first method, calcination treatment was conducted for the prepared mixture of Fe_3O_4 and ZnO at 400°C for 1 h. In the HT method, $ZnCl_2$ and

**Corresponding author*: tmalika@kku.ac.th

Sudheesh K. Shukla, Chaudhery Mustansar Hussain, Santanu Patra, Meenakshi Choudhary (eds.) Electrocatalytic Materials for Renewable Energy, (27–48) © 2024 Scrivener Publishing LLC

$FeCl_3.6H_2O$ were used as chemical precursors and heated in an autoclave at 200°C for 24 h to obtain Fe_3O_4/ZnO-HT. The developed synthetic photocatalyst material properties were analyzed using X-ray diffraction (XRD), Fourier-transform infrared spectroscopy (FTIR), scanning electron microscopy energy-dispersive X-ray analysis (SEM-EDX), accelerated surface area and porosimetry analyzer (ASAP-2460), thermogravimetric analysis (TGA), and UV–Visible spectrophotometry (UV–Vis). The resulting Fe_3O_4/ZnO-HT attained a lower crystalline peak intensity and a higher amount of Fe than Fe_3O_4/ZnO-PC. Notably, the HT method obtained a nanospherical shape of the composite with a higher specific surface area of 17.63 m^2/g and a narrow pore size of 7.27 nm. It was also found that Fe_3O_4/ZnO-HT had a higher photoresponse in UV–Vis light with less energy band gap (2.15 eV), and it was a suitable photocatalyst for DSSC.

Keywords: Fe_3O_4/ZnO, hydrothermal, photocatalyst, UV–Vis

2.1 Introduction

The energy crisis is rapidly growing around the world with the adoption of modernized lifestyles and technology. At present, conventional energy production methods are widely used and likely increase environmental pollution [1–3]. To address energy stabilization and less environmental impact, renewable energy-based power production is the most efficient method [4–6]. Among other renewable energy sources, the solar energy conversion technique earned greater attention especially dye-sensitized solar cells (DSSCs) which are reliable for tropical climatic conditions having low cost and no moving parts, being maintenance free, and requiring no specific land for installation, unlike silicon solar cells [7]. DSSC is the replacement for conventional silicon solar cells as it can be installed indoors to generate power using room lighting [8]. DSSC operates at wavelengths of 100–400 nm and at 400–750 nm of visible light ranges or an infrared range from 750 to 2,500 nm. As DSSC requires less irradiance, in most cases, it is replaced with conventional building walls. Moreover, as it minimizes the building material cost, natural lighting reduces conventional power consumption, and finally, it can produce electrical power under indoor/ambient conditions.

In recent years, ZnO- and Fe_3O_4-based photocatalysts have been widely examined for DSSC as they are readily available, non-toxic, low cost, and chemically and thermally stable [9]. Wang *et al.* synthesized Fe_3O_4/ZnO-based photocatalyst using a simple two-step chemical process (coprecipitation and HT methods) and achieved the same performance after conducting the degradation test six times. A magnetic field is applied to the synthesized photocatalyst to minimize/remove the magnetization resulting in surface oxygenation presenting in higher order, and Fe_3O_4/ZnO performed effectively [10]. Elshypany *et al.* developed a photocatalyst to perform under the blue visible range using Fe_3O_4/ZnO. Different mass ratios of Fe_3O_4/ZnO show that an increase in magnetite compounds reduced reflectance, and 0.8 wt.% obtained a lower reflectance than 1 wt.%. Comparatively, ZnO in the presence of magnetite increased the photocatalyst performance and degradation efficiency shifting from 13.5% to 88.5%. Beneficially synthesized Fe_3O_4/ZnO-based photocatalyst operates at higher efficiency after consecutive cycles [11]. Nikazar *et al.* processed Fe_3O_4 using a standard chemical treatment: ferric chloride hexahydrate ($FeCl_3.6H_2O$), ferrous sulfate heptahydrate ($FeSO_4.7H_2O$), and ammonium hydroxide (NH_4OH). The prepared Fe_3O_4 was then mixed with ZnO for further calcination process at high temperature (550°C for 2 h). It was found that 1:10 of Fe_3O_4:ZnO efficiently removed 88% of phenol and turned into a recyclable catalyst in phenol removal [12].

A one-step synthesis of Zn-doped Fe_3O_4 composites has been performed using the hydrothermal method. The selected photocatalyst material underwent chemical treatment of the one-step synthesis process by mixing zinc chloride ($ZnCl_2$), $FeCl_3.6H_2O$, and ammonium acetate (NH_4Ac) with ethylene glycol ($C_2H_6O_2$) and then heated to 200°C (by hydrothermal method). The photo-Fenton process is used to degrade Rhodamine B (RhB) and cephalexin in hydrogen peroxide (H_2O_2), which contains water under visible light irradiation. It is found that the synthesized catalyst becomes highly stable and recycled by an external magnetic field [13].

In the study of Wang *et al.* [14], Fe_3O_4-based photocatalyst is synthesized by the process of adding iron(III) nitrate nonahydrate ($Fe(NO_3)_3.9H_2O$) and sodium hydroxide (NaOH) to glycol and mixed it well by heating it in an autoclave for 12 h using 200°C to dry the distilled water. Furthermore, $ZnCl_2$ and NaOH were added, stirred, and heated to 80°C for 12 h. It is found that the prepared photocatalyst performed effectively under low UV light with a large specific surface area, strong light absorption, and high adsorption capacity and current density. The synthesized Fe_3O_4/ZnO photocatalyst photocurrent density reached 18.4 $\mu A/cm^2$. From the above literature study, it was found that magnetite and zinc oxide were excellent materials for synthesizing the photocatalyst and several synthesis techniques were studied. However, there have not been any reports comparing the properties of the composites obtained from the two synthetics above. Furthermore, in this study, Fe_3O_4/ZnO-based photocatalyst is prepared using the two-step method, i.e., PC, and the one-step method, i.e., HT synthesis, to find the effectiveness of light absorption for DSSC. Comparative studies are performed for both the PC and HT methods by analyzing the synthesized photocatalyst using XRD, FTIR, SEM/EDX, BET, TGA, and UV–Vis.

2.2 Materials and Methods

2.2.1 Materials

To prepare the synthetic photocatalyst, $FeCl_3.6H_2O$, ferrous chloride tetrahydrate ($FeCl_2.4H_2O$), $ZnCl_2$, NH_4Ac, demineralized water (DI water), NaOH, $C_2H_6O_2$, ethanol (C_2H_5OH), and ZnO chemicals were purchased from Sigma-Aldrich Co. LLC and Elago Enterprises Private Limited with 99% purity.

2.2.2 Photocatalyst Synthesis Methods

Method 1 (the PC Method): Initially, 3.244 g of $FeCl_3.6H_2O$, 1.194 g of $FeCl_2.4H_2O$, and 200 ml of DI water were mixed by a magnetic stirrer at 60°C for 40 min. Next, 200 ml of NaOH was mixed with the prepared solution and stirred for 1.5 h at 60°C and then washed with 95% ethanol to remove the unwanted materials from the solution. After washing the solution, oven drying was done to dry the solution for 12 h at 100°C. The oven-drying process removed the liquid particles from the prepared solution and produced dry Fe_3O_4. The second process of the synthesis was carried out with ZnO, and the prepared Fe_3O_4 and commercial ZnO were mixed at a ratio of 1:1 (0.6 g:0.6 g) using the ultrasonic method by adding 3 ml of ethanol for 30 min. The final stage of the synthesis was calcination, and the mixed Fe_3O_4/ZnO solution was placed for the calcination process for 1 h at 400°C. After the calcination process, the final product Fe_3O_4/ZnO-PC was obtained as shown in Figure 2.1, which can be utilized for DSSC as a photocatalyst.

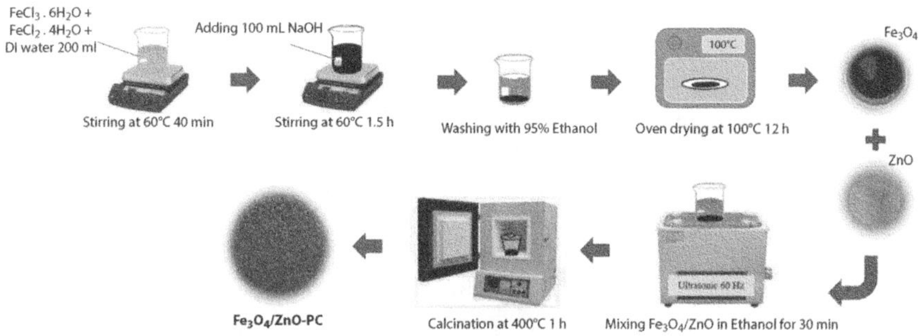

Figure 2.1 Synthesis of the Fe_3O_4/ZnO-PC photocatalyst by the precipitation–calcination method (PC).

Method 2 (the HT Method): Comparatively, the HT method is easier than PC and other synthesis methods. In this method, the Fe_3O_4/ZnO composite was prepared through chemical processing. At first, 0.544 g of $ZnCl_2$, 2.16 g of $FeCl_3.6H_2O$, 4.624 g of NH_4AC, and 200 ml of ethylene glycol were mixed by a magnetic stirrer for 30 min and then the sonication process was performed for 30 min to enhance the chemical reaction. Furthermore, the autoclave process was performed at 200°C for 24 h, and the materials were washed with 95% ethanol. The final step of the HT method was oven drying in order to remove the trace ethanol. The prepared material was dried at 100°C for 12 h, and the final product Fe_3O_4/ZnO-HT was produced by the HT method. The procedure of this method is illustrated in Figure 2.2.

2.2.3 Prepared Sample Characterization

To find the intensity and 2θ angle of the prepared photocatalysts, namely, ZnO, Fe_3O_4, Fe_3O_4/ZnO-PC, and Fe_3O_4/ZnO-HT, XRD (PANalytical, Empyrean) was performed using CuKα radiations (λ = 1.5406 Å), and the scanning range (2θ) was selected from 20° to 100° angles. The morphology of the material was analyzed using a Hitachi SU3800 (SEM) with SEM/EDX. FTIR was also performed using a Bruker ALPHA-II spectrophotometer within the range of 4,000–550 cm^{-1}. To find the thermal stability of the photocatalyst, TGA was conducted with a heating rate of 10°C min^{-1}, starting from 25°C to 700°C using a Shimadzu DTG-60H. Furthermore, the

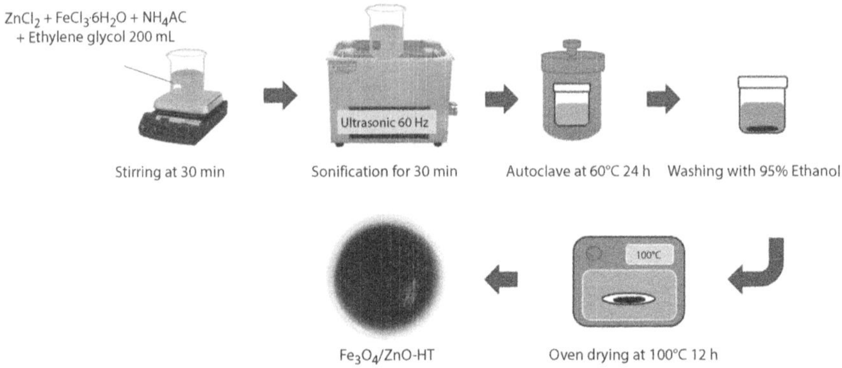

Figure 2.2 Synthesis of Fe_3O_4/ZnO-HT photocatalysts by the hydrothermal method (HT).

specific surface area and UV–Vis absorption spectra were measured using an ASAP-2460 instrument by the N_2 adsorption–desorption method at 77 K and a Shimadzu UV-3101PC ultraviolet spectrophotometer (range = 200–800 nm), respectively.

2.3 Results and Discussion

2.3.1 XRD Pattern

Figure 2.3 shows the crystalline structures of ZnO, Fe_3O_4, Fe_3O_4/ZnO-PC, and Fe_3O_4/ZnO-HT. ZnO exhibits a prominent crystal structure and meets the standard pattern of JCPDS No. 36-1451. Notably, ZnO attained a wurtzite structure [10], with a strong diffraction peak at 2θ = 31.89°, 34.56°, 36.30°, 47.59°, 56.68°, 62.96°, 66.58°, 68.00°, 69.13°, 72.73°, and 77.04° for (100), (002), (101), (102), (110), (103), (200), (112), (201), (004), and (202) respectively. Fe_3O_4 shows a crystal structure with different diffraction peaks at 2θ = 30.43°, 35.67°, 43.23°, 53.64°, 57.14°, and 62.83° for (220), (222), (400), (442), (511), and (440), respectively. Comparatively, these diffraction peaks are in good agreement with magnetite standard data (JCPDS No. 16-0629) [10]. Fe_3O_4/ZnO-PC (method 1) has a mixed crystal structure between ZnO and Fe_3O_4. Overlapping peaks of the crystalline ZnO and Fe_3O_4 can be seen. It indicates that this method influences the phase shift of ZnO, and minor Fe_3O_4 peaks appear at 2θ = 30.31°, 31.97°,

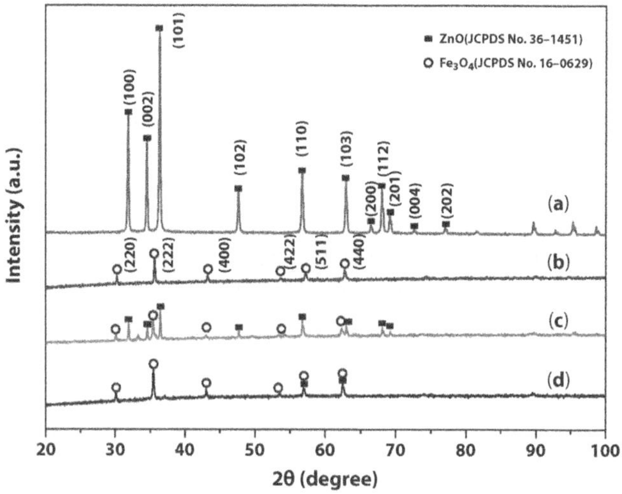

Figure 2.3 XRD spectra of ZnO (a), Fe_3O_4 (b), Fe_3O_4/ZnO-PC (c), and Fe_3O_4/ZnO-HT (d).

34.63°, 35.50°, 36.46°, 42.91°, 47.71°, 53.36°, 56.88°, 62.40°, 63.08°, 68.15°, and 69.23°. Next, the composite material of Fe_3O_4/ZnO synthesized using the HT method (method 2) shows a more predominant crystal structure of Fe_3O_4 than ZnO. The peaks of Fe_3O_4/ZnO-HT appear at $2\theta = 30.16°$, 35.55°, 43.10°, 53.44°, 56.89°, and 62.50°. The reaction mechanism of the prepared sample is characterized by the replacement of Fe^{3+} ions with Zn^{2+} ions [13]. The hybridization of Fe_3O_4/ZnO-HT shows that a ZnO shell grows on the surface of the Fe_3O_4 core [15].

To calculate the crystallite size of the prepared samples, the Debye–Scherrer method was used as expressed in Equation 2.1.

$$D = K\lambda / \beta \cos\theta \tag{2.1}$$

where,
D = sample average crystallite size (nm)
K = Scherrer constant (0.9)
λ = sample wavelength of radiation (CuKα = 1.5406 Å)
β = full-width at half-maximum of peak (FWHM) in radians
θ = diffraction angle (radians)

Also, the crystalline index was evaluated as the percentage of the area of crystalline peaks in the whole area of the crystalline and amorphous peaks as depicted in Table 2.1. Notably, Fe_3O_4 has a crystal size of 33.96 nm, which is larger than ZnO. Moreover, ZnO has a higher percentage of crystallinity, whereas Fe_3O_4/ZnO-PC attained crystallite size and crystalline index between Fe_3O_4 and ZnO. On the other hand, lower crystallite size and crystalline index were attained for Fe_3O_4/ZnO-HT as compared with Fe_3O_4/ZnO-PC following its crystal structure as shown in Figure 2.3. The elemental composition analysis with EDX depicts the Fe_3O_4/ZnO composite attaining high purity. Fe_3O_4/ZnO-HT contains Fe amount (wt.%) which is 1.9 times higher than Fe_3O_4/ZnO-PC. However, Fe_3O_4/ZnO-PC contains an Fe/Zn ratio of only 0.89 by weight or an atomic ratio of 1.04. Thus, Fe_3O_4/ZnO-HT shows better hybridization between Fe_3O_4 and ZnO.

2.3.2 Dispersion of the Particle

Figure 2.4 shows the morphology of the prepared photocatalyst samples analyzed by scanning electron microscopy at ×5k magnification. ZnO particles (Figure 2.4a) are tiny granules that form in clusters scattered across the surface similar to Fe_3O_4 particles (Figure 2.4b); however, its particle size is larger than that of ZnO. As seen in Figure 2.4c, Fe_3O_4/ZnO-PC shows a spherical particle which is the combination of Fe_3O_4 and ZnO crystalline particles bonded together in clusters. Comparatively, Fe_3O_4/ZnO-HT has a spherical structure (Figure 2.4d) similar to Fe_3O_4/ZnO-PC and the inset image at ×30k revealed that the particle had a uniform submicrosphere shape consistent with the report of Nguyen *et al.* [13]. However, the particle size of Fe_3O_4/ZnO-HT is smaller (300 nm) due to different concentrations of the synthesized substance and the technique, more likely a minor range

Table 2.1 Comparison of the crystalline properties and components of the samples.

Properties	ZnO	Fe_3O_4	Fe_3O_4/ ZnO-PC	Fe_3O_4/ ZnO-HT
Crystallite size (nm)	21.99	33.96	32.46	24.45
Crystallinity index (%)	64.66	26.49	44.98	29.2
Elemental composition (wt.%)	Zn (80.93), O (19.07)	Fe (81.3), O (18.3)	Fe (39.01), Zn (43.86), O (17.13)	Fe (74.02), Zn (3.57), O (22.42)

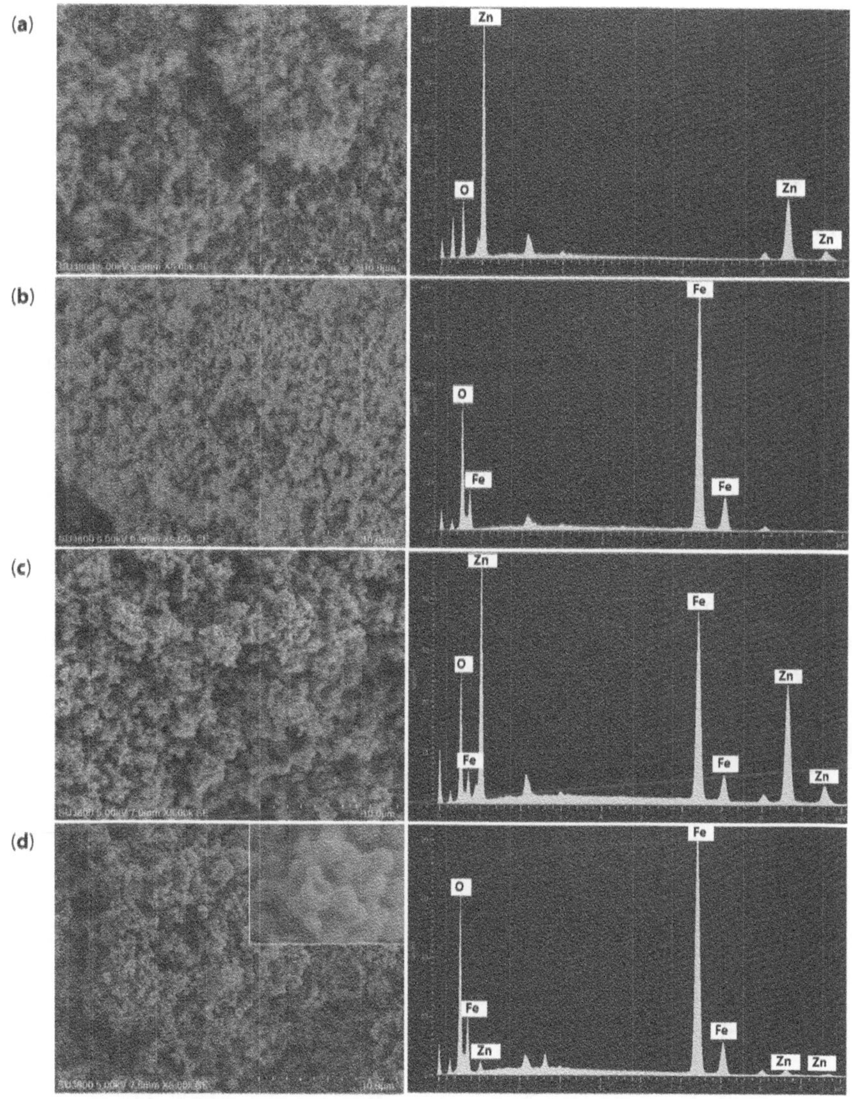

Figure 2.4 Morphology (×5k) and EDX analysis of (a) ZnO, (b) Fe_3O_4, (c) Fe_3O_4/ZnO-PC, and (d) Fe_3O_4/ZnO-HT (inset: SEM image ×30k).

of ZnO on Fe_3O_4. The reason behind this is a hydrothermal method and it composited a minor portion of ZnO present within the Fe_3O_4 structure.

2.3.3 FTIR Analysis

The FTIR spectroscopy method was used to characterize the surface sites of the prepared catalysts. From the FTIR spectra as seen in Figure 2.5, the characteristic bands of ZnO were observed at 873.56 cm^{-1} (tetrahedral coordination of Zn appeared) and 669.59 cm^{-1} (ZnO is stretched due to vibration), which corresponded to the report of da Silva-Neto *et al.* [16]. For Fe_3O_4, transmission was observed at 1,613.21 cm^{-1} which was favorable to –OH group bending vibrations and stretching vibration at 1,120.80 cm^{-1} of the COO–Fe bond. It was noted that Fe–O bond characteristics were observed around 684.01 cm^{-1}. The Fe_3O_4/ZnO composites from both synthesis methods showed peaks at the low wavenumber, indicating the interaction between Fe_3O_4 and ZnO phases, giving a change in the vibration bands corresponding to the metal–O bonds. After ZnO was mixed with Fe_3O_4, the characteristic bands of ZnO and Fe_3O_4 were found at the wavenumber of 1,149.65 cm^{-1} and assigned to the COO–Fe bond. The stretching vibrations of Zn–O and Fe–O bonds were revealed at 869.44 cm^{-1} and 684.01 cm^{-1}, respectively. Comparatively, Fe_3O_4/ZnO-HT attained a higher

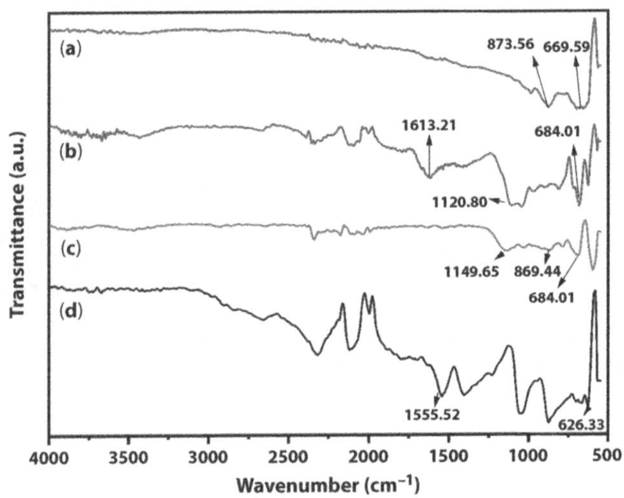

Figure 2.5 FTIR spectra of ZnO (a), Fe_3O_4 (b), Fe_3O_4/ZnO-PC (c), and Fe_3O_4/ZnO-HT (d).

intensity of peaks than Fe_3O_4/ZnO-PC and had characteristic bands of ZnO. The functional group of Zn–O was found in the range of 679.89 cm^{-1} and absorption at 1,555.52 cm^{-1} and 1,061.05 cm^{-1} in the presence of a COO–Fe bond. This shows that Fe_3O_4/ZnO-HT attained strong hybridization between Fe_3O_4 and ZnO phases in the composites.

2.3.4 TGA and DTG

TGA and DTG analyses were conducted using the operating temperature of 30°C–700°C. It was found that ZnO semiconductor has high thermal stability over the operating temperature (Figure 2.6a). Notably, high decomposition was observed for Fe_3O_4 with a pronounced loss rate of –0.22%/min at 49°C (Figure 2.6b) as the water molecules evaporated. Secondly, Fe_3O_4 was less thermally stable than ZnO but at different temperatures. There was no decomposition observed for Fe_3O_4/ZnO-PC (Figure 2.6c), with almost no apparent max peak loss. However, Fe_3O_4/ZnO-HT has high-temperature sensitivity and is unstable. There were three degradation regions with the highest rates at 282.3°C, 413.79°C, and 507.12°C for –0.20, –0.07, and

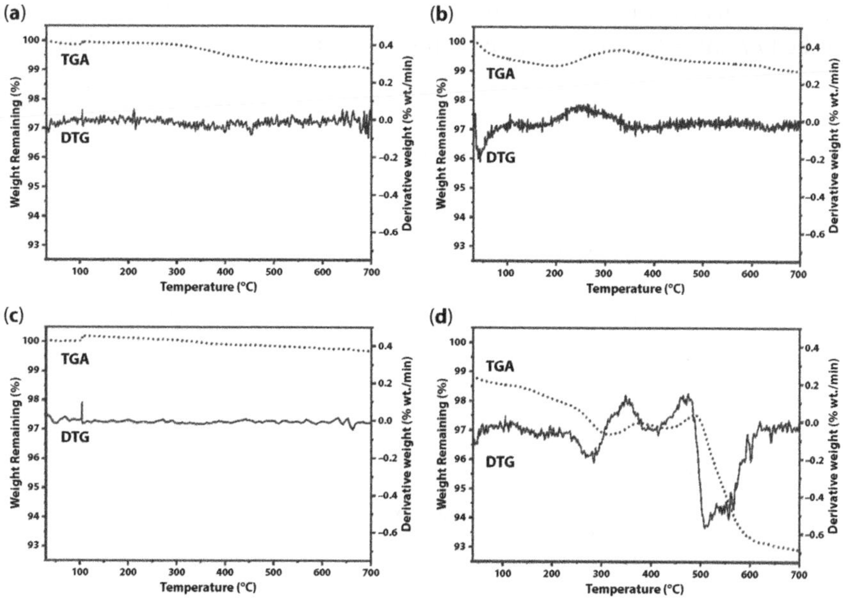

Figure 2.6 TGA and DTG thermal curves of ZnO (a), Fe_3O_4 (b), Fe_3O_4/ZnO-PC (c), and Fe_3O_4/ZnO-HT (d).

−0.55%/min, respectively. More importantly, it is not suggested to be used at temperatures above 480°C. The total remaining weight percentages at 700°C of ZnO, Fe_3O_4, Fe_3O_4/ZnO-PC, and Fe_3O_4/ZnO-HT were 99.08%, 98.99%, 99.69%, and 92.93%, respectively.

2.3.5 BET Analysis

Figure 2.7 shows the adsorption and desorption isotherms of ZnO, Fe_3O_4, Fe_3O_4/ZnO-PC, and Fe_3O_4/ZnO-HT. It shows that the prepared materials attained type IV isotherms as per the IUPAC, and there is a hysteresis loop of type H3. These characteristics reveal that desorption happens at a lower partial pressure than adsorption and the limiting uptake over a range of high P/P_0 associated with capillary condensation in mesopores [17], and it is related to stemming from slit-shaped pores with a combination of meso-/macropores. It is noticed that ZnO has a lower N_2 adsorption and desorption capacity than Fe_3O_4, and Fe_3O_4/ZnO-PC has a lower N_2 adsorption and desorption capacity than Fe_3O_4/ZnO-HT. From the conclusion of their porous structures as presented in Table 2.2, ZnO has a pore size of 55.29 nm, and the specific surface area of the samples is lower (1.55 m^2/g). Fe_3O_4 attained a high pore volume and a smaller pore size resulting in a higher specific surface area. It is worth mentioning that the specific surface area of Fe_3O_4/ZnO-PC is close to Fe_3O_4 but the pore size is smaller at 9.38 nm,

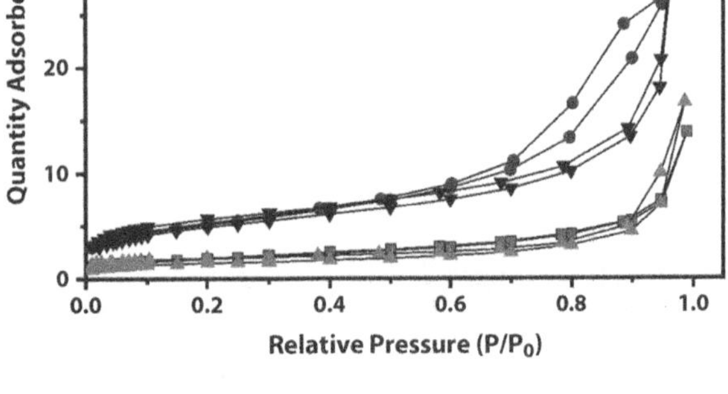

Figure 2.7 Nitrogen adsorption/desorption isotherm of ZnO, Fe_3O_4, Fe_3O_4/ZnO-PC, and Fe_3O_4/ZnO-HT.

Table 2.2 Porous property of the prepared samples.

Sample	BET surface area (m^2/g)	Pore volume (cm^3/g)	Pore size (nm)
ZnO	1.55	0.02	55.29
Fe_3O_4	5.42	0.04	36.19
ZnO/Fe_3O_4-PC	5.20	0.01	9.38
ZnO/Fe_3O_4-HT	17.63	0.03	7.27

which is caused by small crystals of ZnO depositing around the surface of Fe_3O_4 as shown in Figure 2.4c. Fe_3O_4/ZnO-HT (Figure 2.4d) has a remarkable result as it has a higher specific surface area than Fe_3O_4/ZnO-PC and a relatively smaller pore size and a pore volume; therefore, it is more suitable for catalyst application.

2.3.6 UV–Vis Spectra

UV–Vis absorption spectroscopy was carried out to analyze the energy band gap of the prepared samples. Figure 2.8 shows the optical properties of ZnO, synthesized Fe_3O_4, and Fe_3O_4/ZnO composites. It is found that

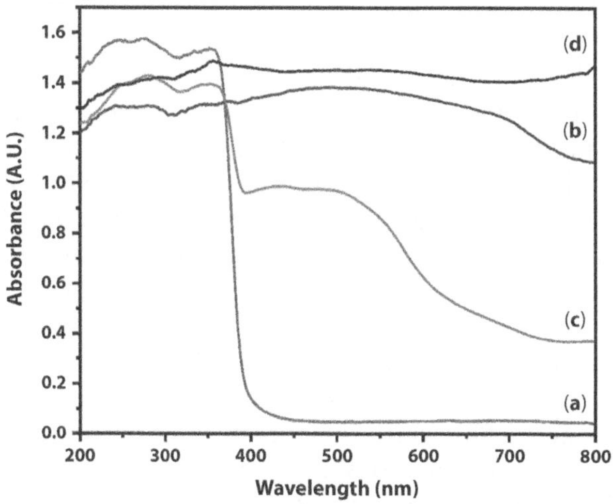

Figure 2.8 UV–Vis light absorption spectrum of ZnO (a), Fe_3O_4 (b), Fe_3O_4/ZnO-PC (c), and Fe_3O_4/ZnO-HT (d).

ZnO has a high absorption ability in the UV range and a strong wavelength at 358 nm. The absorption spectra showed a shoulder in the region from 350 nm to 400 nm and had low absorption within the visible light. Comparatively, Fe_3O_4 has better absorbency than ZnO in both the UV and Vis ranges. The formation of Fe_3O_4/ZnO favors altering the electronic band structure. The composite material of Fe_3O_4/ZnO-PC has an absorbency in both the UV and Vis ranges, midway between ZnO and Fe_3O_4, which is better than ZnO but still lower than Fe_3O_4. There are two shoulders of its absorption spectra resulting in double absorption edges corresponding to ZnO absorption and the presence of magnetic Fe_3O_4. Beneficially, Fe_3O_4/ZnO-HT has outstanding agility in all materials as it has a high absorbency and ability to absorb both UV and Vis light.

The optical absorption strength is determined using Equation 2.2 as it depends on the photon energy and the band gap.

$$(\alpha h\nu)^2 = A(h\nu - E_g) \tag{2.2}$$

where,

h = Planck's constant
ν = photon's frequency
α = absorption coefficient
A = proportionality constant
E_g = band gap energy

Based on the optical absorbance data, Tauc plots of the absorption coefficient times are performed between the photon energy and second power to the total incident photon energy (Tauc plots) as shown in Figure 2.9. During low photon, the prepared photocatalyst absorption rate reached zero, while the absorption got stronger at higher photon energies with linearity. The band gap energy is determined from the extrapolation of the linear region to the *X*-axis. The lowest energy gap is obtained for Fe_3O_4/ZnO-HT as compared with Fe_3O_4, Fe_3O_4/ZnO-PC, and ZnO. Herein, ZnO has an energy band gap of 3.22 eV (Figure 2.9a), similar to the existing study of Ma *et al.* [18]. The synthesized Fe_3O_4 attained an energy band gap of 2.27 eV (Figure 2.9b), which is higher than the investigation of Millaty Abadiah *et al.* [19]. Figure 2.9c shows that Fe_3O_4/ZnO-PC has a value of 3.04 eV which is halfway between ZnO and Fe_3O_4 due to the influence of Fe_3O_4 having a lower band gap energy. Thus, Fe_3O_4/ZnO is applied in

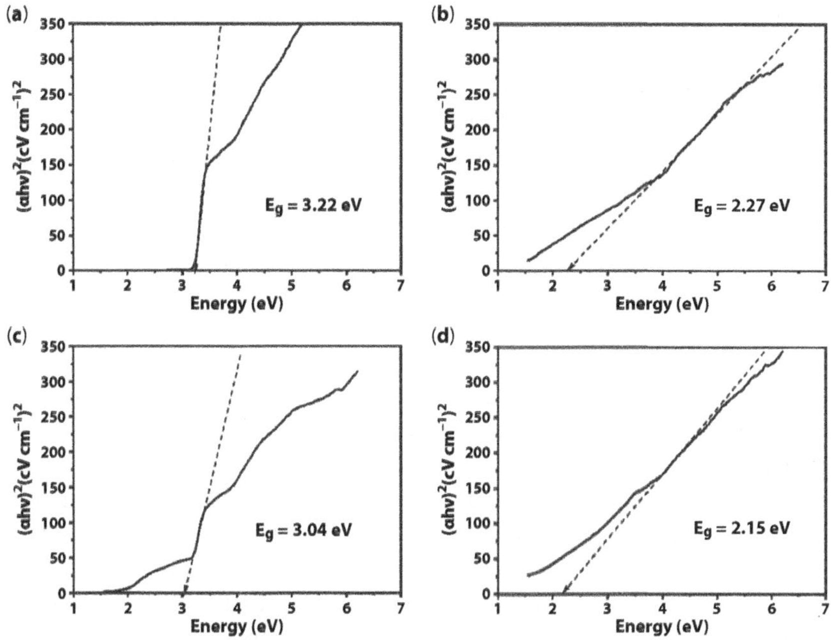

Figure 2.9 Band gap energy (E_g) determination from the Tauc plot of ZnO (a), Fe_3O_4 (b), Fe_3O_4/ZnO-PC (c), and Fe_3O_4/ZnO-HT (d).

the active photocatalytic solar cells and photodiodes. As mentioned earlier, Fe_3O_4/ZnO-HT has the lowest band gap energy (Figure 2.9d) since it has a higher specific surface area as well as a more prominent structure, leading to higher support of the active sites and better charge accumulation. Consequently, grain boundary is limited to electron trapping, slowing down the combination process of photogenerated electron/hole pairs [13]. A low band gap energy will result in a low activation energy for electron activation between the valence band and the conduction band, favoring an easier and more effective optical reaction.

Table 2.3 shows the comparative study of the Fe_3O_4/ZnO structure with other works, and Fe_3O_4 has excellent properties as a catalyst since it has a high specific surface area, high biocompatibility, high electronic conductivity, and low cost and is environmentally friendly. Moreover, Fe^{3+} of Fe_3O_4 can increase light absorption from the UV range to the visible light range; in addition to the properties mentioned above, Fe_3O_4 has a low energy gap band. Therefore, it is used to modulate the wide band gap semiconductors to effectively shift the edge of the absorbance band to the visible light region in order to enhance their photovoltaic properties. The synthesized Fe_3O_4/ZnO nanocomposite attained a lower band gap energy than existing studies as listed in Table 2.3, showing that ZnO conduction bands are formed at lower energies due to the fine coupling between Fe and Zn 4s orbitals [31]. The photocomposition with small band gap energy has better performance in solar utilization. When the incident light falls on the band gap, the catalyst absorbs the photon with minimum energy, allowing the photon to excite electrons from the valence band to the conduction band. This phenomenon produces pairs of electron holes, which can either recombine or migrate to different surface reaction sites to drive chemical reactions. Therefore, composite catalysts have been synthesized to delay the recombination of electron/hole pairs, allowing electrons to continuously migrate to the catalyst surface, resulting in better performance in various applications. The Fe_3O_4/ZnO composite catalyst exhibits a difference in energy levels between Fe_3O_4 and ZnO, resulting in photo-excitation electron transfer from the surface of ZnO to Fe_3O_4. The photogenerated electrons transfer from the conduction bands of ZnO to the conduction bands of Fe_3O_4, while holes transfer from the valence band of ZnO to the valence band of Fe_3O_4. Therefore, the recombination of photogenerated electrons and holes is reduced, extending the material life and enhancing the photocatalytic performance. Thus, the Fe_3O_4/ZnO composite photocatalyst with low band gap energy has attracted attention to be used as nanomaterials for photocatalysis, sensors, photodetector solar cells, supercapacitors, and optoelectronic devices. As mentioned above, using this low-energy band gap material as a photocatalyst can acquire higher energy which could be favorable in minimizing the conventional energy source. The innovation of a higher efficiency renewable energy harvester can predominantly be utilized in an urban area to minimize the on-peak demand and environmental pollution. Switching urban power consumption from non-renewable to renewable energy can directly affect the well-being and daily routine life of people in a productive way by controlling pollution. On the other hand, Fe_3O_4/ZnO will help promote the United Nations Sustainable Development Goals (SDGs), also known as the Global Goals, providing the present and future generations a better life by minimizing fossil fuel consumption.

Table 2.3 Comparative analysis of the Fe_3O_4/ZnO structure with recent studies.

Catalyst structure	**Synthesis method**	**Band gap energy (eV)**	**Application reported**	**Reference**
Nanocomposite	Two-step liquid chemical method	2.9	Photocatalysis	[14]
Nanoparticles	A simple low-temperature chemical synthesis process	2.88	Photocatalytic degradation of methyl orange	[20]
Nanoparticle	One-step chemical bath deposition method	2.82	Decolorization of acid blue 113	[21]
Nanocomposite	Wet chemical route	3.27	None	[22]
Nanocomposite	Chemical method	3.1	Photocatalytic degradation of RhB	[23]
Nanocomposite	Microwave-sonochemical	2.47	Antibacterial performance	[24]
Nanoparticles	Solvent thermal technique	3.21	TOC degradation	[25]
Nanoparticles	Sol–gel process	2.95	RhB degradation	[26]
Nanocomposite	Co-precipitation method	2.3	Dye degradation	[27]
Nanocomposite	Co-precipitation method	2.75	Photocatalytic degradation of metalaxyl	[28]
Nanoparticles	Wet-chemical technique	3.2	Detection of L-aspartic acid and glycine	[29]
Nanocomposite	Sol–gel and hydrothermal method	3.1	Degradation of MO and MB	[30]
Nanosphere composite	***Hydrothermal method***	***2.15***	***DSSC***	***Present study***

2.4 Conclusion

The composite material Fe_3O_4/ZnO was synthesized by the PC and HT methods. XRD and FTIR analysis shows that Fe_3O_4/ZnO-HT has a lower crystalline peak intensity and higher Fe content in structure than Fe_3O_4/ZnO-PC as the reaction mechanism is characterized by ion replacement. SEM depicts that Fe_3O_4/ZnO-HT attained a uniform nanosphere shape and EDX indicates a higher Fe/Zn ratio than Fe_3O_4/ZnO-PC. Beneficially, Fe_3O_4/ZnO-HT has a narrow pore size (7.27 nm), higher surface area (17.63 m^2/g), and larger pore volume (0.03 cm^3/g). The effective performance of Fe_3O_4/ZnO-HT was found up to 480°C and achieved a good light response in the UV–Vis range with less band gap (2.15 eV) that favors incorporation with DSSC. This study provides a method for synthesizing such materials with outstanding chemical and photoresponse properties. Therefore, it will be used as a photoanode material instead of conventional semiconductors in our future work soon. The light harvesting efficiency in the DSSC efficiency term will be compared with that of the other materials.

Acknowledgments

This study was financially supported by the Research Funding for Postgraduate KKU Engineering Student, Khon Kaen University. Also, the authors would like to thank the Center for Alternative Energy Research and Development, Khon Kaen University for technical support.

References

1. Yashwantrao, G. and Saha, S., Perspective on the rational design strategies of quinoxaline derived organic sensitizers for dye-sensitized solar cells (DSSC). *Dyes Pigm.*, 199, 110093, 2022.
2. Velmurugan, K., Karthikeyan, V., Kumarasamy, S., Wongwuttanasatian, T., Sa-Ngiamsak, C., Thermal mapping of photovoltaic module cooling via radiation-based phase change material matrix: A case study of a large-scale solar farm in Thailand. *J. Energy Storage*, 55, 105805, 2022.
3. Velmurugan, K., Karthikeyan, V., Sharma, K., t.b. Korukonda, V.K., Balasubramanian, D., Wongwuttanasatian, T., Contactless phase change material based photovoltaic module cooling: A statistical approach by clustering and correlation algorithm. *J. Energy Storage*, 53, 105139, 2022.

4. Roslan, N., Ya'acob, M.E., Radzi, M.A.M., Hashimoto, Y., Jamaludin, D., Chen, G., Dye Sensitized Solar Cell (DSSC) greenhouse shading: New insights for solar radiation manipulation. *Renewable Sustain. Energy Rev.*, 92, 171–186, 2018.
5. Velmurugan, K., Madurai Elavarasan, R., De, P., Karthikeyan, V., Korukonda, T.B., Arockia Dhanraj, J., Emsaeng, K., Chowdhury, M., Techato, K., Khier, B., Attia, E.-A., A review of heat batteries based PV module cooling—Case studies on performance enhancement of large-scale solar PV system. *Sustainability*, 14, 1–65, 2022.
6. Velmurugan, K., Kumarasamy, S., Wongwuttanasatian, T., Seithtanabutara, V., Review of PCM types and suggestions for an applicable cascaded PCM for passive PV module cooling under tropical climate conditions. *J. Cleaner Prod.*, 293, 126065, 2021.
7. Kusumawati, Y., Hutama, A.S., Wellia, D.V., Subagyo, R., Natural resources for dye-sensitized solar cells. *Heliyon*, 7, 12, 08436, 2021.
8. Alizadeh, A., Roudgar-Amoli, M., Bonyad-Shekalgourabi, S.-M., Shariatinia, Z., Mahmoudi, M., Saadat, F., Dye sensitized solar cells go beyond using perovskite and spinel inorganic materials: A review. *Renewable Sustain. Energy Rev.*, 157, 112047, 2022.
9. Elshypany, R., Selim, H., Zakaria, K., Moustafa, A.H., Sadeek, S.A., Sharaa, S.I., Raynaud, P., Nada, A.A., Magnetic ZnO crystal nanoparticle growth on reduced graphene oxide for enhanced photocatalytic performance under visible light irradiation. *Molecules*, 26, 8, 2269, 2021.
10. Wang, J., Yang, J., Li, X., Wang, D., Wei, B., Song, H., Li, X., Fu, S., Preparation and photocatalytic properties of magnetically reusable Fe_3O_4@ZnO core/shell nanoparticles. *Physica E*, 75, 66–71, 2016.
11. Elshypany, R., Selim, H., Zakaria, K., Moustafa, A.H., Sadeek, S.A., Sharaa, S.I., Raynaud, P., Nada, A.A., Elaboration of Fe_3O_4/ZnO nanocomposite with highly performance photo catalytic activity for degradation methylene blue under visible light irradiation. *Environ. Technol. Innovation*, 23, 101710, 2021.
12. Nikazar, M., Alizadeh, M., Lalavi, R., Rostami, M.H., The optimum conditions for synthesis of Fe_3O_4/ZnO core/shell magnetic nanoparticles for photodegradation of phenol. *J. Environ. Health Sci. Eng.*, 12, 1–6, 2014.
13. Nguyen, X.S., Zhang, G., Yang, X., Mesocrystalline Zn-Doped Fe_3O_4 hollow submicrospheres: Formation mechanism and enhanced photo-fenton catalytic performance. *ACS Appl. Mater. Interfaces*, 9, 10, 8900–8909, 2017.
14. Wang, Y., Gao, J., Liu, Y., Li, M., Zhang, M., He, G., Sun, Z., Facile fabrication of ZnO nanorods modified Fe_3O_4 nanoparticles with enhanced magnetic, photoelectrochemical and photocatalytic properties. *Optical Materials,* 111, 110608, 2021.
15. Pham, H.L., Nguyen, V.D., Nguyen, V.K., Le, T.H.P., Ta, N.B., Pham, D.C., Tran, Q.T., Dang, V.T., Rational design of magnetically separable core/ shell

Fe_3O_4/ZnO heterostructures for enhanced visible-light photo degradation performance. *RSC Adv.*, 11, 36, 22317–22326, 2021.

16. da Trindade, L.G., Hata, G.Y., Souza, J.C., Soares, M.R.S., Leite, E.R., Pereira, E.C., Longo, E., Mazzo, T.M., Preparation and characterization of hematite nanoparticles-decorated zinc oxide particles (ZnO/Fe_2O_3) as photo electrodes for solar cell applications. *J. Mater. Sci.*, 55, 7, 2923–2936, 2020.
17. Ojuva, A., Pulsed current processing of zeolites: consolidation behaviour, porosity and crystallinity, Department of physical, inorganic and structural chemistry, Stockholm University, 2009.
18. Ma, S., Huang, Y., Hong, R., Lu, X., Li, J., Zheng, Y., Enhancing photo catalytic activity of ZnO nanoparticles in a circulating fluidized bed with plasma jets. *Catalysts*, 11, 1, 77, 2021.
19. Millaty Abadiah, N., Yuliantika, D., Amalia Hariyanto, Y., Eko Saputro, R., Masruroh, Taufiq, A., Soontaranoon, S., Nanostructure, Band Gap, and Antibacterial Activity of Spinel Fe_2Mo_4/OO Magnetic Fluids. *IOP Conference Series: Earth and Environmental Science*, vol. 276, p. 012064, 2019.
20. Feng, Q., Li, S., Ma, W., Fan, H.-J., Wan, X., Lei, Y., Chen, Z., Yang, J., Qin, B., Synthesis and characterization of Fe_3O_4/ZnO-GO nanocomposites with improved photo catalytic degradation methyl orange under visible light irradiation. *J. Alloys Compd.*, 737, 197–206, 2018.
21. Yilmaz, M., Mengelizadeh, N., Saloot, M.k., shahbaksh, S., Balarak, D., Facile synthesis of Fe_3O_4/ZnO/GO photo catalysts for decolorization of acid blue 113 under solar, visible and UV lights. *Mater. Sci. Semicond. Process.*, 144, 106593, 2022.
22. Roychowdhury, A., Mishra, A., Pati, S., Das, D., Synthesis and characterization of multifunctional Fe_3O_4-ZnO nanocomposite. *AIP Conf. Proc.*, 1447, 283–284, 2012.
23. Yu, R., Shang, Y., Zhang, X., Liu, J., Zhang, F., Du, X., Sun, H., Zeng, J., Self-templated synthesis of core-shell Fe_3O_4@ZnO@ZIF-8 as an efficient visible-light-driven photocatalyst. *Catal. Commun.*, 174, 106583, 2023.
24. Vijayalakshmi, K. and Noor Ul Haq, L., Microwave-sonochemical synergistically assisted synthesis of hybrid Ni-Fe_3O_4/ZnO nanocomposite for enhanced antibacterial performance. *Mater. Today Commun.*, 26, 101835, 2021.
25. Liu, K., Qin, Y., Muhammad, Y., Zhu, Y., Tang, R., Chen, N., Shi, H., Zhang, H., Tong, Z., Yu, B., Effect of Fe_3O_4 content and microwave reaction time on the properties of Fe_3O_4/ZnO magnetic nanoparticles. *J. Alloys Compd.*, 781, 790–799, 2019.
26. Pham, H.L., Hao Nguyen, V., Vu, T.T., Khien Nguyen, V., Ta, N.B., Le, T.T., Nguyen, V.D., Phung, T.O., Novel 3D hierarchical nanostructure of Fe_3O_4/ZnO hybrid composites for enhanced solar light photo catalytic performance. *J. Mater. Sci.: Mater. Electron.*, 32, 20, 25018–25032, 2021.
27. Shashank, M., Naik, H.S.B., Nagaraju, G., Keri, R.S., Naik, M.M., Lingaraju, K., Facile synthesis of Fe_3O_4/ZnO nanocomposite: Applications to photo catalytic and antibacterial activities. *J. Electronic Mater.*, 50, 6, 3557–3568, 2021.

28. Dehghan, S., Jafari, A.J., FarzadKia, M., Esrafili, A., Kalantary, R.R., Visible-light-driven photocatalytic degradation of Metalaxyl by reduced graphene oxide/Fe_3O_4/ZnO ternary nanohybrid: Influential factors, mechanism and toxicity bioassay. *J. Photochem. Photobiol., A*, 375, 280–292, 2019.
29. Hussain, M.M., Asiri, A.M., Rahman, M.M., Simultaneous detection of l-aspartic acid and glycine using wet-chemically prepared Fe_3O_4@ ZnO nanoparticles: real sample analysis. *RSC Adv.*, 10, 33, 19276–19289, 2020.
30. Wei, P., Yu, X., Li, Y., Preparation of Fe_3O_4/ZnO/ZnS composites with enhanced photo performance under solar irradiation. *J. Electron. Mater.*, 48, 8, 4877–4885, 2019.
31. Abdullah Mirzaie, R., Kamrani, F., Firooz, A.A., Khodadadi, A.A., Effect of α-Fe_2O_3 addition on the morphological, optical and decolorization properties of ZnO nanostructures. *Mater. Chem. Phys.*, 133, 1, 311–316, 2012.

3

Study and Analysis of Hybrid Nanofluid-Based Heat Pipes for Renewable Energy Applications

Ramkumar Venkatasamy[1], Joshuva Arockia Dhanraj[1,2], Nadanakumar Vinayagam[3], Chatchai Sirisamphanwong[4], Karthikeyan Velmurugan[4], Rattaporn Ngoenmeesri[5] and Chattariya Sirisamphanwong[6]*

[1]*Centre for Automation and Robotics (ANRO), Department of Mechatronics Engineering, Hindustan Institute of Technology and Science, Chennai, Tamil Nadu, India*
[2]*University Centre for Research & Development (UCRD), Chandigarh University, Mohali, Punjab, India*
[3]*Department of Automobile Engineering, Hindustan Institute of Technology and Science, Chennai, India*
[4]*Smart Energy System Integration Research Unit, Department of Physics, Faculty of Science, Naresuan University, Phitsanulok, Thailand*
[5]*Research and Academic Service, Faculty of Science, Naresuan University, Phitsanulok, Thailand*
[6]*Department of Physics and General Science, Faculty of Science, Nakhon Sawan Rajabhat University, Nakhon Sawan, Thailand*

Abstract
A deep convolution neural network (DCNN) was used to study and simulate a heat pipe using a hybrid nanofluid for renewable energy applications. Python was used to create the DCNN model. Several input factors, such as flow rate, inclinations, heat input, flow rates, hybrid nanofluid ratios, water inlet temperature, and ambient temperature, are employed for the analysis. The hybrid nanofluids used for this study are silver nanofluid (Ag) and alumina (Al_2O_3). The heat pipe specification used for this investigation is a heat pipe with a length of 1 m, an outer diameter of 19 mm, and a thickness of 2 mm of copper material. The heat

**Corresponding author*: chattariya.s@nsru.ac.th

Sudheesh K. Shukla, Chaudhery Mustansar Hussain, Santanu Patra, Meenakshi Choudhary (eds.) Electrocatalytic Materials for Renewable Energy, (49–70) © 2024 Scrivener Publishing LLC

pipe's wick structure is built with two layers of stainless steel screen mesh with a mesh size of 2,000. The two-hybrid nanofluids were made with the combinations of Ag and Al_2O_3 nanoparticles in DI water as the base fluid (Ag 80%, Al_2O_3 20%, and Ag 60% $A1_2O_3$ 40%). Heat inputs vary from 40 W to 100 W in steps of 10 W at different inclination angles (0°, 30°, and 45°). Two different flow rates of 0.2 kg/min and 0.3 kg/min have been investigated. A comparison study on heat inputs under ideal working circumstances was also given. The rectified linear unit function activated the convolution layer in the DCNN model, which was trained and tested using the experimental result data. The dense layer, in comparison, uses the linear function. A network model for DCNN was created. Three hundred epochs were determined to match best the experimental data set after the model was built for various epochs ranging from 100 to 1,000. As a result, weights corresponding to 300 epochs were frozen, and a simulation for a fresh set of data (not experimented) was created and assessed to obtain the ideal value of water outlet temperature. Based on its performance, the hybrid nanofluid had an excellent heat input of 95 W, a flow velocity of 0.3 kg/min, and an inclination of 40°.

Keywords: DCNN, nanofluids, inclinations, flow rate, temperature

3.1 Introduction

A heat pipe is a handy tool that employs a two-phase fluid flow and has a high thermal conductivity. High thermal conductivity and phase change can effectively transfer heat between two components. Heat may be efficiently transferred from one place to another using heat pipes with very little heat loss. In place of sensible heat, the latent heat of the working fluid is used to conduct heat. As a result, the heat pipe's thermal conductivity is significantly greater than that of solid conductors made of the same material. Machine learning constructed a model for data analysis to automatically learn from data, make judgments, and identify various patterns. The procedure starts with feeding the data and then training the machines by building machine learning models using the data. Deep learning is considered a division of machine learning where the ANN learns from data. Deep learning permits solving complicated problems even when the data are extremely diverse, assorted, and unstructured. Deep learning instructs the computer to understand the process of prediction and classification of information through layers. The authors discussed the heat transfer enhancement of nanofluids and the technique for preparing a nanofluid, nanophase powders, and a base liquid consisting of a suspension. In heating or cooling fluids, they suspend nanophase particles. The dispersed nanoparticles enhance the surface area of the fluid, as well as its heat capacity and thermal conductivity. The thermal conductivity of nanofluids was explored theoretically and tested with a hot wire setup using suspended copper nanophase particles. Finally, they demonstrated how a nanofluid moving through a tube transfers heat [1]. In this paper, the authors explain the circular tube in which the convective heat transfer of Al_2O_3 + water nanofluid takes place, and various concentrations were taken for this investigation. As a result, the heat transfer coefficient increases based on the rise of nanoparticle concentration and Peclet number [2].

This chapter explains the investigation of the rheological behavior and thermal properties of nanofluids (Al_2O_3 + water), and this result reveals that adding Al_2O_3 nanoparticles increases the thermal conductivity of water [3]. In a circular pipe, an aqueous solution of gold nanoparticles of different sizes is used as a nanofluid and compared with DI water. As a result, the thermal resistance of DI water is greater than that of the nanoparticle solution [4]. This paper uses an aqueous solution of silver nanoparticles as a nanofluid with an average diameter of 10 nm. The thermal resistance result of the heat pipe with nanofluid is less than pure water [5]. The aqueous solution of silver nanoparticles with diameters of 10 nm and 35 nm at various concentrations of 1 mg/l, 10 mg/l, 50 mg/l, and 100 mg/l was taken and compared with DI water, which shows that an increase in the number of silver nanoparticles in the heat pipe wall temperature is smaller compared with pure water [6]. A flat copper heat pipe transient thermal and hydrodynamic model was used for cooling electronics components. The wick and vapor core fluid fluxes were calculated using a transient 2D hydrodynamic model linked with a transient 3D thermal model. Pressure drops in the phases determine the capillary limit. It signifies that the heat pipe operates correctly during the transient operation [7]. The thermal performance of the heat pipes uses nanoparticles such as Al_2O_3, CuO, and TiO_2, and this investigation shows that temperature, velocity, pressure, and maximum heat transfer were attained with different nanofluid concentrations of the heat pipe [8]. The screen mesh wick heat pipe is used with water-based Al_2O_3 nanofluids for investigating the thermal properties. This method shows that water-based Al_2O_3 nanofluids improve the thermal properties of the heat pipe compared with water [9].

In this paper, the authors used a horizontal heat pipe and investigated two base fluids of water and ethylene glycol mixed with Al_2O_3

nanoparticles using the computational method. The results show that ethylene glycol base fluid has a higher heat transfer coefficient than water [10]. This paper reveals the study of convective heat transfer and the rheological behavior of copper and titanium oxide in nanometers. The experimental result shows that increased volume concentrations (0.1% to 2.0%) increase the heat transfer coefficient [11]. This investigation is based on numerical analysis by comparing the water-based Al_2O_3 and Al_2O_3–Cu hybrid nanofluid in the effect of forced convective heat transfer [12]. The thermal conductivity and dynamic viscosity of mixed nanofluids of various volume fractions, such as Ag–MgO/water, improve thermal conductivity properties [13].

The authors developed an artificial neural network model that matches the experimental values of thermal conductivity of hybrid nanofluids such as Cu/TiO_2–water/EG hybrid [14]. The thermal conductivity property is increased in plate heat exchangers using nanofluids such as alumina and multiwalled carbon nanotubes [15]. This paper deals with the nanoparticle concentration and the temperature effect based on the rheological behavior of hybrid nanofluids Fe_3O_4–Ag/EG. Adding a volume fraction below 0.3% obeys the Newtonian law, and more than 1.2% had non-Newtonian behavior [16]. The deep learning algorithm performs SHW in solar thermal systems in different meteorological conditions. Techniques like artificial neural networks, recurrent neural networks, long short-term memory, and software TRNSYS are used [17]. The authors also used various machine learning methods such as support vector regression, process regression, and convolution neural network to predict the thermal conductivity of the composite material. These methods are useful for predicting the thermal conductivity of the composite material faster with the data set and training set of the model [18]. The authors created an online monitoring system for SiCf–SiCm composite deterioration. They used AE data as the raw input to develop a deep learning-based end-to-end convolutional neural network model for online monitoring of the SiCf–SiCm composite tube damage progression. Compared with 74% of the RF models, the CNN model with single events may achieve an average prediction accuracy of 84.4% [19].

3.2 Materials and Methods

3.2.1 Heat Pipe

A heat pipe contains a vacuum-sealed pipe or tube, generally known as a container. The container material and working fluid must be completely compatible with each other. Copper and water are compatible, as well as aluminum. A vacuum pump is used to create a vacuum [20]. The heat pipe is filled with a measured volume of a working fluid and vacuum-sealed. Working fluids are chosen based on the temperature at which the heat pipes are about to operate, with extremely low and extremely high temperatures [21]. The majority of the heat pipes used for room temperature applications employ ammonia as the fluid for an operating temperature range between 213 K and 373 K, alcohol in the temperature range of 283–403 K, water in the temperature range of 298–573 K, or ethanol in the temperature range of 273–403 K [22]. Heat pipes are made up of copper material, and water is used as a working fluid while operating at the temperature range of 293–423 K. To fill the water in the heat pipe, it is heated until the water boils. After that, the air is displaced and sealed. It contains the saturated liquid; the vapor is ready for the function of the heat pipe, and this liquid goes toward the condenser and is cooled [23]. Then, it becomes

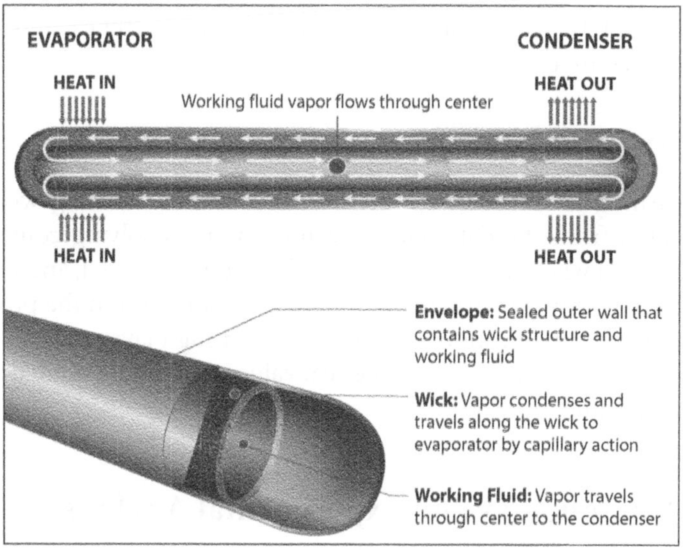

Figure 3.1 Heat pipe.

a saturated liquid state and returns to the heat pipe evaporator [24]. The capillary action uses a wick structure in the working liquid phase [25]. The heat pipe is shown in Figure 3.1.

3.2.2 Container

A heat pipe has a container, a wick structure, and a working fluid. Heat pipes are constructed in such a way that no maintenance is required even after a long run, so the heat pipe container, wick, and working fluid must be compatible [26]. Copper and water working fluid, aluminum and ammonia, steel and refrigerant R134a, and copper and refrigerant R134a are the most widely used container fluid combinations. Containers made of super alloy (alkali metals of cesium, potassium, sodium), hydrogen, neon, nitrogen, oxygen, and helium are all compatible with stainless steel. Copper and methanol are both compatible [27]. A combination of aluminum and ethanol heat pipes is commonly employed in spaceship thermal management.

3.2.3 Nanofluids

A nanofluid is a fluid that comprises particles sized to nanometers. Nanometer-scale particles are dispersed in a base fluid such as water, ethylene glycol, or oil [28]. In the preparation of nanofluids, nanoparticles of materials like metallic oxides, semiconductors, carbide ceramics, nitride ceramics, metals, single and multiwalled carbon nanotubes, and alloyed nanoparticles are used [29].

3.2.4 Deep Convolutional Neural Network

Deep learning is considered a division of machine learning where the ANN learns from the data. Deep learning permits solving complicated problems even when the data are extremely diverse, assorted, and unstructured. Deep learning instructs the computer to understand the process of prediction and classification of information through layers. Data sets can be in images, text, sound, and numerical values and can be trained using a machine learning algorithm.

3.3 Methodology and Experimental Analysis

The paper's objective is to predict and simulate the results of hybrid nanofluids of heat pipes using the DCNN model. The training data and testing

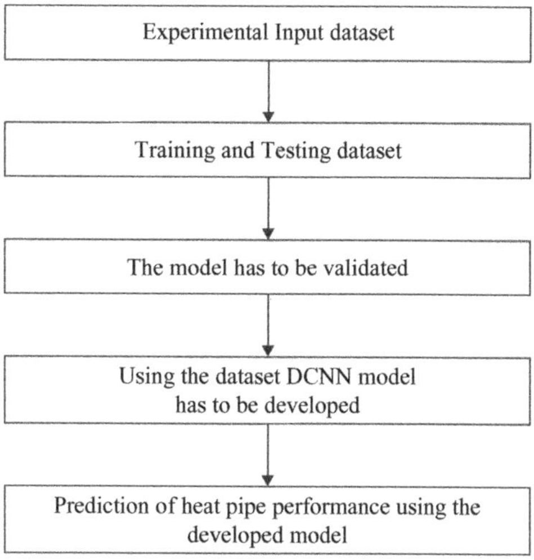

Figure 3.2 Methodology.

data were obtained from the experimental analysis of the heat pipe and should be considered a data set. The methodology is shown in Figure 3.2.

3.3.1 DCNN Model of a Heat Pipe with Hybrid Nanofluid

This section describes the development procedure of the DCNN heat pipe model with a hybrid nanofluid. The experimental data obtained were used to train and test the model.

Figure 3.3 depicts the DCNN model utilized in the proposed work. The input layer of this analysis consists of the following parameters: heat input in watts, inclination of the heat pipe in degrees, flow rate in kg/min, fluid ratio, and atmosphere in °C.

The water outlet temperature is the expected output of the network. ReLu was used as an activation function for the convolution layer, and linear was used as an activation function for the dense (fully connected) layer. There were 2,055 samples used for training and 78 samples for testing the model. The DCNN model is shown in Figure 3.3.

The model is created in Python using Kera's Jupyter Notebook. The data set for the input parameters is recorded in CSV format. The model has two 1D CNN layers, according to the suggested project. CNN layers are often created in two groups to give the model a decent capacity to learn parameters from input data. The known features have been flattened to a long vector and fed through a fully linked layer after the convolution layer.

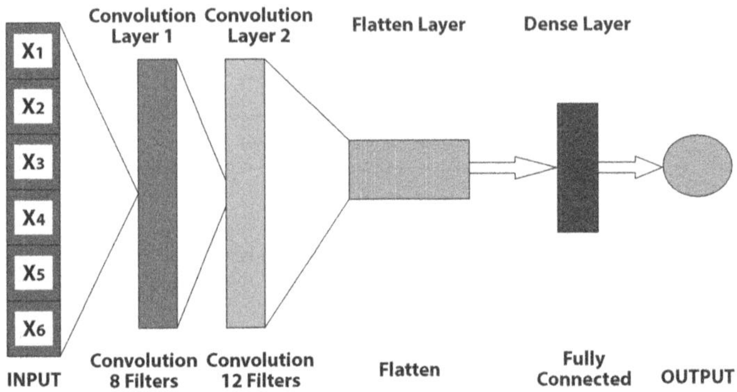

Figure 3.3 The DCNN model.

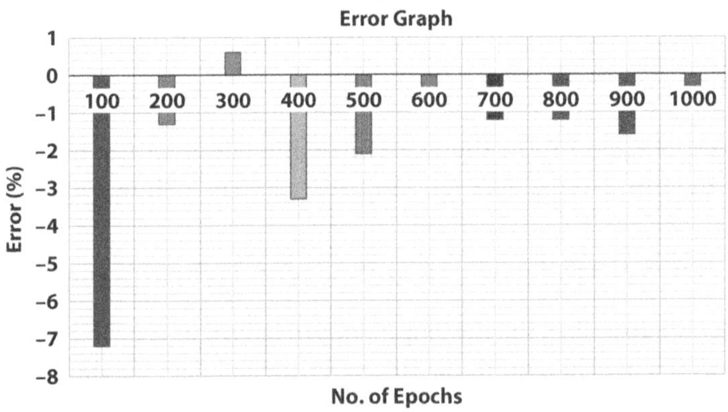

Figure 3.4 Percentage deviations of the predicted values.

The output layer makes a forecast. The output layer's goal is to understand the learned parameters before producing a prediction, and the fully connected layer acts as a buffer between the known parameters and the output. The model is in good shape. The percentage of predicted value is shown in Figure 3.4.

3.4 Results and Discussion

The results obtained by the heat pipe DCNN model are presented in this section. Figure 3.4 gives a comparison of the predicted and measured water outlet temperature. Figure 3.5 shows a correlation coefficient (R^2) of 0.991

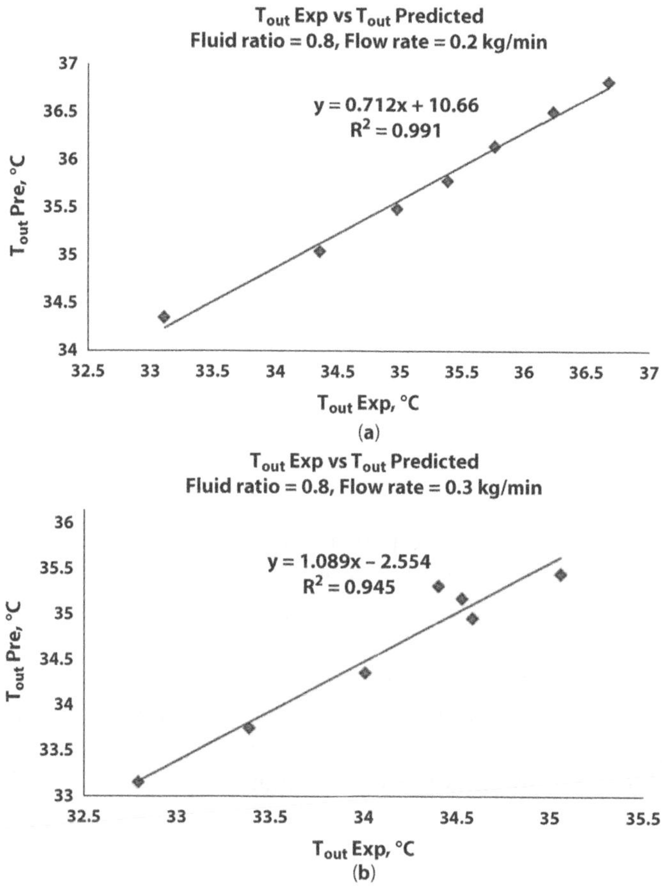

Figure 3.5 (a, b) Comparison of the predicted and measured water outlet temperature.

for a fluid ratio of 0.8 at a flow rate of 0.2 kg/min at 0° inclination of the heat pipe. Similarly, the correlation coefficient for 0.3 kg/min at the same fluid ratio has been obtained as 0.945 as shown in Figure 3.5. A correlation coefficient of 0.912 was accepted for 0.3 kg/min flow rate at a fluid ratio of 0.6 as evident from Figures 3.5a and b.

As represented in Figure 3.6, the graph follows $y = x$, which implies the model generated with DCNN can predict with high precision. The low-data neural network may result in overfitting problems. This overfit phenomenon makes the model unfit to simulate the system. Hence, one must establish that the overfit phenomenon has not occurred in our analysis. The correlation coefficient obtained in the figure indicates clearly that the overfitting phenomenon has not happened in the present DCNN model.

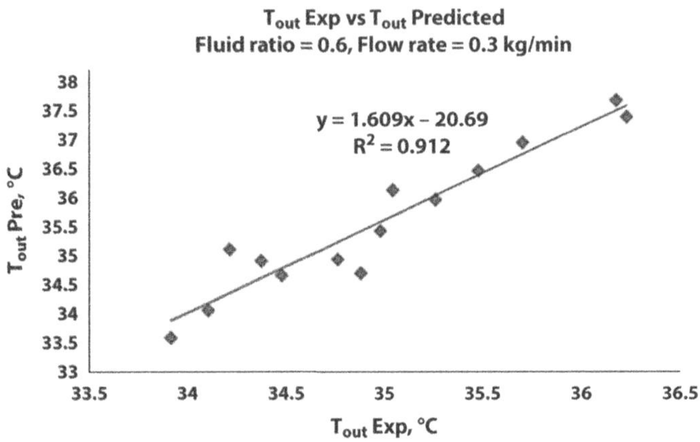

Figure 3.6 Comparison of the predicted and measured water outlet temperature.

It has been found that the percentage deviation of the predicted values is well within the acceptable limit of 5%. Hence, this model has been used to simulate the heat pipe with a new set of unknown data, as given in Table 3.1.

A heat pipe output parameter simulation was run for various combinations of tabulated input parameters, and the results are presented here. Figure 3.7 gives a comparison between the outlet temperature and the heat input. Figure 3.7 provides the water outlet temperature for a fluid ratio of 0.65 at varying water inlet temperatures such as 28°C, 29°C, 30°C, and 31°C. Figure 3.8 shows that the output temperature of around 32°C was obtained for the inlet temperature of 28°C. Similarly, Figure 3.9 gives the water outlet temperature for the fluid ratio of 0.7 at different water inlet temperatures. Here, the obtained maximum output temperature was

Table 3.1 Input parameters and values considered for the simulation.

S. no.	Parameters	Input values
1	Heat input	45,55, 65, 75, 85, 95 (W)
2	Inclination	40 (°)
3	Fluid ratio	0.65, 0.675, 0.7, 0.75
4	Flow rate	0.2, 0.3 (kg/min)
5	Atmospheric temperature	30 (°C)

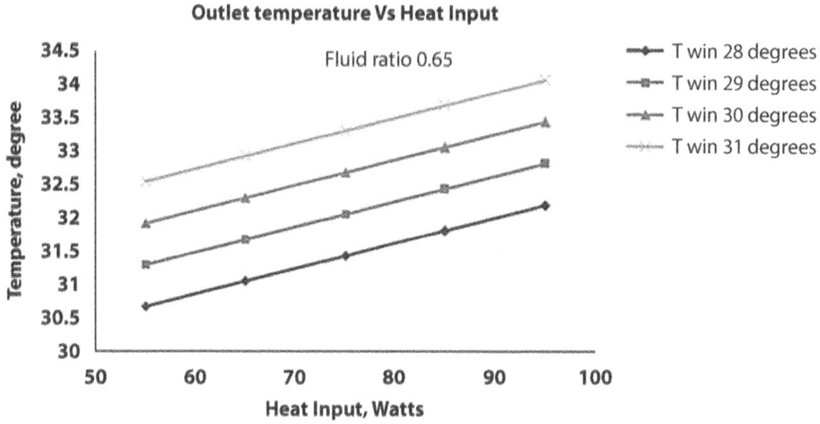

Figure 3.7 Outlet temperature vs. heat input for a fluid ratio of 0.65.

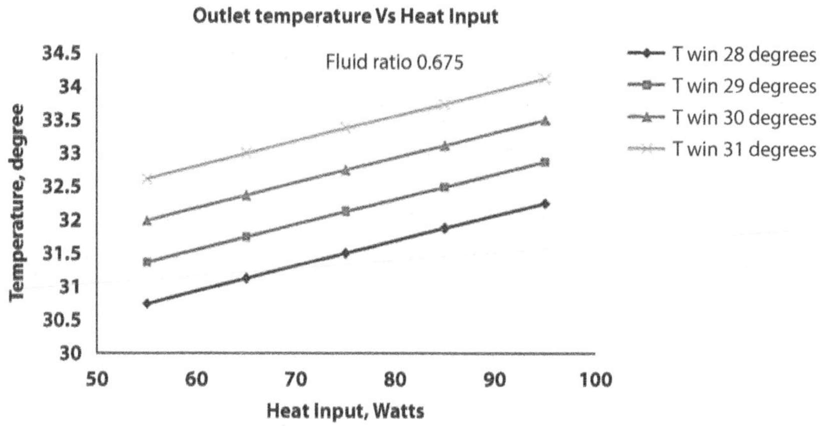

Figure 3.8 Outlet temperature vs. heat input for a fluid ratio of 0.675.

32.5°C for the inlet temperature of 29°C. Figure 3.10 gives the outlet temperature for a fluid ratio of 0.75, and here, the maximum outlet temperature was 33°C for the inlet temperature of 30°C. Table 3.2 shows the outlet temperature vs. heat input for a fluid ratio of 0.65, and Table 3.3 shows the outlet temperature vs. heat input for a fluid ratio of 0.675. Figure 3.10 gives the outlet temperature for a fluid ratio of 0.75 at different water inlet temperatures; here, the maximum outlet temperature was 34°C for the inlet temperature of 31°C.

Table 3.4 shows the outlet temperature vs. heat input for a fluid ratio of 0.7, while Table 3.5 shows the outlet temperature vs. heat input for a

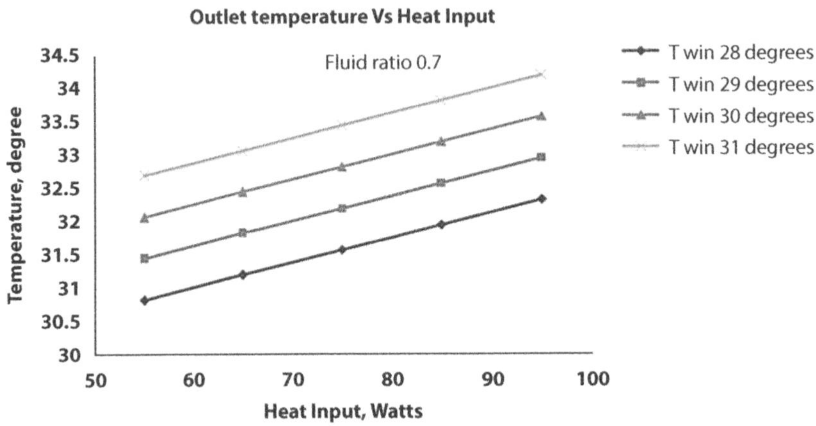

Figure 3.9 Outlet temperature vs. heat input for a fluid ratio of 0.7.

Table 3.2 Outlet temperature vs. heat input for a fluid ratio of 0.65.

	Heat input (W)				
Outlet temperature	**55**	**65**	**75**	**85**	**95**
Twin 28	30.1	31.05	31.43	31.81	32.191
Twin 29	31.30	31.68	32.06	32.43	32.83
Twin 30	31.92	32.30	32.68	33.06	33.44
Twin 31	32.55	32.93	33.30	33.68	34.06

Table 3.3 Outlet temperature vs. heat input for a fluid ratio of 0.675.

	Heat input (W)				
Outlet temperature	**55**	**65**	**75**	**85**	**95**
Twin 28	30.74	31.12	31.49	31.87	32.25
Twin 29	31.36	31.74	32.12	32.50	32.87
Twin 30	31.99	32.37	32.74	33.12	33.50
Twin 31	32.61	32.99	33.37	33.75	34.12

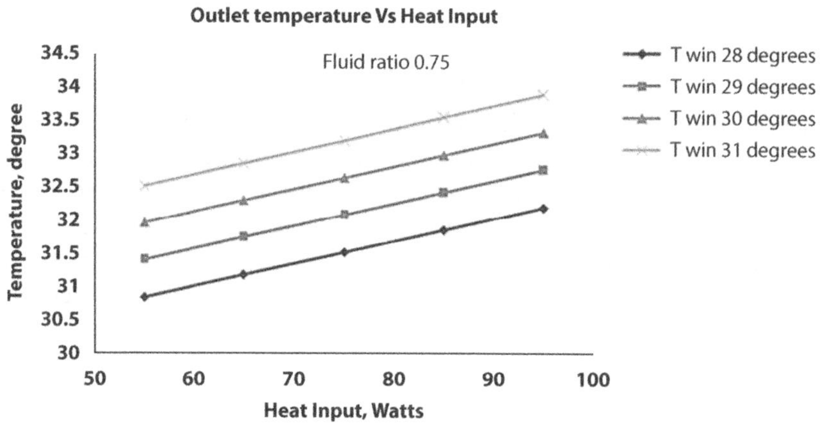

Figure 3.10 Outlet temperature vs. heat input for a fluid ratio of 0.75.

Table 3.4 Outlet temperature vs. heat input for a fluid ratio of 0.7.

	Heat input (W)				
Outlet temperature	**55**	**65**	**75**	**85**	**95**
Twin 28	30.08	31.1	31.56	31.94	32.31
Twin 29	31.43	31.80	32.18	32.56	32.94
Twin 30	32.05	32.43	32.81	33.19	33.56
Twin 31	33.68	33.05	33.43	33.81	34.19

Table 3.5 Outlet temperature vs. heat input for a fluid ratio of 0.75.

	Heat input (W)				
Outlet temperature	**55**	**65**	**75**	**85**	**95**
Twin 28	30.93	31.31	31.69	32.06	32.44
Twin 29	31.55	31.93	32.31	32.69	33.07
Twin 30	32.18	32.56	32.94	.33.31	33.69
Twin 31	32.80	33.18	33.56	33.94	34.32

fluid ratio of 0.75. The estimated outlet temperatures of water, simulated using the DCNN model, were used to calculate the heat pipe performance. Figures 3.11, 3.12, 3.13, and 3.14 give the version of the heat pipe. A comparison of efficiency for a fluid ratio of 0.65 is shown in Figure 3.11. The same trend was obtained for all fluid balances of the hybrid nanofluid. Figure 3.12 shows a comparison of efficiency for a fluid ratio of 0.675. The efficiency decreases with an increase in water inlet temperature. Figure 3.13 compares the efficiency for a fluid ratio of 0.7, and Figure 3.14 compares the efficiency for a fluid ratio of 0.75. While 28°C and 29°C inlet temperatures decrease the efficiency as the heat input increases, the efficiency

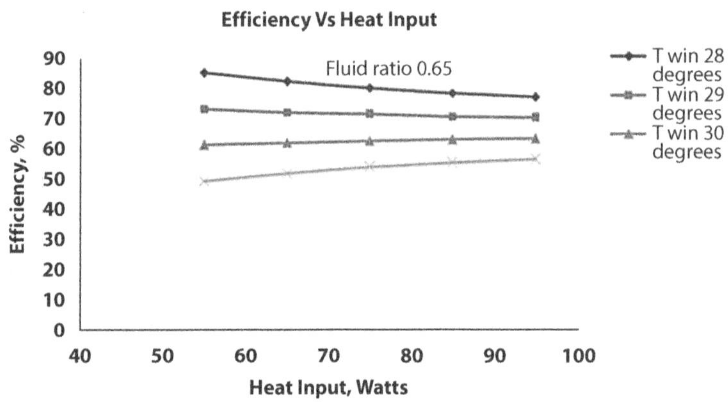

Figure 3.11 Comparison of efficiency for a fluid ratio of 0.65.

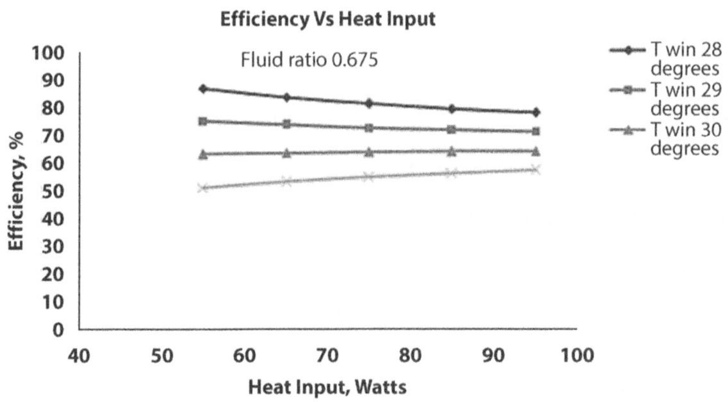

Figure 3.12 Comparison of efficiency for a fluid ratio of 0.675.

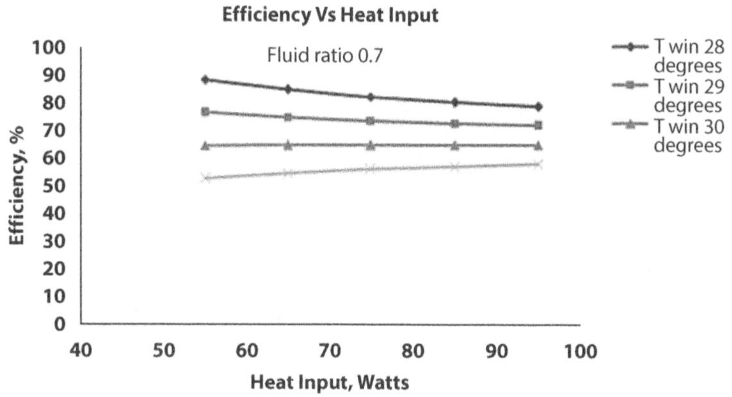

Figure 3.13 Comparison of efficiency for a fluid ratio of 0.7.

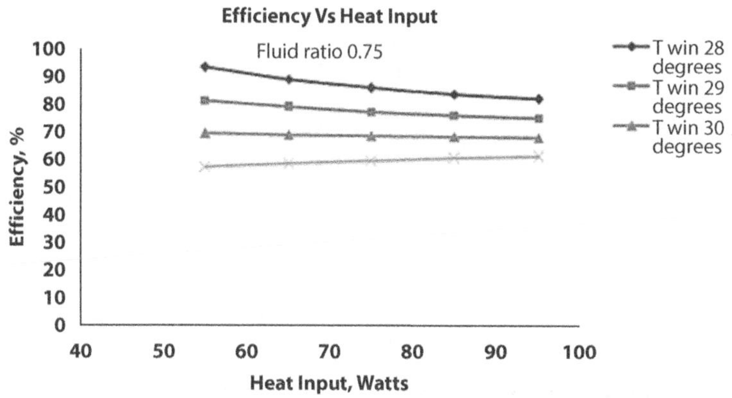

Figure 3.14 Comparison of efficiency for a fluid ratio of 0.75.

marginally increases with an increase in heat input for inlet temperatures of 30°C and 31°C.

Table 3.6 shows the efficiency for a fluid ratio of 0.65, Table 3.7 shows the efficiency for a fluid ratio of 0.675, and Table 3.8 shows the efficiency for a fluid ratio of 0.7.

Figure 3.15 gives the variation of water outlet temperature with water inlet temperature for different heat inputs. Table 3.9 shows the efficiency

Table 3.6 Efficiency for a fluid ratio of 0.65.

Efficiency	Heat input (W)				
	55	**65**	**75**	**85**	**95**
Twin 28	84.94	82.01	79.86	78.22	76.93
Twin 29	73.04	71.95	71.146	70.53	70.04
Twin 30	61.15	61.88	62.42	62.83	63.15
Twin 31	49.26	51.82	53.70	55.13	56.27

Table 3.7 Efficiency for a fluid ratio of 0.675.

Efficiency	Heat input (W)				
	55	**65**	**75**	**85**	**95**
Twin 28	86.96	83.72	81.35	79.53	78.10
Twin 29	75.07	73.66	72.63	71.84	71.21
Twin 30	63.17	63.59	63.90	64.14	64.33
Twin 31	51.28	52.52	55.18	56.44	52.44

Table 3.8 Efficiency for a fluid ratio of 0.7.

Efficiency	Heat input (W)				
	55	**65**	**75**	**85**	**95**
Twin 28	88.98	85.44	82.83	80.84	79.27
Twin 29	77.09	75.37	74.113	73.149	72.38
Twin 30	65.20	65.31	65.39	65.45	65.50
Twin 31	53.30	55.24	56.66	57.75	58.61

for a fluid ratio of 0.75. The outlet temperature increases with increased heat input and water inlet temperature. Figure 3.16 gives the variation of water outlet temperature for various fluid ratios of the hybrid nanofluid. It grows with fluid ratio and heat input.

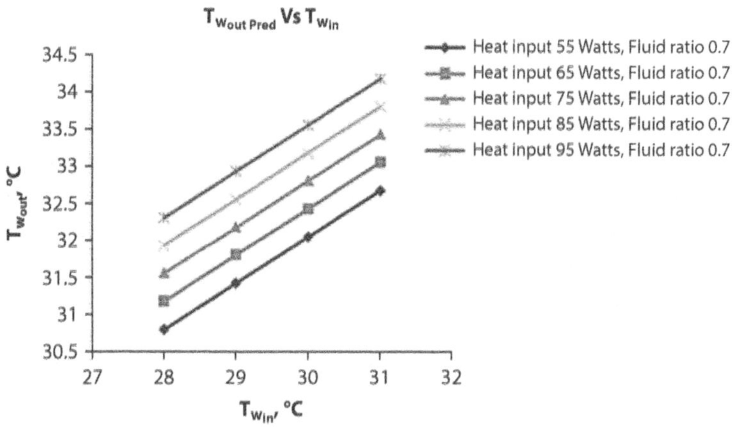

Figure 3.15 Variation of water outlet temperature for various heat inputs.

Table 3.9 Efficiency for a fluid ratio of 0.75.

	Heat input (W)				
Efficiency	**55**	**65**	**75**	**85**	**95**
Twin 28	93.03	88.86	85.80	83.46	81.61
Twin 29	81.14	78.79	77.08	75.76	74.73
Twin 30	69.24	68.73	68.35	68.07	67.84
Twin 31	57.35	58.66	59.63	60.37	60.95

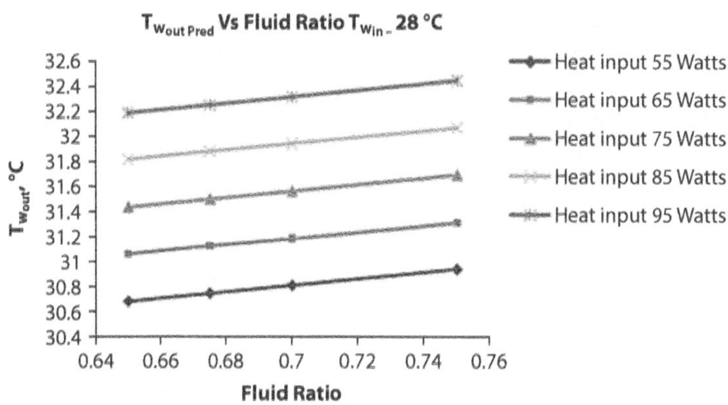

Figure 3.16 Variation of water outlet temperature for various fluid ratios.

3.5 Conclusion

This study uses the neural network analysis of heat pipes using hybrid nanofluids. The deep convolutional neural network (DCNN) was adopted for the examination. The experimental research reveals that while the surface temperature increases with an increase in input, it decreases gradually while moving away from the evaporator. Using the experimental data, a hybrid nanofluid generated a DCNN model for the heat pipe. The model created was tested against a set of experimental data. The DCNN model, with a precision of 0.991, predicts the outlet temperature of the heat pipe using a hybrid nanofluid consisting of Ag and Al_2O_3 for different inclinations of the heat pipe, flow rate, fluid ratio, and heat input. The predicted temperature has been used to estimate the efficiency of the heat pipe. The correlation coefficient of 0.991 obtained in the DCNN model clearly shows that the DCNN approach can be extended to predict the heat pipe performance. Heat pipe efficiency increases with a decrease in water inlet temperature and decreases with an increase in heat input. Furthermore, it has been evident from the analysis that the water output temperature rises as the fluid ratio increases, as does the temperature of the water inflow. The future directions of this study are to fine-tune the DCNN model algorithm using more data sets, compare the simulation results of other models, and implement this model in real time for renewable energy applications.

References

1. Xuan, Y. and Li, Q., Heat transfer enhancement of nanofluids. *Int. J. Heat Fluid Flow*, 21, 58–64, 2000.
2. Heris, S.Z., Esfahany, M.N., Etemad, S.G., Experimental investigation of convective heat transfer of Al2O3/water nanofluid in circular tube. *Int. J. Heat Fluid Flow*, 28, 203–210, 2007.
3. Ghanbarpour, M., Bitaraf Haghigi, E., Khodabandeh, R., Thermal properties and rheological behavior of water based Al_2O_3 nanofluid as a heat transfer fluid. *Exp. Thermal Fluid Sci.*, 53, 227–235, 2014.
4. Ramesh, C. *et al.*, Performance enhancement of selective layer coated on solar absorber panel with reflector for water heater by response surface method: A case study. *Case Stud. Thermal Eng.*, 36, 102093, 2022.
5. Wei, W.C., Tsai, S.H., Yang, S.Y., Kang, S.W., Effect of nanofluid on heat pipe thermal performance. *3rd IASME/WSEAS Int. Conf. on Heat Transfer, Thermal Engineering and Environment*, pp. 115–117, 2005.

6. Khimsuriya, Y.D. *et al.*, Artificially roughened solar air heating technology-A comprehensive review. *Appl. Thermal Eng.*, 214, 118817, 2022.
7. Sonan, R., Harmand, S., Pellé, J., Leger, D., Fakès, M., Transient thermal and hydrodynamic model of flat heat pipe for the cooling of electronics components. *Int. J. Heat Mass Transfer*, 51, 6006–6017, 2008.
8. Shafahi, M., Bianco, V., Vafai, K., Manca, O., An investigation of the thermal performance of cylindrical heat pipes using nanofluids. *Int. J. Heat Mass Transfer*, 53, 376–383, 2010.
9. Do, K.H., Ha, H.J., Jang, S.P., Thermal resistance of screen mesh wick heat pipes using the water-based Al2O3 nanofluids. *Int. J. Heat Mass Transfer*, 53, 5888–5894, 2010.
10. Nuim Labib, M., Nine, J., Afrianto, H., Chung, H., Jeong, H., Numerical investigation on effect of base fluids and hybrid nanofluid in forced convective heat transfer. *Int. J. Thermal Sci.*, 71, 163–171, 2013.
11. Madhesh, D., Parameshwaran, R., Kalaiselvam, S., Experimental investigation on convectiveheat transfer and rheological characteristics of Cu TiO2 hybrid nanofluids. *Exp. Thermal Fluid Sci.*, 52, 104–115, 2014.
12. Moghadassi, A., Ghomi, E., Parvizian, F., A numerical study of water based Al2O3 and Al2O3 Cu hybrid nanofluid effect on forced convective heat transfer. *Int. J. Thermal Sci.*, 92, 50–57, 2015.
13. Esfe, M.H., Akbar A, A., Rezaie, M., Yan, W.M., Karimipour, A., Experimental determination of thermal conductivity and dynamic viscosity of Ag MgO/water hybrid nanofluid. *Int. Commun. Heat Mass Transfer*, 66, 189–195, 2015.
14. Esfe, M.H., Wongwises, S., Naderi, A., Asadi, A., Safaei, M.R., Rostamian, H., Dahari, M., Karimipour, A., Thermal conductivity of Cu/TiO2-water/EG hybrid nanofluid: Experimental data and modeling using artificial neural network and correlation. *Int. Commun. Heat Mass Transfer*, 66, 100–104, 2015.
15. Huang, D., Wub, Z., Sunden, B., Effects of hybrid nanofluid mixture in plate heat exchangers. *Exp. Thermal Fluid Sci.*, 72, 190–196, 2016.
16. Afrand, M., Toghraie, D., Ruhani, B., Effects of temperature and nanoparticles concentration on rheological behavior of Fe3O4-Ag/EG hybrid nanofluid. An experimental study. *Exp. Thermal Fluid Sci.*, 77, 38–44, 2016.
17. Correa-Jullian, C., Cardemil, J.M., Droguett, E.L., Behzad, M., Assessment of deep learning techniques for prognosis of solar thermal systems. *Renew. Energy*, 145, 2178–2191, 2020.
18. Wei, H., Zhao, S., Rong, Q., Bao, H., Predicting the effective thermal conductivities of composite materials and porous media by machine learning methods. *Int. J. Heat Mass Transfer*, 127, 908–916, 2018.

19. Nasiri, A., Bao, J., Mccleeary, D., Louis, S.-Y.M., Huang, X., Hu, J., Online damage monitoring of SiCf-SiCm composite materials using acoustic emission and deep learning. *IEEE Access*, 7, 140534–140541, 2019.
20. Miandoab, A.R., Bagherzadeh, S.A., Isfahani, A.H., Numerical study of the effects of twisted-tape inserts on heat transfer parameters and pressure drop across a tube carrying graphene oxide nanofluid: An optimization by implementation of artificial neural network and genetic algorithm. *Eng. Anal. Boundary Elements*, 140, 1–1, 2022 Jul 1.
21. Esfe, M.H., Hajian, M., Toghraie, D., Rahmanian, A., Pirmoradian, M., Rostamian, H., Prediction the dynamic viscosity of MWCNT-Al2O3 (30: 70)/Oil 5W50 hybrid nano-lubricant using principal component analysis (PCA) with Artificial Neural Network (ANN). *Egyptian Inf. J.*, 23, 427–436, 2022 Apr 12.
22. Li, Y., Liu, Z., Wang, Y., Liu, T., Xie, Y., Real-time physical field reconstruction for nanofluids convection using deep Learning with auxiliary tasks. *Numer. Heat Transfer, Part A*, 18, 1–24, 2022 Jun.
23. Ullah, H., Shoaib, M., Akbar, A., Raja, M.A., Islam, S., Nisar, K.S., Neurocomputing for hall current and MHD effects on the flow of micro-polar nanofluid between two parallel rotating plates. *Arabian J. Sci. Eng.*, 24, 1–21, 2022.
24. Mesgarpour, M., Mir, M., Alizadeh, R., Abad, J.M., Borj, E.P., An evaluation of the thermal behaviour of a lithium-ion battery pack with a combination of pattern-based artificial neural networks (PBANN) and numerical simulation. *J. Energy Storage*, 47, 103920, 2022 Mar 1.
25. Kumararaja, K., Khiran Kumar, C.S., Sivaraman, B., A convolutional neural network analysis of a heat pipe with hybrid nanofluids. *Int. J. Ambient Energy*, 29, 1–3, 2021 Dec.
26. Modi, K.V., Patel, P.R., Patel, S.K., Applicability of mono-nanofluid and hybrid-nanofluid as a technique to improve the performance of solar still: A critical review. *J. Cleaner Prod.*, 386, 135875, 2023.
27. Yasmin, H., Giwa, S.O., Noor, S., Sharifpur, M., Experimental exploration of hybrid nanofluids as energy-efficient fluids in solar and thermal energy storage applications. *Nanomaterials*, 13, 2, 278, 2023.
28. Priyanka, Kumar, S., Kumar, A., Effect of mono/hybrid nanofluids and passive techniques on thermal performance of parabolic trough solar collector: A review. *Energy Sources, Part A*, 45, 1, 1686–1709, 2023.

29. Aghakhani, S., Pordanjani, A.H., Afrand, M., Farsani, A.K., Karimi, N., Sharifpur, M., Entropy generation and exergy analysis of Ag–MgO/ water hybrid nanofluid within a circular heatsink with different number of outputs. *Int. J. Thermal Sci.*, 184, 107891, 2023.

4

Nanosilver-Based Electrocatalytic Materials

Ahmed Mourtada Elseman* **and Sabah M. Abdelbasir**†

Central Metallurgical Research and Development Institute, Helwan, Cairo, Egypt

Abstract

Silver metal-based nanomaterials (SMNs) have sparked considerable interest due to their multifunctional and interactive properties, which allow for a wide range of catalytic applications. Combining experimental and theoretical research has developed various strategies for tuning electrocatalytic performance by controlling the nanoparticles' size, composition, morphology, and crystal structure to balance the performance and financial aspects appropriately. Despite significant progress, the chemical synthesis of SMNs with paragon catalytic efficiencies remains difficult. This chapter provides an overview of structure-based sorting and general SMN electrocatalyst synthesis. Furthermore, the current state of technological advancements, post-synthetic treatments to enhance activity, and the incorporation of SMNs into proper support for practical electrocatalytic applications are highlighted. Finally, future directions and challenges related to the electrocatalysis of SMNs are covered.

Keywords**:** Electrocatalysis, silver metal, electrocatalysts, synthesis, applications

4.1 Introduction

A new era in nanoscience and nanotechnology was launched in 1889 with the publication of M. C. Lea's key paper on producing silver nanoparticles (Ag NPs). Ag NPs typically range in size from 1 to 100 nm. These nanometric-sized Ag particles' versatility and unique characteristics fueled contemporary civilization's invention [1]. Maximum electrical and thermal

**Corresponding author*: AhmedMourtada5555@yahoo.com

†*Corresponding author*: sfoda20@hotmail.com

Sudheesh K. Shukla, Chaudhery Mustansar Hussain, Santanu Patra, Meenakshi Choudhary (eds.) Electrocatalytic Materials for Renewable Energy, (71–110) © 2024 Scrivener Publishing LLC

conductivity and excellent brightness are three different features of silver. Nanosilver has a wide range of academic applications and provides several critical benefits to health and agriculture; scientific interest is sparked by industries like electronics, robotics, the food industry, the textile business, and others [2–5]. As a result, a lot of effort has gone into designing and synthesizing Ag NPs in various sizes in addition to forms like rods, cubes, wires, prisms, tubes, and plates [6–13]. The chemical, optical, and electromagnetic characteristics of Ag NPs vary greatly depending on their shape and size [14, 15]. Ag NPs have been synthesized using various synthetic techniques and assemblies to suit a variety of purposes [16–20].

Furthermore, researchers have focused on developing more hybrid or complicated Ag-based nanomaterials (NMs), whose elastic properties enable significant applications in technology-driven domains [21–23]. It is worth noting that Ag NPs have aided in the improvement of catalytic processes in particular [24–27]. Because of their enormous surface area and abundance of coordinative unsaturated atoms (placed at various steps, kinks, or terraces), these nanosized particles have a distinctive proliferation that improves overall catalytic effectiveness by coordinating with substrates [28–30].

It is crucial to consider the reducing agents used to create Ag NPs in determining cytotoxicity [31, 32]. These outstanding characteristics of Ag NPs have prompted scientists to develop potent catalysts that may be used in industrial operations. In addition, the continual growth in the number of papers dedicated to Materials based on silver can be directly associated with the remarkable success in this domain. Nanosilver and nanosilver-based materials are now being studied in depth. Much of the discussion is based on the assumption that nanosilver is a versatile material with emerging uses in catalysis and biomedicine, thanks to developments in nanotechnology. According to Nowack and colleagues, nanosilver, which has been available as colloidal silver for over a century, is more effective than regular bulk silver [33]. Bigioni and colleagues released a paper detailing the synthesis, structure, and reactivity of Ag molecular NPs, their genesis, and history [34]. Kang *et al.* researched the techniques for stabilizing gold and silver nanoparticles (NPs) and their use in enhancing plasmonic characteristics [35]. In another review, He *et al.* examined the kinetic and thermodynamic perspectives for the creation and transition of silver sulfide nanoparticles and their cytotoxicity [36]. To synthesize Ag NMs, batch reactors are frequently employed; nevertheless, spatial variation in Ag NMs arises owing to incorrect mixing and differences in reaction conditions. He *et al.* examined the relevance of flow microreactors and the different synthesis procedures used to create Ag NMs under flow reaction conditions [37]. Several reviews on the synthesis techniques and uses of Ag NPs in antibacterial activity and catalysis have been published so far. Among these, few informative evaluations highlighted the advancement of silver-based NM manufacturing processes [38–40]. Cinelli and colleagues presented a classification model for Ag NM syntheses based on green chemistry concepts [41]. He *et al.* emphasized the chemical and biological syntheses of silver graphene-based nanocatalysts, as well as their potential as antibacterial agents, catalysts, and sensors in biomedicine [40]. Liao *et al.* provided an overview of the production techniques for Ag-based nanocomposites and their applications in catalysis, including oxidation and reduction processes and photo- and electrocatalysis [17].

In this chapter, light is shed on various synthesis approaches of Ag-based nanomaterials as well as their catalytic applications in electrocatalysis. We also briefly reviewed the favorable function of Ag NM-stimulated catalytic processes, concluding with future research directions in Ag NM engineering and catalysis. We are confident that the in-depth examination of the many synthetic procedures described here for the creation of supported Ag catalysts and their practical implementation in a number of applications will be of great interest to a vast readership.

4.2 Synthesis Methodologies of Silver-Based Nanomaterials

Silver-based nanomaterials are easy to produce and have a significant potential for usage due to their superior electric conductivity and catalytic activities in electrochemical devices. The ionic conductivity of silver is the greatest of any metal, measuring 6.301 10^7 S m^{-1} and 429 W m/K^{-1}, respectively. Electrical and thermal characteristics can be improved when silver is composited into active materials with a variety of morphologies [42, 43]. Due to its lower price, better availability, and stronger oxygen reduction reaction (ORR) activity than other catalysts, silver has drawn considerable interest as a potential replacement for the Pt catalyst in electrocatalysis [44]. Silver nanowires (Ag NWs) have received a lot of attention among SNMs because of their tiny high aspect ratio and diameter, which allows for efficient interaction with electrolytes and provokes quick electron transport in the process of electrochemical processes. Furthermore, because of the nanoscale size effect, they have high photon transmission and resistance to elastic bending [45]. Compared with silver particles as fillers, percolation channels may be easily constructed with a relatively small loading of Ag NWs.

As a result, many studies have been done on the properties and electrochemical uses of Ag NWs.

4.2.1 Physical Techniques

Evaporation/condensation, electric or gamma irradiation, laser ablation, and lithography are examples of physical methods for producing SMNs that do not use chemicals, such as polymers, redox reagents, and colloid stabilizer electrolytes [45]. The most crucial processes are evaporation/condensation and laser ablation of these procedures. The apparent lack of adulteration in the created thin films caused by solvents and the uniformity of particle dispersion are the benefits of physical techniques over chemical procedures [46, 47].

Atmospheric pressure physical synthesis of SNMs has significant limitations. The tube furnace, for instance, requires a lot of room, uses a lot of energy to raise the temperature surrounding the source material, and takes a while to reach thermal stability. Additionally, a typical tube furnace takes several kilowatts of electricity and many tens of minutes for preheating before it can attain a consistent working temperature [48, 49]. It was demonstrated that a tiny ceramic heater with a confined heating region might be used to create silver nanoparticles [50]. To evaporate the source materials, a small ceramic heater was used. It can cool at an appropriate rate due to the temperature gradient in the region of the evaporated vapor. This allows for the creation of tiny NPs at high concentrations. The design of particles is comparatively steady since the heater's surface temperature does not fluctuate over time. This physical method can be utilized as a calibration device for nanoparticle measuring equipment as well as a nanoparticle producer for studies on long-term inhalation toxicity. It can cool at an appropriate rate due to the temperature gradient in the region of the evaporated vapor. This allows for the creation of tiny NPs at high concentrations. Because the heater surface temperature does not change over time, particle production is relatively stable. The aforementioned method could be used to create nanoparticles for long-term inhalation toxicological research as well as a calibration tool for nanoparticle measuring devices [50]. The results revealed that when the surface temperature of the heater rises, the geometric characteristics and total concentration of NPs increase. Even at high concentrations and high heater surface temperatures, spherical non-agglomerated NPs were found. The diameter of silver NP ranged from 6.2 to 21.5 nm, with a standard variation of 1.23–1.88 nm. Laser ablation might produce SNMs from metallic bulk materials [51–53].

The ablation efficiency and characteristics of produced nanosilver particles are affected by the wavelength of the laser impinging on the metallic target, the length of the laser pulses, the laser fluence, the duration of the ablation time, and the effective liquid medium, with or without surfactants [54, 55]. The absence of chemical reagents in solutions significantly benefits laser ablation over other methods for producing metal colloids. As a result, this process may be used to create metal colloids that are clean and pure for future applications [56]. Femtosecond laser pulses synthesized silver nanospheroids (ranging from 20 to 50 nm) at 800 nm in water [57]. Colloidal particles generated with nanosecond laser pulses were compared with particle creation efficiency and size. As a result, manufacturing femtosecond pulses was substantially less efficient than producing nanosecond pulses. When compared with nanosecond pulses, femtosecond pulses had smaller, less dispersed colloids. Additionally, femtosecond ablation efficiency in water was lower than in air, although nanosecond ablation efficiency was the same in both environments. Table 4.1 shows various physical techniques used for synthesis of SNM.

Table 4.1 SNM synthesis using physical techniques.

Technique	Morphology and size	Conditions	Ref.
Laser ablation	Spherical shape, 12–31 nm	A 532–1,064-nm laser with a relatively high fluence of 36 J/cm^2 is used.	[56, 58]
	Irregular shape, 15–20 nm	A 750-m wire diameter laser ablation efficiency on a small silver wire in a water flow reactor.	[59]
	Spherical shape, 5–13 nm	A pulsed Nd:YAG laser operating at 1,064 nm was used for laser ablation.	[59]
	Spherical shape, 27–120 nm	Nd:YAG laser 1,064 nm, 532 nm for laser ablation, and fragmentation in liquid monomer isodecyl acrylate (IDA).	[60]

(Continued)

Table 4.1 SNM synthesis using physical techniques. (*Continued*)

Technique	Morphology and size	Conditions	Ref.
	Irregular shape, 27–41 nm	Laser ablation in water using 800 nm femtosecond laser pulses.	[57]
	Spherical particles, 5–50	Laser ablation with a pulsed Nd:YAG laser operating at 1,064 nm for 5 min at 10 Hz.	[61]
	Spherical particles, 50 nm	Laser source of ns diode-pumped Nd:YVO_4 laser system at 532 nm with a pulse duration of 50 ns.	[62]
γ-Irradiation	Spherical particles, 9–15 nm	γ-Irradiation (from 2 to 50 kGy) of aqueous solutions containing $AgNO_3$ and gelatin as a silver source and stabilizer.	[63]
Thermal decomposition	Spherical particles, 4–7 nm	Silver oxalate thermal degradation in water and ethylene glycol at 140°C.	[64]
	Spherical particles, 4.7 nm	Decomposition of a silver–fatty acid complex at 250°C in a nitrogen atmosphere.	[65]
	Spherical particles, 8.0 ± 1.3 nm	Thermal decomposition of the prepared Ag–oleate complex at 300°C in an autoclave for almost 4 h.	[66]
	Spherical particles, 40–50 nm	Pyrrole was thermally polymerized with silver nitrate before being converted to polypyrrole (PPy)/Ag nanocomposites without using any external oxidizing agent or solvent.	[67]

(*Continued*)

Table 4.1 SNM synthesis using physical techniques. (*Continued*)

Technique	Morphology and size	Conditions	Ref.
	Spherical particles, 9.5 ± 0.7 nm	Ethylene glycol (EG) reduced $AgNO_3$ in the presence of Pt seeds and polyvinylpyrrolidone with a long chain length.	[68]
	Spherical shape 3.1 ± 0.7 to 4.5 ± 0.8 nm	Microwave irradiation produced polypyrrole (PPy)/Ag nanocomposites based on interfacial polymerization at a water/chloroform interface.	[69]
	Spherical shape, 24–132 nm	Silver nitrate aqueous solution and trisodium citrate were microwave-irradiated using formaldehyde as a reducing agent.	[70]
Small ceramic heater	Agglomerated spherical shape, 6–21.5 nm	A tiny ceramic heater is used in the synthesis to provide a limited heating area where source metals can be evaporated. At a small heating area of 5 × 10 mm^2, the heater may achieve 1,500°C in around 10 s.	[50]

4.2.2 Chemical Techniques

Depending on the preparation procedure, multiple chemical approaches are used to synthesize SNMs of varied shapes and sizes (Figure 4.1). Chemical diminution (using a range of organic or inorganic agents), sonochemical procedures, microwave (MW)-assisted synthesis, radiolysis, and physical–chemical reduction are the most frequent of these approaches. The bulk of these techniques is still in the early phases of research, with NP stability

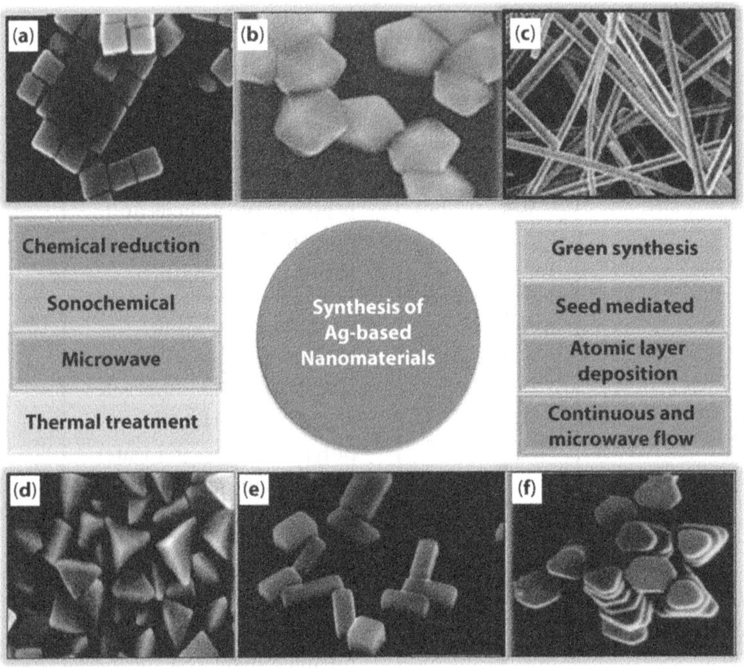

Figure 4.1 Primary synthesis techniques for producing SNMs. Reproduced from ref. [60].

and aggregation, crystal growth control, form, size, and size distribution being the most frequent challenges. Furthermore, the purification and separation of the generated NPs for future use are still significant problems [58, 59].

4.2.2.1 *Chemical Reduction Technique*

One of the most popular processes for creating nanostructured Ag is chemical reduction. The precursor Ag salt, reducing agents, and stabilizing or capping agents are the three essential elements that govern or impact the synthesis of NPs. The most significant silver salt is $AgNO_3$, which may be converted to silver nanoparticles (Ag NPs) by reducing it with a variety of reductants, including sodium borohydride ($NaBH_4$), sodium citrate, ascorbic acid, hydrazines, alcohols, hydroquinone, dimethylformamide (DMF), glucose, and many more [61–64]. $NaBH_4$, a powerful reductant, produces smaller NPs. To control the NP growth and form the looked-for morphologies, appropriate stabilizers or capping agents such as polymers, ligands, or surfactants such as polyvinylpyrrolidone (PVP), polymethacrylic acid (PMAA), polyethylene glycol (PEG), collagen, and sodium oleate are added (Figure 4.2) [65, 66].

Several techniques for the synthesis of SNMs are being developed daily. These techniques are mainly classified into two categories: physical

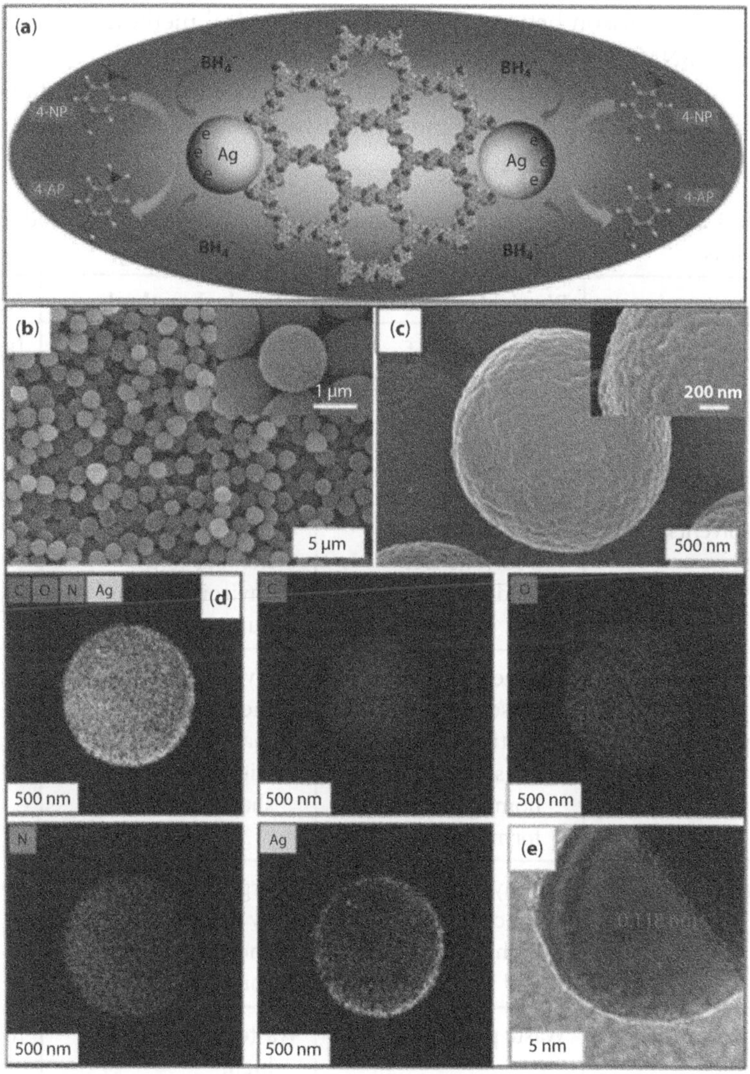

Figure 4.2 (a) a schematic of the 4-NP reduction with $NaBH_4$ in the Ag NPs@SCOF microspheres. (b) and (c) SEM images of the SCOF and Ag NPs@SCOF microspheres. (d) STEM-EDX elemental mapping images for C, O, N, and Ag. (e) HR-TEM image taken at SCOF of Ag nanoparticles. Reproduced from ref. [60] with permission.

methods and chemical methods. Capping and reducing agents are critical in the synthesis of nanoparticles. Harmful and highly poisonous chemicals are used in the chemical and physical processes of nanoparticle synthesis, causing environmental defects [67]. Chemical and physical methods use expensive reducing or capping agents. Chemicals developed biologically are used in biological ways that are not detrimental to our environment [68]. A comparison between physical and chemical methods for the synthesis of SNMs is revealed in Table 4.2.

The citrate reduction process, in which citric ions function as a stabilizer and a reductant, has been the most extensively used method for reducing

Table 4.2 Comparison of the physical and chemical strategies for the production of SNMs.

Aspect	Physical methods	Chemical methods	Ref.
Production cost	Several types of equipment required for physical methods are costly.	Reliant on the procedure used, the cost of fabrication may be high.	[85]
Chemicals or solvents used	Typically, the reaction takes place in a solvent-free medium.	Some chemicals and solvents used are noxious to the environment and human health.	[86]
Nanomaterial yield and properties	The yield is reasonable, and the particles produced are stable and homogeneous in form and size.	The yield is high, and the synthesized materials are stable with well-defined shapes and sizes.	[83]
Energy required	All physical synthesis procedures need high temperatures and pressures.	Energy is determined by the reactants used and the synthesis conditions.	[84]
Waste produced	Waste produced is low.	Waste produced is high.	[87]
Possibility of large-scale production	It can be used to produce SNMs on a large scale.	It can be used to produce SNMs on a large scale.	[88]

$AgNO_3$ to colloidal Ag [69, 70, 89, 90]. Furthermore, this synthetic technique yields Ag NPs of various forms and sizes. Since then, a variety of artificial styles utilizing hydrogen, ascorbic acid, and dextrose as reductants for the synthesis of SNMs have been reported [71, 72]. The blend of two reductants, sodium citrate and tannic acid, was used to synthesize monodispersed particles with a size ranging from 10 to 200 nm [73]. This approach also yielded Ag NPs with a limited range of measures that can be modified and functionalized to be employed in biomedicine and catalysis [74, 75].

4.2.2.2 Sonochemical Synthesis

Sonochemical, also known as an ultrasound-mediated synthesis of SNMs, is a unique technique that has lately piqued scientists' interest. This method employs strong, powerful, and safe ultrasound waves (20 kHz to 10 MHz) to fabricate various nanomaterials with varying morphologies. Sonochemistry emerges due to ultrasound's severe transitory circumstances, which create custom hot spots that can generate regional pressures and temperatures of 1,000 atmospheres and about 5000 K, respectively [76]. A sonochemical operation has no direct contact between chemical species and ultrasonic waves. When a liquid is subjected to intensified ultrasound, alternating expanding and compressive acoustic waves are produced, causing the density of the liquid layer to fluctuate [76]. Sonochemical production of Ag NPs has made significant progress, yielding a range of morphologies, nanorods, nanotriangles, nanowires, and fascinating fluorescence characteristics. A straightforward preparative approach was described for manufacturing blue-emitting Ag nanoclusters by combining glutathione and ultrasonic irradiations as stabilizing agents [77]. Suslick and colleagues used a sonochemical technique to create nanosized Ag clusters with polymethacrylic acid (PMAA) as a stabilizer; they were discovered to be fluorescent and highly stable, even when exposed to water [78].

4.2.2.3 Microwave-Aided Synthesis

Fabricating Ag-nanostructured materials using microwave irradiations is a promising strategy that has piqued academic and industrial research [79, 80]. In 1986, Gedye *et al.* aimed to develop organic compounds using microwave (MW) irradiation, and it has since been commonly employed for NP fabrication [81]. The MW chemistry is based on the interaction of irradiation with molecules present in reaction media. It is based on a dielectric heating mechanism that involves dipolar polarization and ionic conduction processes. MW irradiation heats materials with moveable electrical charges, such as polar molecules or ions. Polar molecules are activated by MW irradiation because of their inherent bipolarity. The fluctuating electromagnetic field also causes the polar molecules and ions to realign themselves with the electric field. Rotation, molecular collisions, molecular friction, and dielectric loss are all brought on by the reorientation process, which results in a rapid internal heating event. Ions bounce back and forth across the solution under the applied electrical field, causing endless variations due to friction and collision and increasing total temperature [82]. The MW chemistry is based on a dielectric heating mechanism that includes the processes of dipolar polarization and ionic conduction. It is based on the interaction of irradiation with molecules present in reaction media. Polar molecules or ions or other materials with moveable electrical charges are warmed by MW irradiation. Polar molecules are activated by MW irradiation because they are naturally bipolar.

The pulsating electromagnetic field simultaneously realigns the polar molecules and ions with the electric field. As a result of rotation, molecular collisions, molecular friction, and dielectric loss brought on by the reorientation process, a rapid internal heating event results. As an electrical field is produced, ions bounce back and forth across the solution, causing constant variations due to friction and collision, raising the entire system's temperature [83]. Silver nanomaterials have also been created from silver nitrate with the help of Pt seeds, ethylene glycol, and polyvinyl pyrrolidine as a stabilizer by applying MW radiation at temperatures of 100°C to 200°C [84]. When polyvinyl pyrrolidone was used, the diameter of the NPs ranged from 62 to 78 nm. Furthermore, MV heating and a silver salt dissolved in oil could create Fe–Ag bimetallic NPs [85]. The produced spherical Ag NPs were 30 nm in size.

4.3 Electrocatalysis

Excessive reliance on non-renewable energy sources, particularly fossil fuels, has led to resource depletion and a devastating energy crisis. Designing and developing renewable energy sources in order to avoid this calamity and meet the world's energy needs for a sustainable environment, harnessing the power of the sun, wind, and tides has become essential [86]. Many transition metal-based nanocatalysts have so far demonstrated exceptional catalytic activity in the electroreduction of CO_2, oxygen, nitrogen, hydrogen, and methanol [87, 88].

Due to the competitive co-evolution of hydrogen, these catalysts displayed low faradaic efficiency (FE), substantial overpotential, and poor product selectivity during the electrochemical conversion [89]. Ag NMs have various physicochemical features, including customizable size and shape, electrical behavior, long-term operating stability, and minimal toxicity. They are less costly than Pd, Pt, and Au, which has boosted their utility for electrocatalysis and other applications [17, 90, 91]. Ag has a strong tolerance for methanol poisoning and membrane damage, as well as a similar mechanism and kinetics for electrochemical conversions to Pt deterioration, i.e., short hydroxide ion to hydrogen peroxide decomposition [92]. As a result, Ag NMs are widely used as an energy source for the electrochemical conversion of fuels and chemicals. Controlling the shape, size, and content of these NMs may readily alter their electrocatalytic effectiveness [93]. Compared with aggregates, distributed Ag NPs offer more catalytically active sites [94]. The alcohol oxidation process (MOR), the ethanol oxidation reaction, the oxygen evolution reaction (OER), the oxygen reduction reaction (ORR), the carbon dioxide reduction reaction (CO_2RR), the nitrogen reduction reaction (NRR), and the hydrogen evolution reaction (HER) have all been investigated for electrocatalytic applications over the last two decades (Figure 4.3) [95–97].

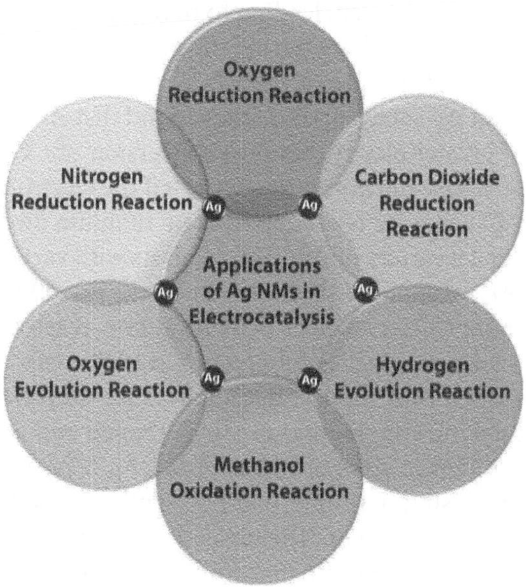

Figure 4.3 The electrocatalytic uses of silver nanocatalysts are depicted schematically.

4.3.1 CO_2 Reduction Reaction

Growing industrialization, urbanization, and global warming pose serious environmental threats. Since CO_2 has been recognized as one of the leading causes of the greenhouse effect, its sequestration is necessary. Therefore, as a potential energy source in the future, researchers are focusing on the catalytic conversion of CO_2 to liquid fuels and chemical feedstock [98, 99]. One of the burgeoning topics is renewable power for the electrochemical conversion of CO_2 to C_1/C_2 compounds such as carbon monoxide, formic acid, methanol, methane, ethanol, and ethylene [100, 101]. Silver is an attractive electrocatalyst because of its potential to catalyze the electrochemical conversion of CO_2 to carbon monoxide at ambient temperature [102]. Shao and associates performed X-ray absorption spectroscopy in the immediate operand to trace the composition and structural evolution of an Ag_2O precatalyst for CO_2 electrochemical reduction (XAS) [103]. The Ag-D (Ag nanocrystals with rich defects) catalyst demonstrated around 100% FE for CO_2 to CO conversion with negligible deteriorating efficiency after 120 h in an H-cell reactor. As pioneered by Hori *et al.*, silver has been acknowledged as the best metal for CO production via electrocatalytic CO_2RR (EC-CO_2RR) in an aqueous medium, with 81.5% FE at 1.14 V [104].

Since Ag NMs with large surface areas provide many active sites and well-defined morphologies, electrocatalytic efficiency has increased to the most significant level ever seen, leading to remarkable advancement in the field of nano-Ag electrocatalysts during the past 20 years [105]. Masel and colleagues showed that the size of Ag NPs directly impacts the electrocatalytic efficacy of an Ag catalyst [106]. They found that Ag NPs (5 nm) promoted a 10-fold greater electrochemical conversion of CO_2 to CO than a bulk Ag catalyst. Kenis and colleagues reported the electrocatalytic reduction of CO_2 to CO using Ag NPs supported on titanium dioxide (Ag/TiO_2) [107]. They created a variety of Ag/TiO_2 catalysts with varying Ag loadings on TiO_2 and 40% Ag supported on carbon black. The 40% Ag/TiO_2 catalyst produced more CO in an electrochemical flow reactor [108] through EC-CO_2RR. Various approaches for creating Ag NPs with decreased sizes have been developed over the last decade because they play a significant role in electrocatalytic CO_2 reduction [109]. Hwang and coworkers [110] generated three different sizes of immobilized Ag NPs spread on carbon (nanosized Ag/C) using a simple one-pot synthesis method using cysteamine as an anchoring agent. The created nanosized Ag/C revealed an excellent potential for CO production from electrochemical CO_2RR in an aqueous electrolyte. Among the samples, the 5-nm Ag particles exhibited a high FE of 84.4%, long-term operational stability for electrochemical CO_2 reduction, and excellent electrocatalytic activity. Compared with Ag foil, nanosized Ag/C (5 nm) demonstrated less overpotential, or 300 mV, at a current density of 1 mA/cm^2. The anchoring substance, cysteamine, caused the Ag–S interaction. According to computer simulations, the Ag–S contacts that formed following cysteamine anchoring produced surface localization of the unpaired electrons, which further stabilized the intermediates and contributed to the high catalytic efficiency of the electrocatalytic CO_2 reduction process.

During CO_2 electrolysis, a sizable quantity of overpotential is needed to prevent the hydrogen evolution side reaction [111]. To stop hydrogen co-evolution, Hsieh *et al.* created coral-like Ag nanostructures with broad surfaces for selective CO synthesis from electrochemical CO_2 reduction at the right overpotential [112]. An Ag nanocoral-based electrocatalyst was designed using the oxidation–reduction process in the presence of aqueous chloride anions, and it demonstrated a current efficiency of 95% at 0.37 V of overpotential. This coral-like Ag nanostructure showed a 32-fold

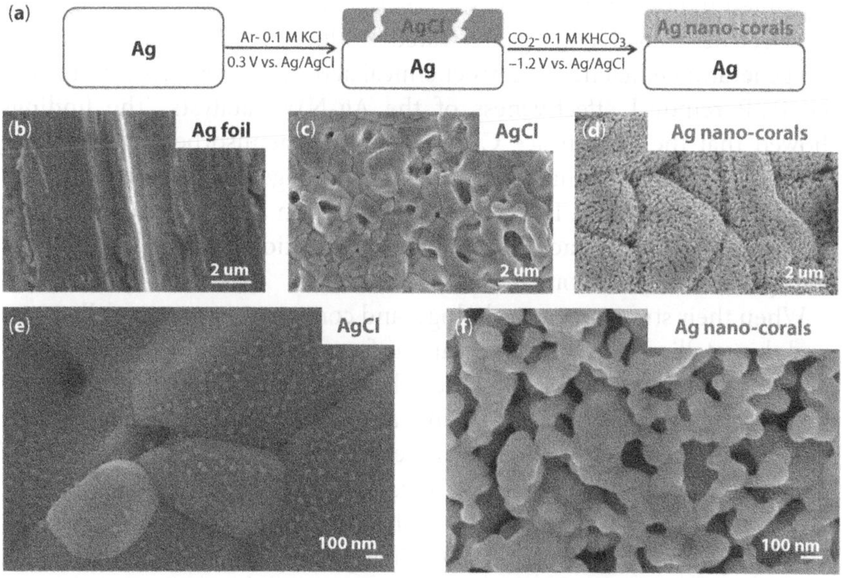

Figure 4.4 (a) Schematic representation of the production of AgCl and Ag nanocorals in their prepared state. SEM pictures of (b) untreated silver foil, (c) produced silver chloride after a 12-h oxidation period, and (d) silver nanocorals. SEM pictures at high magnification of (e) the as-prepared AgCl after a 12-h oxidation period and (f) Ag nanocorals. Reproduced from ref. [112].

higher catalytic efficiency than its Ag metallic foil, based on a normalized surface area and a 0.49-V overpotential. Ag nanocorals were shown to have a higher electrocatalytic activity when chloride ions were adsorbed on the catalyst's surface, creating a greater surface area (Figure 4.4). The adsorbed chloride anions reduced hydrogen generation while enhancing CO_2 electroreduction [112].

Because of their outstanding conductivity and film formability, silver nanowires (Ag-NWs) produced in a polyol solution were used to create an Ag NW-based electrode using a straightforward drop-casting method, increasing the electrocatalyst's surface area [113]. With a low working potential of 0.9 V vs. a reversible hydrogen electrode, an Ag NW-based electrode was used to electrochemically reduce CO_2 to CO (FE = 80%) and formic acid (FE = 2.6%–5.6%) (vs. RHE). Broekmann and coworkers described Ag-NWs for synthesizing the electrical reduction of CO of CO_2 on a similar premise. A PVP-assisted polyol synthesis process was used [114]. For Ag-NWs, PVP acts as a capping agent. After that, the Ag-NWs were coated with glassy and highly porous carbon support, significantly enhancing the electrocatalytic performance of the naked Ag NWs. The three-dimensional and loosely packed Ag NWs were evenly dispersed and implanted in both supports, as seen by the SEM images (Figures 4.5a–d). The electrochemical looping-pretreated produced Ag-NWs showed a 94% CO generation efficiency. Electrochemical activation treatments improved the PVP removal effectiveness of the Ag-NW catalysts. The findings showed that the EC surface CO_2RR without chemisorbed chloride and PVP performed better than the latter, which changed the selectivity toward the HER (Figure 4.5e). The CO_2RR can function more electrochemically by having high-index facets or active edge locations responsible for CO_2 adsorption and activation.

When their structure, morphology, and composition are carefully managed, bimetallic electrocatalysts can perform better than their monometallic counterparts in operation and selectivity [17, 115, 116]. To improve the performance of the electrochemical CO_2RR, a second metal atom is included in the outer layer. This increases the activity and selectivity of bimetallic nanocomposites by significantly expanding the outer layer. A structure-controlled model for the polyol synthesis of bimetallic Ag@Cu NPs was put out in this respect by Chang *et al.* For the electrocatalytic reduction of CO_2, a binary Ag@Cu nanocomposite was used. An inner Ag core and an outer Cu shell were present on the surface of Ag@Cu NPs, as described by the architectonic characterization. Period-dependent bimetallic Ag/Cu NP samples such as Ag@Cu-5, Ag@Cu-7, Ag@Cu-10, Ag@Cu-15, Ag@Cu-20, and Ag@Cu-25 were created by altering the heating

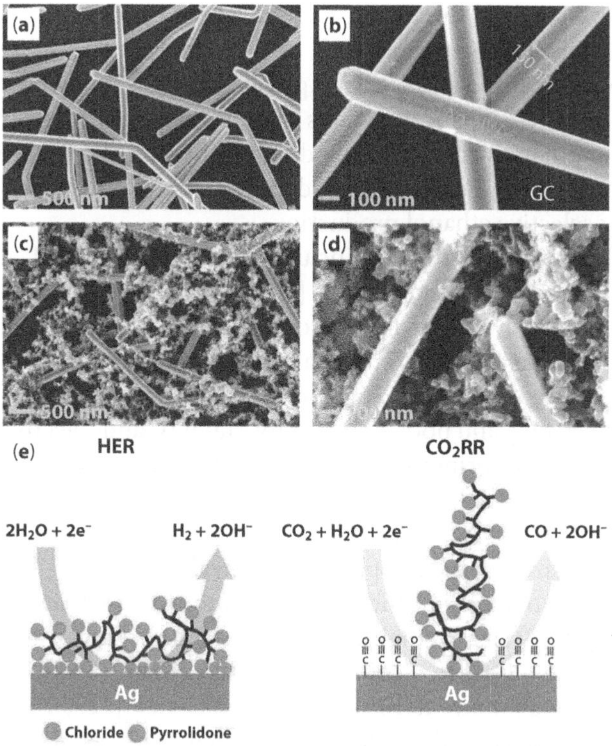

Figure 4.5 Images captured using a scanning electron microscope (SEM) depict Ag-NWs supported with a glassy carbon (GC) electrode (a and b), Ag-NWs reinforced with carbon (c and d), and schematics of the PVP and Cl terminated Ag surface (e). Reproduced from ref. [114] with permission.

time from 0 to 25 min. Ag@Cu-7 and Ag@Cu-20 samples had greater FEs toward CO and hydrocarbon production or ethylene formation. The Ag-dominant Ag@Cu-7 NPs and the Cu-dominant Ag@Cu-20 NPs significantly reduced hydrogen evolution. At a potential of –0.5 V, they boosted CO_2RR activity. Due to the "dilution" effects between Ag and Cu, the selectivity of the Ag-based electrocatalyst for CO generation was more excellent. Still, adding Cu altered the selectivity in favor of the hydrocarbon. The component mixing patterns, which are critical in defining the activity and selectivity of each catalyst, are primarily responsible for the synergistic or diluting impact of Cu and Ag in these Ag@Cu bimetallic electrocatalysts. Ethylene is produced by the electrochemical reduction of CO_2 over an Ag–Cu bimetallic catalyst, according to Wang *et al.* [117]. The nanocomposite's Ag and Cu components had a priority, making it easier to synthesize ethylene selectively. For ethylene formation, they reached an FE of 42%, which was twice as high as the pure Cu catalyst. Yeo and colleagues found that the surfaces of Ag NPs and the Cu–Ag boundaries on the surface of Cu–Ag nanostructures were in charge of the electrochemical synthesis of ethanol from CO_2 in a different investigation [118]. The easy availability of CO at the Cu^-+Ag borders aided the C–C interaction on Cu, permitting selective ethanol production.

Other NMs with an Ag base include AgIn neurons [119], Ag with Cu dots [120], an AgCo surface alloy [121], Ag-adorned copper [122], an active Cu/Ag catalyst with bifunctionality [123], Cu–Ag shell-core NPs [124], a catalyst made of mixed metal oxides, or $Ag_2Cu_2O_3$ [125] and Ag@ZnO@rGO electrocatalysts [126], and Ag NPs can be combined with other elements to change and boost the activity, stability, and selectivity of active catalysts while also bringing down the total cost. This has been demonstrated in the approaches discussed above. Adjusting and improving the catalytic activities of Ag-based nanocomposites is equally as important as the materials' size, shape, and physicochemical properties.

Ag-based nanocatalyst studies have demonstrated that reducing the nanoparticle (NP) size to single atoms (below 5 nm), known as single-atom catalysts (SACs), leads to a considerable increase in surface area, thereby boosting the maximum atom utilization efficiency [127, 128]. However, few atomically dispersed Ag-supported metal oxide catalysts have been found due to the limited stability of Ag SACs [129–131].

Ethanol production requires a high overpotential and several proton–electron exchanges to convert CO_2 to C^{2+} compounds such as ethanol. A fuel cell uses ethanol as a fuel for transportation and a substitute energy source, making it more valuable to the industry. For the first time, ethanol was produced from CO_2 electroreduction utilizing a low-cost catalyst: Ag supported on N-doped carbon foam with integrated graphene, according to Zhang and coworkers' findings (Ag-G-NCF) [132]. It outperformed other catalysts, including NCF, G-NCF, Ag NCF, and Ag-G-NCF, converting CO_2 to ethanol with a maximum yield of 85.2% of FE in ethanol at overpotentials as low as 0.62 V (vs. RHE). The Ag NPs converted CO* to the intermediate of the OC–COH ethanol. It was discovered that the pyridinic N atoms of the Ag-G-NCF catalyst were more able to connect to CO* intermediates than other N species. Due to the hierarchical porous structure and strong electrical conductivity, the high N content in the pyridinic architecture and the Ag NPs increased the electrocatalytic activity of CO_2RR to ethanol. Wang *et al.* created CuAg electrocatalysts for acetaldehyde synthesis by electrochemical CO reduction and EC-CO_2RR [133]. Several Ag-based NMs, including TiO_2-supported Ag NPs, have shown good promise for the synthesis of liquid fuels and hydrocarbons [107], (Cu)m(Ag)n NPs [134], porous sponge-like Ag nanocubes [135], MOFs encased in Ag NPs [136], MnO_2/Ag(111) nanocatalysts [137], and Ag/tert-GO heterogeneous catalysts [138].

4.3.2 Oxygen Evolution Reaction (OER)

The four-electron/four-proton transport channel is used in the anodic process known as the OER during the electrochemical water oxidation reaction [139]. Like metal–air batteries and alkaline fuel cells, the OER is essential for electrochemical conversion and energy storage [140]. A number of noble and non-noble metal-based electrocatalysts have been described for the OER [141, 142]. Ag NMs showed good electrocatalytic performance toward the electrochemical OER due to their high activity, greater tolerance, long-term operational stability, and extended performance. Because combining two or more different materials has a synergistic impact, Ag nanocomposite catalysts have become a feasible option for electrocatalytic OER. An Ag@CoxP core@shell nanoelectrocatalyst for OER was created by Hou *et al.* [143]. Core–shell nanostructures of Ag@Co_xP were produced by growing phosphides from transitional metals on an Ag seed *in situ*, which included manufacturing Ag seeds, developing the Co shell, and phosphating. These nanoparticles (NPs) were found to have a round shape, with the metallic Ag in the core having lattice fringes on (111) facets on TEM and HRTEM images. Significant electrical interactions exist between the Ag core and the Co_xP shell because of the peculiar structure of the Ag core. This nanostructured core–shell Ag@Co_xP electrocatalyst displayed important electrocatalytic OER performance and long-term catalytic stability at a current density of 10 mA cm^{-2} and a moderate overpotential of 310 mV. The catalytic activity of Ag@Co_xP nanocomposites was eight times larger than that of pure Co_2P NPs. According to the study of Acharya and colleagues [144], multifunctional Ag NPs demonstrated decreased power density, improved activity, and durability for the OER under environmentally friendly circumstances when supported by reduced graphene oxide (rGO)-coated electrocatalysts (AgNP@rGO/MIL-88B(Fe)). At a potential of 395 mV and a current density of 1 mA cm^{-2} at pH neutral, the Ag NPs@rGO/MIL-88B(Fe) hybrid increased the OER activity of the glassy carbon (GC) electrode. The Ag catalyst's 30-day operating stability was just 2.6% compared with the original run. The electrocatalyst's conductivity gradually improved OER's performance. $Co(OH)_2$ nanosheets containing Ag nanoparticle inclusions were made by OER researchers using a particular reduction–oxidation method [145]. As illustrated in Figure 4.6, it was

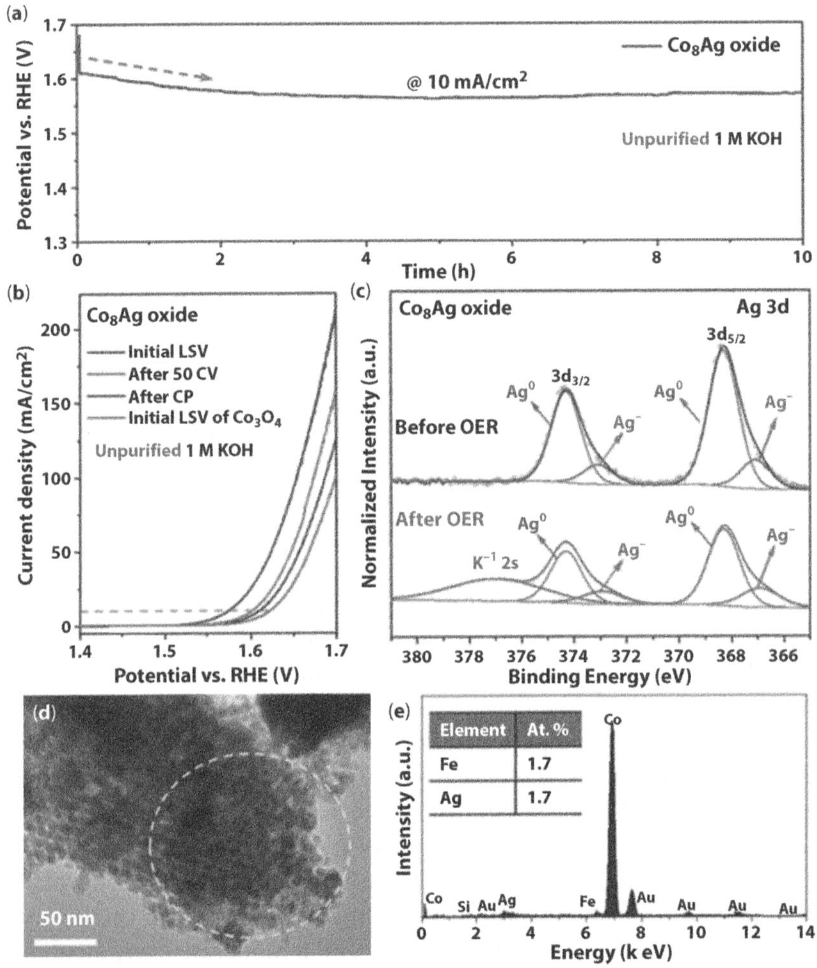

Figure 4.6 (a) Chronopotentiometric (CP) curve at a current density of 10 mA cm^{-2}. (b) Initial LSV curves after 50 CV and CP measurements of Co_8Ag oxide. (c) XPS spectra of Ag 3d on Co_8Ag oxide/carbon fiber before and after CP test. (d) TEM image of Co_8Ag oxide after CP measurement. (e) Local EDX analysis according to the yellow circle in (d). Reproduced from ref. [146].

discovered that the Ag NPs were uniformly dispersed throughout the $Co(OH)_2$ nanosheets. The electrical conductivity of $Co(OH)_2$ nanosheets was dramatically improved by doping Ag into the nanosheets. Ag was applied to the $Co(OH)_2$ nanosheets in loadings ranging from 0% to 20%. Using linear sweep voltammetry (LSV) with a scan rate of 5 mV s^{-1}, the OER performance of a 10% Ag-decorated $Co(OH)_2$ nanocomposite was examined.

That is what these findings taught us, and it is essential to note that the overpotential shrank when Ag was introduced. A 270-mV overpotential was visible in Ag-decorated $Co(OH)_2$ nanosheets at a current density of 10 mA/cm^2, a result of their high electrical conductivity and active catalytic surfaces. Yu *et al.* also developed a mesoporous Ag–Co oxide nanocomposite

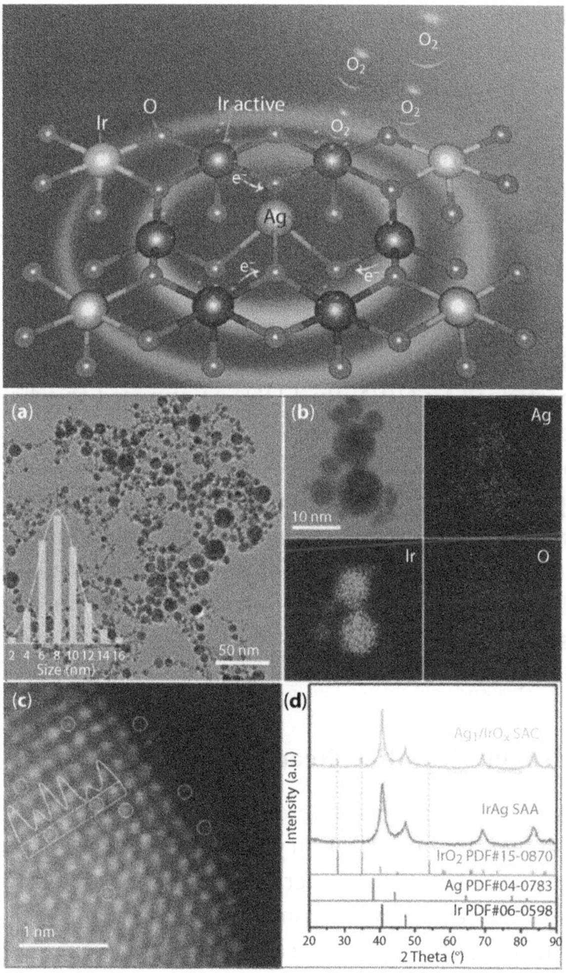

Figure 4.7 The Ag1/IrO_x SAC for OER electrolyte performance. Scan rate 5 mV/s for LSV polarization curves (iR-corrected). For example, consider the following: (a) geometric electrode area normalized current density (Jgeo) is represented by this value. (b) The polarization curves are visible in the panel and are plotted on a table. (c) Detailed actions (j represents the current density normalized by ECSA). The excess potential is 10 mA cm^2 ECSA, as shown in (d) (without carbon black). Reproduced from ref. [150].

for the OER [146]. OER performance was more than quadrupled when Ag was applied to the Co_3O_4 matrix. Co_8Ag oxide was synthesized by mixing Co_3O_4 (88%) with metallic Ag (12%). Nanocasting, a hard template procedure, was used to introduce In-mesostructured Co_3O_4, and metallic Ag NPs and Ag_2O clusters are present.

Zhang *et al.* recently presented an innovative method for dispersing Ag NP clusters across metal oxide [147]. To manufacture Ag NPs assisted by $Co(OH)_2$ nanohybrid ($Ag@Co(OH)_x$), a spontaneous redox reaction was employed at ambient temperature. According to experimental and theoretical analyses, a robust electronic interaction between Ag and $Co(OH)_2$ resulted in the activation of the OER catalytically active sites, such as Ag clusters, which increased the binding energies of the reaction intermediates. Bifunctional electrocatalysts are made of Ag/Co_3O_4 [148]. The OER and ORR of electrochemistry have been described. According to an electrochemical investigation, the synergistic impact of Ag and Co on catalytic activity was identified by accelerating the reduction process and breaking the O–O bond in molecular oxygen more quickly. They utilized nickel–ruthenium-layered double hydroxide nanosheets (NiRu-LDH) coated with silver nanoparticles (Ag NP/NiRu-LDH) as opposed to what Chala *et al.* used for the OER and ORR under electrochemical circumstances [149]. These approaches significantly increased catalytic performance by altering the surface properties and improving the electrical conductivity of metal oxide nanostructures. Recently, Zhang *et al.* presented for electrochemical OER a single-atom catalyst (SAC) contained in an IrO_2 matrix with single Ag atoms (Figure 4.7). With a long-term business plan in place, Zhang *et al.* [150] showed a low overpotential of 224 mV at 10 mA cm^{-2} current density using Ag-based SACs.

4.3.3 Hydrogen Evolution Reaction (HER)

Recent years have seen a lot of interest in hydrogen as a possible replacement for fossil fuels in the 21st century. Hydrogen is a clean, renewable, and sustainable resource with a high energy density. Sustainable hydrogen production using water electrolysis is one of the most promising methods for addressing the steadily growing need for renewable energy storage [151, 152]. Future hydrogen fuel cells and energy sources will need environmentally friendly hydrogen production [153]. In the HER process, electrochemical proton reduction at the metal electrode surface results in hydrogen fuel. Platinum and platinum-based electrocatalysts are challenging to employ due to their high cost, lack of availability, and restricted availability [154]. Many research teams have created a range of metal-based, metal-free, or noble metal-free electrocatalytic devices in the quest for efficient hydrogen generation electrocatalysts. Because Ag has a higher electrical and thermal conductivity than other Pt group metals, is more affordable and readily available, and is nanostructured, Ag and Ag-based catalysts are effective for electrocatalytic hydrogen evolution [155]. For electrochemical hydrogen evolution, the use of an Ag NP array has been described by Campbell *et al.* [95]. With the help of citric acid, they reduced $AgNO_3$ and employed colloidal suspension to form an array on a basal electrode. Because of the NPs' shape, electrocatalytic efficiency was improved. Ag–Pt nanoplates with hexagonal, circular, and triangular shapes were made by Lee and colleagues using a simple galvanic displacement approach [156]. When it came to electrocatalytic HER, Ag–Pt nanoplates were employed. Hydrogen generation was most efficient in the triangular Ag–Pt nanoplates. Ag_2O has received a lot of interest because of its high catalytic effectiveness and selectivity as a catalyst. Dendritic MoO_3/Ag catalyst with high crystallinity was developed by Wang and colleagues for electrochemical hydrogen evolution [157]. HER efficiency was improved thanks to the heterojunction of Ag_2O and MoO_3, which raised the composite's crystallinity. An improvement in electrocatalytic efficiency was achieved using binary and ternary catalyst systems. The HER (Ag_2S/MoS_2/RGO) was synthesized in one pot by Solomon *et al.* using Ag_2S, MoS_2, and reduced graphene oxide (RGO) [158]. Hydrothermal dispersion of Ag_2S NPs on MoS_2 nanosheets stabilized by RGO was used to fabricate the ternary Ag_2S/MoS_2/RGO system. With a Tafel slope of 56 mV dec^{-1} and a current density of −10 mA cm^{-2} at a −190 mV vs. RHE overpotential, the Ag_2S/MoS_2/RGO sample shows higher electrocatalytic HER activity than the bare binary MoS_2/RGO and MoS_2 composites. Other researchers have shown electrochemical hydrogen production using Pd and Ag bimetallic surface alloys devoid of Pt in acidic and alkaline settings. In the study of Yao *et al.* [159], as a result of Ag's reduction of the hydrogen bonding energy of the surrounding surface Pd atoms, the HER's inherent behavior and the interconnective metal skeleton were enhanced. In electrolytes containing 0.5% H_2SO_4 and 1.0% KOH, these electrodes demonstrated overpotential onset velocities of 26 mV dec^{-1} and ultralow Tafel slope values of 63 and 108 mV, respectively. Preliminary studies show that the electrocatalytic activity of surface atoms at both the top and bottom facets of nanosheets is improved, as are the edge and corner sites. As a surfactant template and a reducing agent, Cheng and colleagues used octadecyl trimethylammonium chloride and L-ascorbic acid to create trimetallic dendritic PdAgPt nanoalloys [160]. Catalytically active sites may be found in this ternary nanoalloy in various crystal defects, such as twin boundaries, steps, and atomic holes. There are multiple holes in the center and nanodendrites on the edges of $Pd_{43}Ag_{21}Pt_{36}$ nanoalloys. According to the geometry of the catalyst, mass transfer occurs through diffusion channels. Center holes and nanodendrites contribute to the high specific area of Ag nanocomposites, while many step sites enhance electrocatalytic activity [161, 162]. The $Pd_{43}Ag_{21}Pt_{36}$ nanowheels showed the best electrocatalytic hydrogen generation capability and operational stability among the several catalysts examined (binary $Pd_{68}Ag_{32}$ nanodendrites, $Pd_{62}Pt_{38}$ nanospheres, $Pd_{43}Ag_{21}Pt_{36}$ nanowheels, and Pt/C). The ternary $Pd_{43}Ag_{21}Pt_{36}$ nanowheels with optimized composition and shape demonstrated maximum mass/specific activity and stability in the oxidation of alcohol (methanol and ethanol). Silver has the highest electrical conductivity among transition metals and is well-documented. Hydrogen adsorptive energy is minimal because of D10's electronic structure [163]. According to the literature, the change in the catalyst's ability to adsorb and interact with intermediates was caused by small coordination numbers; surface characteristics such as a corner, edge, or step; and significant surface strain [164]. Li *et al.* claimed that the behavior of an Ag catalyst

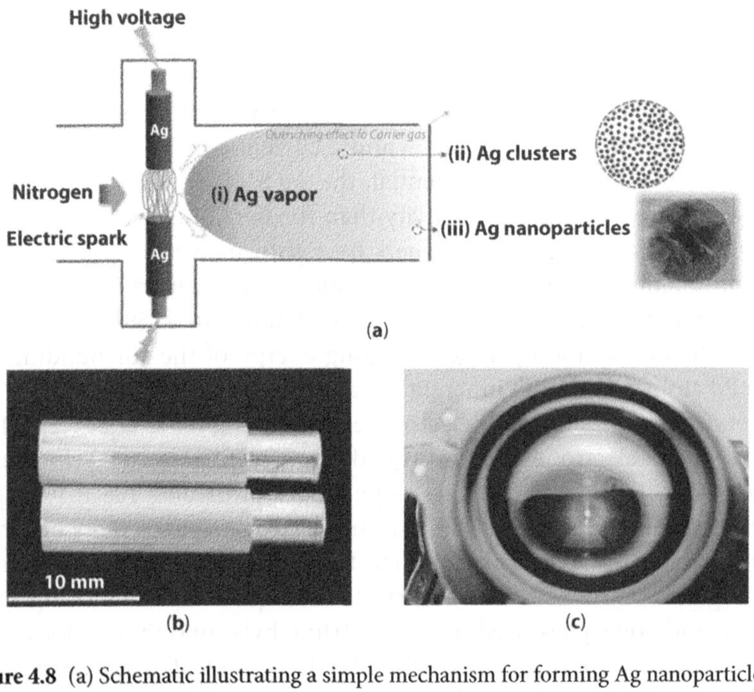

Figure 4.8 (a) Schematic illustrating a simple mechanism for forming Ag nanoparticles on a carbon fiber. Photographs of (b) a pair of pure Ag electrodes and (c) the electric spark machining process. Reproduced from ref. [167].

might be changed to improve the HER performance by utilizing features including a low number of coordinated atoms and a high surface strain [165]. They used laser ablation to induce multiple stacking flaws in Ag NPs by first forming Ag nanoparticles that heated quickly and became cold liquid (L-Ag NPs), resulting in low CN and high tensional strain. In an aqueous acid electrolyte, it was found that the L-Ag NPs exhibited a low overpotential of about 32 mV, with a Tafel slope of 31 mV per decade (dec). The long-term stability of L-Ag NPs was also demonstrated, with an operational time of 160 h. Kang and colleagues also employed laser ablation in the liquid approach to generate HER Ag NPs of varied sizes [166]. Another study used spark ablation in gas approach to deposit Ag NPs on a higher-purity carbon fiber and create a clean surface for HER (Figure 4.8) [167]. Other intriguing methods for the electrochemical generation of hydrogen have been presented, including nanoporous Ag fluff [155], Ag/WO_{3-x} heterostructures [168], Ag/Au NP-supported MoS_2 (1T-MoS_2@Ag/AuNPs) [169], heterogeneous Ag_xNi_{1-x} alloy nanocrystal [170], and SiO_2–Ag_0 NPs [171]. Laser ablation can be an effective method to create stacking flaws in metallic NPs, as seen above. Many different transition metal-based NPs may be synthesized using this approach.

4.3.4 Nitrogen Reduction Reaction (NRR)

A revolutionary way to generate carbon-free hydrogen sources is electrochemical NRR in an ambient environment [172]. Ammonia production from NRR is a sustainable and clean approach that has gotten a lot of interest as a potential energy alternative and storage car fuel [173, 174]. However, the sluggish reaction rate continues to limit NRR's catalytic selectivity and activity [175]. Electrocatalysts must be sensibly chosen to achieve a high electrochemical NRR in terms of performance and selectivity developed to maximize mass transport and modify chemical and physical parameters alongside proton/electron transfer [176]. An Ag electrode catalyst displayed significant catalytic activity among the reported non-noble metal-based electrocatalyst ammonia production [177, 178]. As proven by Sun and colleagues, Ag nanosheets showed an important NRR activity for synthesizing ammonia at ambient temperature and pressure [179]. With an NH_3 production rate of 4.62×10^{-11} mol s^{-1} cm^{-2} and an overpotential of −0.60 V vs. RHE, an Ag-based electrocatalyst demonstrated 4.8% FE. Nazemi *et al.* created bimetallic Au–Ag nanocages with adjustable localized surface plasmon resonance (LSPR) peak locations to circumvent the slow kinetics of the NRR [97]. When Ag NPs are present, the wall and corner pores of porous Au–Ag nanocages vary in size and density, which causes the LSPR peak to move from 635 to 715 nm. Oxygen treatment at LSPR peak locations of 644 and 719 nm resulted in Ag_2O–Au nanocages (Ag_2O–Au-644 and Ag_2O–Au-719) from the galvanic replacement process used to produce porous Au–Ag bimetallic nanocages. An NH_3 generation rate of 2.14 mg cm^{-2} h^{-1} was achieved by Ag_2O–Au-719, which had a slightly better electrocatalytic NRR efficiency than RHE. Li *et al.* employed a reduction process to synthesize boron-doped Ag nanosponges (B-AgNSs) for electrocatalytic nitrogen reduction to ammonia [180]. The B-AgNSs' linear sweep voltammetry (LSV) curves are measured in a 0.1-M HCl solution saturated with Ar and N_2 (Figure 4.9a). To evaluate the electrocatalytic NRR performance, chronoamperometry tests are performed at various potentials (Figure 4.9b). The electrolytes formed following electrolysis had a greater absorbance than the background electrolyte, indicating that the B-AgNSs can manufacture ammonia from the

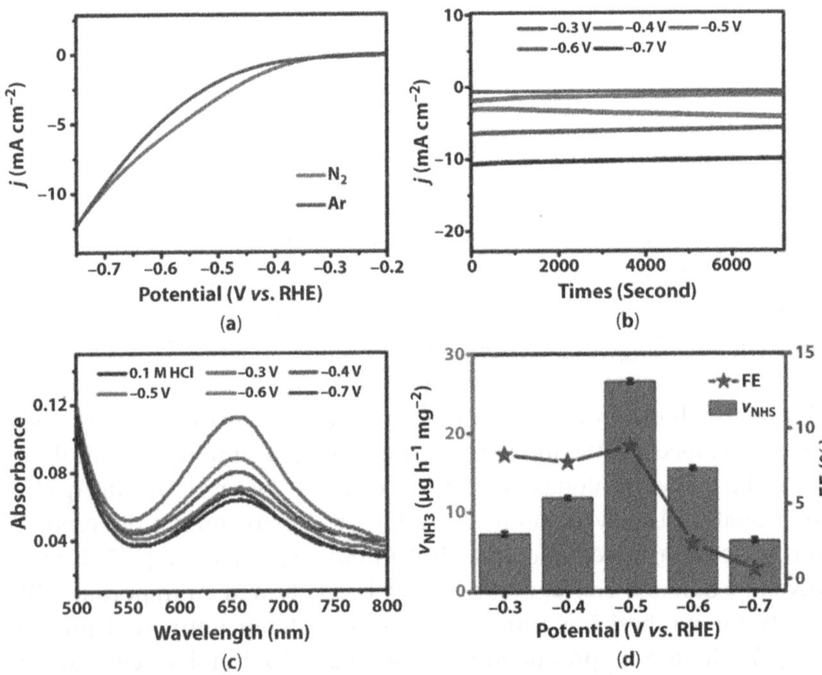

Figure 4.9 Recorded at 10 mV s^{1} in N_2- and Ar-saturated, 0.1 M HCl, (a) LSV curves in N_2 and Ar. For each potential value, (b) chronoamperometry curves are shown. Photomicrographs of electrolyte UV–Vis absorption spectra stained with indophenol indicator after 2 h of charging at various potentials (c). Different possibilities of NH_3 production and the FE that goes with it (d) [180].

electrocatalytic NRR (Figure 4.9c). At 0.5 V, the ideal vNH_3 and FE of 26.48 μg h^{-1} $mgcat.^{-1}$ and 8.86% are obtained, respectively (Figure 4.9d).

4.4 Conclusions

These findings will shed light on the rapidly expanding area of electrocatalysis' influence on current advances in manufacturing silver and silver-based nanomaterials (NMs). Thermal treatment, microwave therapy, chemical reduction, and sonochemical treatment were all discussed in the synthetic section. In a vain attempt to establish long-term processes for producing agricultural nanoparticles (Ag NPs), new, greener methods and techniques for continuous flow are being developed and then deployed. Ag-based nanomaterials can now catalyze a wide range of industrially critical organic reactions, as well as the developing disciplines of photocatalysis (H_2 evolution, CO_2 reduction, organic pollutants degradation), electrocatalysis (H_2 evolution, H_2 reduction), linking (Heck, Suzuki-Miyaura, Sonogashira, etc.), adding, and C–H activation (fuel cell reactions, N reduction). They are creating a new class of hybrid materials with enhanced capabilities for diverse chemical processes by combining Ag NPs with extra functionality (stable support material) or linking with other metals.

The scientific community continues to encounter considerable obstacles in this field, despite the great progress that has been accomplished. Making nanomaterials more stable by using neoteric manufacturing methods is being investigated to speed up the process.

Scientists in the field of materials science are still unable to precisely regulate the atomic size and form of the resulting Ag NPs. Catalytic efficiency may be improved by modifying the structure and morphology of the catalyst, which in turn affects its physical and chemical properties. To maintain a consistent catalyst nanostructure while achieving critical goals in catalysis, such as high activity, good selectivity, and exceptional recyclability, it is necessary to understand the relationship between the statics and dynamics of active catalytic centers at the nanoscale, which is still a difficult task.

Scientists are being forced to move away from material-driven approaches to catalysis. In this shape, selective catalysts or catalytic sites can be explicitly developed for targeted purposes based on knowledge gained through surface science and quantum mechanistic studies to meet the current energy and sustainability challenges. Organic synthesis for the medical and pharmaceutical sectors necessitates the development of catalysts with high stability and selectivity, i.e., enantio-, chemo-, or regioselectivity. We also looked at the most recent advancements in the rapidly growing field of Ag-based electrocatalytic devices. In this grueling field of study, there is much room for improvement and overcoming several obstacles.

To ensure the long-term expansion of the "green" energy supply, electrochemical technologies must be employed to convert CO_2, H_2O, and N_2 into chemical energy and raw material sources. Silver is a viable alternative to Pt due to its similar electrocatalytic activities and kinetics. Silver is also more resistant to methanol toxicity than Pt. The degradation of the proton exchange membrane in fuel cells is caused by hydrogen peroxide, which is produced by Ag NMs breaking down hydroxide ions. Silver-based nanocatalysts appear essential in catalytic activity and sustainability for various electrocatalytic processes in fuel cells and electrolyzers.

Single Ag NPs have many limitations compared with other metal NPs, such as rapid oxidation and aggregation, poor stability, and limited catalytic activity. It is necessary to alter the morphology of silver with other metal atoms and a supporting medium to increase its electrical conductivity, catalytic selectivity, and stability for long-term operations. It is necessary to investigate the synergistic effects between Ag NPs and other transition metals, metal oxides, or carbon supports to increase the electrocatalytic activity, selectivity, and stability of Ag-based nanocomposites for critical electrochemical processes such as the HER, ORR, OER, CO_2RR, and NRR. Models that can describe the complex existence of catalysts under specified reaction conditions and connected tandem processes in the electrochemical double layer will be necessary for further improvement in this field. The electrochemical double layer and the transitory nature of electrocatalysts are essential for a comprehensive understanding of reaction processes and for developing better catalysts and electrochemical interfaces. Another important step toward a cleaner future is coordinating electrochemical devices with electricity supplied by renewable energy sources; nevertheless, the cost of renewable energy generation must be considered when estimating the carbon footprint. An overview of the electrocatalytic applications of Ag NMs has included the CO_2RR and HER and the OER and ORR. Since they offer various catalytic features, such as a significant electron transport coefficient and a low overpotential, Ag-based NMs have emerged as low-cost technology for energy production that will last into the future. The mechanistic role of Ag species in photocatalysis must also be further studied and understood. The application of Ag nanoparticles in the industry requires continuous and scalable synthetic design. For the scientific community, we believe this study will not only provide new paths but also can be used to learn from future discoveries in this developing field.

References

1. Lea, M.C., ART. L.–On allotropic forms of silver. *Am. J. Sci.*, 37, 222, 476, 1880-1910.
2. Beyene, H.D. *et al.*, Synthesis paradigm and applications of silver nanoparticles (AgNPs), a review. *Sustain. Mater. Technol.*, **13**, 18–23, 2017.
3. Gunawan, C. *et al.*, Widespread and indiscriminate nanosilver use: genuine potential for microbial resistance. *ACS Nano*, **11**, 4, 3438–3445, 2017.
4. Fernando, A. *et al.*, Quantum mechanical studies of large metal, metal oxide, and metal chalcogenide nanoparticles and clusters. *Chem. Rev.*, **115**, 12, 6112–6216, 2015.
5. Zhang, X.-F. *et al.*, Silver nanoparticles: synthesis, characterization, properties, applications, and therapeutic approaches. *Int. J. Mol. Sci.*, 17, 9, 1534, 20161534.
6. Jana, N.R., Gearheart, L., Murphy, C.J., Wet chemical synthesis of silver nanorods and nanowires of controllable aspect ratioElectronic supplementary information (ESI) available: UV–VIS spectra of silver nanorods. *Chem. Commun.*, 2001, 7, 617–618, 2001. http://www.rsc.org/suppdata/cc/b1/b100521i.
7. Personick, M.L. *et al.*, Plasmon-mediated synthesis of silver cubes with unusual twinning structures using short wavelength excitation. *Small*, 9, 11, 1947–1953, 2013.
8. Wiley, B., Sun, Y., Xia, Y., Synthesis of silver nanostructures with controlled shapes and properties. *Accounts Chem. Res.*, 40, 10, 1067–1076, 2007.
9. Zhang, S.-H. *et al.*, Synthesis of silver nanotubes by electroless deposition in porous anodic aluminium oxide templates. *Chem. Commun.*, 9, 1106–1107, 2004.
10. Li, B. *et al.*, Stress-induced phase transformation and optical coupling of silver nanoparticle superlattices into mechanically stable nanowires. *Nat. Commun.*, 5, 1, 1–7, 2014.
11. Gao, Y. *et al.*, Studies on silver nanodecahedrons synthesized by PVP-assisted N, N-dimethylformamide (DMF) reduction. *J. Crystal Growth*, **289**, 1, 376–380, 2006.
12. Tang, S. *et al.*, Rapid synthesis of pentagonal silver nanowires with diameter-dependent tensile yield strength. *Mater. Chem. Phys.*, **142**, 1, 17–26, 2013.
13. Wiley, B.J. *et al.*, Right bipyramids of silver: a new shape derived from single twinned seeds. *Nano Lett.*, 6, 4, 765–768, 2006.
14. González, A. *et al.*, Size, shape, stability, and color of plasmonic silver nanoparticles. *J. Phys. Chem. C*, **118**, 17, 9128–9136, 2014.
15. Bilankohi, S.M., Optical scattering and absorption characteristics of silver and silica/silver core/shell nanoparticles. *Oriental J. Chem.*, **31**, 4, 2259–2263, 2015.
16. Heiligtag, F.J. and Niederberger, M., The fascinating world of nanoparticle research. *Mater. Today*, **16**, 7-8, 262–271, 2013.

17. Liao, G. *et al.*, Ag-Based nanocomposites: Synthesis and applications in catalysis. *Nanoscale*, **11**, 15, 7062–7096, 2019.
18. Desireddy, A. *et al.*, Ultrastable silver nanoparticles. *Nature*, **501**, 7467, 399–402, 2013.
19. Naik, R., Stringer, S.J., Agarwal, G., Jones, S.E., Stone, M.0., Biomimetic synthesis and patterning of silver nanoparticles. *Nat. Mater.*, **1**, 3, 169–172, 2002.
20. Pandey, S. *et al.*, Fast and highly efficient catalytic degradation of dyes using κ-carrageenan stabilized silver nanoparticles nanocatalyst. *Carbohydr. Polym.*, **230**, 115597, 2020.
21. Liu, H. *et al.*, Noble metal-based composite nanomaterials fabricated via solution-based approaches. *J. Mater. Chem. A*, 3, 7, 3182–3223, 2015.
22. Divya, K.P. *et al.*, *In situ* synthesis of metal nanoparticle embedded hybrid soft nanomaterials. *Accounts Chem. Res.*, **49**, 9, 1671–1680, 2016.
23. Wu, M. *et al.*, Removal of silver nanoparticles by mussel-inspired Fe_3O_4@ polydopamine core-shell microspheres and its use as efficient catalyst for methylene blue reduction. *Sci. Rep.*, 7, 1, 1–9, 2017.
24. Patel, A.C. *et al.*, Electrospinning of porous silica nanofibers containing silver nanoparticles for catalytic applications. *Chem. Mater.*, **19**, 6, 1231–1238, 2007.
25. Edison, T.J.I. and Sethuraman, M., Instant green synthesis of silver nanoparticles using Terminalia chebula fruit extract and evaluation of their catalytic activity on reduction of methylene blue. *Process Biochem.*, **47**, 9, 1351–1357, 2012.
26. Fang, G. and Bi, X., Silver-catalysed reactions of alkynes: Recent advances. *Chem. Soc. Rev.*, **44**, 22, 8124–8173, 2015.
27. Pellissier, H., Enantioselective silver-catalyzed transformations. *Chem. Rev.*, **116**, 23, 14868–14917, 2016.
28. Zhu, Q.-L. and Xu, Q., Immobilization of ultrafine metal nanoparticles to high-surface-area materials and their catalytic applications. *Chem*, **1**, 2, 220–245, 2016.
29. Song, H. *et al.*, Hydrothermal growth of mesoporous SBA-15 silica in the presence of PVP-stabilized Pt nanoparticles: Synthesis, characterization, and catalytic properties. *J. Am. Chem. Soc.*, **128**, 9, 3027–3037, 2006.
30. Liu, J., Catalysis by supported single metal atoms. *ACS Catalysis*, 7, 1, 34–59, 2017.
31. Carlson, C., Hussain, S., Schrand, A., Braydich-Stolle, L.K., Hess, K., Jones, R.L., Schlager, J.J., Unique cellular interaction of silver nanoparticles: Size-dependent generation of reactive oxygen species. *J. Phys. Chem. B*, **112**, 43, 13608–13619, 2008.
32. Li, Y. *et al.*, Surface coating-dependent dissolution, aggregation, and ROS generation of silver nanoparticles under different irradiation conditions. *Environ. Sci. Technol.*, **47**, 18, 10293–10301, 2013.
33. Nowack, B., Krug, H.F., Height, M., 120 years of nanosilver history: Implications for policy makers. *ACS Publications*, 45, 4, 1177–1183, 2011.

34. Bhattarai, B. *et al.*, Chemistry and structure of silver molecular nanoparticles. *Accounts Chem. Res.*, **51**, 12, 3104–3113, 2018.
35. Kang, H. *et al.*, Stabilization of silver and gold nanoparticles: Preservation and improvement of plasmonic functionalities. *Chem. Rev.*, **119**, 1, 664–699, 2018.
36. He, D. *et al.*, Silver sulfide nanoparticles in aqueous environments: formation, transformation and toxicity. *Environ. Science: Nano*, 6, 6, 1674–1687, 2019.
37. He, X. *et al.*, Overview of the application of flow microreactors in the synthesis of silver nanomaterials. *Nano*, **12**, 11, 1730002, 2017.
38. Guo, Z. *et al.*, Silver nanomaterials in the natural environment: An overview of their biosynthesis and kinetic behavior. *Sci. total Environ.*, **643**, 1325–1336, 2018.
39. Sadovnikov, S. and Gusev, A., Recent progress in nanostructured silver sulfide: From synthesis and nonstoichiometry to properties. *J. Mater. Chem. A*, 5, 34, 17676–17704, 2017.
40. He, K. *et al.*, Advancement of Ag–graphene based nanocomposites: An overview of synthesis and its applications. *Small*, **14**, 32, 1800871, 2018.
41. Cinelli, M. *et al.*, A green chemistry-based classification model for the synthesis of silver nanoparticles. *Green Chem.*, **17**, 5, 2825–2839, 2015.
42. Lima, F.H.B., de Castro, J.F.R., Ticianelli, E.A., Silver-cobalt bimetallic particles for oxygen reduction in alkaline media. *J. Power Sources*, **161**, 2, 806–812, 2006.
43. Lima, F.H.B., Calegaro, M.L., Ticianelli, E.A., Electrocatalytic activity of dispersed platinum and silver alloys and manganese oxides for the oxygen reduction in alkaline electrolyte. *Russian J. Electrochemistry*, **42**, 12, 1283–1290, 20062006.
44. Erikson, H., Sarapuu, A., Tammeveski, K., Oxygen reduction reaction on silver catalysts in alkaline media: A minireview. *ChemElectroChem*, 6, 1, 73–86, 2019.
45. Zhang, L. *et al.*, Recent progress for silver nanowires conducting film for flexible electronics. *J. Nanostructure Chem.*, **11**, 3, 323–341, 2021.
46. Lee, S.-H., Min, D.-R., Lee, K.-M., Sintering characteristics of au and ag nanoparticles prepared by inert gas condensation. *J. Powder Mater.*, **14**, 3, 165–172, 2007.
47. Raffi, M. *et al.*, Studies of the growth parameters for silver nanoparticle synthesis by inert gas condensation. *J. Mater. Res.*, **22**, 12, 3378–3384, 2007.
48. Kmis, F.E., Fissan, H., Rellinghaus, B., Sintering and evaporation characteristics of gas-phase synthesis of size-selected PbS nanoparticles. *Mater. Sci. Eng. B: Solid-State Mater. Advanced Technol.*, **69**, 329–334, 2000.
49. Boies, A.M. *et al.*, Hot-wire synthesis of gold nanoparticles. *Aerosol Sci. Technol.*, **45**, 5, 654–663, 2011.
50. Jung, J.H. *et al.*, Metal nanoparticle generation using a small ceramic heater with a local heating area. *J. Aerosol Sci.*, **37**, 12, 1662–1670, 2006.

51. Mafuné, F. *et al.*, Formation and size control of silver nanoparticles by laser ablation in aqueous solution. *J. Phys. Chem. B*, **104**, 39, 9111–9117, 2000.
52. Kabashin, A.V. and Meunier, M., Synthesis of colloidal nanoparticles during femtosecond laser ablation of gold in water. *J. Appl. Phys.*, **94**, 12, 7941–7941, 2003.
53. Dolgaev, S.I. *et al.*, Nanoparticles produced by laser ablation of solids in liquid environment. *Appl. Surface Sci.*, **186**, 1-4, 546–551, 2002.
54. Link, S. *et al.*, Laser-induced shape changes of colloidal gold nanorods using femtosecond and nanosecond laser pulses. *J. Phys. Chem. B*, **104**, 26, 6152–6163, 2000.
55. Kawasaki, M. *et al.*, 1064-nm laser fragmentation of thin Au and Ag flakes in acetone for highly productive pathway to stable metal nanoparticles. *Appl. Surface Sci.*, **253**, 4, 2208–2216, 2006.
56. Tsuji, T. *et al.*, Preparation of silver nanoparticles by laser ablation in solution: influence of laser wavelength on particle size. *Appl. Surface Sci.*, **202**, 1-2, 80–85, 2002.
57. Tsuji, T., Kakita, T., Tsuji, M., Preparation of nano-size particles of silver with femtosecond laser ablation in water. *Appl. Surface Sci.*, 206, 1–4, 314–320, 2003.
58. Ahmad, A. *et al.*, Extracellular biosynthesis of silver nanoparticles using the fungus Fusarium oxysporum. *Colloids Surfaces B: Biointerfaces*, **28**, 4, 313–318, 2003.
59. Iravani, S., Green synthesis of metal nanoparticles using plants. *Green Chem.*, **13**, 10, 2638–2650, 2011.
60. Sharma, R.K. *et al.*, Silver nanomaterials: synthesis and (electro/photo) catalytic applications. *Chem. Soc. Rev.*, **50**, 20, 11293–11380, 2021.
61. Caswell, K.K., Bender, C.M., Murphy, C.J., Seedless, surfactantless wet chemical synthesis of silver nanowires. *Nano Lett.*, **3**, 5, 667–669, 2003.
62. Ben Moshe, A. and Markovich, G., Synthesis of single crystal hollow silver nanoparticles in a fast reaction-diffusion process. *Chem. Mater.*, **23**, 5, 1239–1245, 2011.
63. Pérez, M.A. *et al.*, Hydroquinone synthesis of silver nanoparticles: A simple model reaction to understand the factors that determine their nucleation and growth. *Crystal Growth Design*, **8**, 4, 1377–1383, 2008.
64. Vinoth, S., Govindasamy, M., Wang, S.F., Solvothermal synthesis of silver tungstate integrated with carbon nitrides matrix composites for highly sensitive electrochemical nitrofuran derivative sensing in biological samples. *Analytica Chimica Acta*, **1192**, 339355–339355, 2022.
65. Isao, W. *et al.*, Reduction by the end groups of poly(vinyl pyrrolidone): A new and versatile route to the kinetically controlled synthesis of ag triangular nanoplates. *Advanced Mater.*, **18**, 13, 1745–1749, 2006.
66. Choi, K.W. *et al.*, Shape- and size-controlled synthesis of noble metal nanoparticles. *Adv. Mater. Res.*, **3**, 4, 199–216, 2014.

67. Ijaz, I. *et al.*, Detail review on chemical, physical and green synthesis, classification, characterizations and applications of nanoparticles. *Green Chem. Lett. Rev.*, **13**, 3, 59–81, 2020.
68. Vishwanath, R. and Negi, B., Conventional and green methods of synthesis of silver nanoparticles and their antimicrobial properties. *Curr. Res. Green Sustain. Chem.*, **4**, 100205–100205, 2021.
69. Rycenga, M. *et al.*, Controlling the synthesis and assembly of silver nanostructures for plasmonic applications. *Chem. Rev.*, **111**, 6, 3669–3712, 2011.
70. Pillai, Z.S. and Kamat, P.V., What factors control the size and shape of silver nanoparticles in the citrate ion reduction method? *J. Phys. Chem. B*, **108**, 3, 945–951, 2003.
71. Nadagouda, M.N. and Varma, R.S., *Risk reduction via greener synthesis of noble metal nanostructures and nanocomposites*, pp. 209–217, 2009.
72. Han, C. *et al.*, Expeditious synthesis of noble metal nanoparticles using Vitamin B12 under microwave irradiation. *Appl. Sci.*, **5**, 415–426, 20152015.
73. Bastús, N.G. *et al.*, Synthesis of highly monodisperse citrate-stabilized silver nanoparticles of up to 200 nm: Kinetic control and catalytic properties. *Chem. Mater.*, **26**, 9, 2836–2846, 2014.
74. Chen, S. *et al.*, Synthesis of lignin-functionalized phenolic nanosphere supported Ag nanoparticles with excellent dispersion stability and catalytic performance. *Green Chem.*, **22**, 9, 2879–2888, 2020.
75. Giner-Casares, J.J. *et al.*, Inorganic nanoparticles for biomedicine: where materials scientists meet medical research. *Mater. Today*, **19**, 1, 19–28, 2016.
76. Dheyab, M.A., Aziz, A.A., Jameel, M.S., Recent advances in inorganic nanomaterials synthesis using sonochemistry: A comprehensive review on iron oxide, gold and iron oxide coated gold nanoparticles. *Molecules*, **26**, 9, 2453, 2021.
77. Zhou, T. *et al.*, Sonochemical synthesis of highly fluorescent glutathione-stabilized Ag nanoclusters and S2– sensing. *Nanoscale*, **4**, 14, 4103–4106, 2012.
78. Xu, H. and Suslick, K.S., Sonochemical synthesis of highly fluorescent Ag nanoclusters. *ACS Nano*, 4, 6, 3209–3214, 2010.
79. Gawande, M.B. *et al.*, Microwave-assisted chemistry: Synthetic applications for rapid assembly of nanomaterials and organics. *Accounts Chem. Res.*, **47**, 4, 1338–1348, 2014.
80. Rathi, A.K. *et al.*, Microwave-assisted synthesis–Catalytic applications in aqueous. *Coordination Chem. Rev.*, 291, 68–94, 2015.
81. Gedye, R. *et al.*, The use of microwave ovens for rapid organic synthesis. *Tetrahedron Lett.*, 27, 3, 279–282, 1986.
82. Kumar, A. *et al.*, Microwave chemistry, recent advancements, and ecofriendly microwave-assisted synthesis of nanoarchitectures and their applications: A review. *Mater. Today Nano*, **11**, 100076–100076, 2020.
83. Chen, J. *et al.*, Microwave-assisted green synthesis of silver nanoparticles by carboxymethyl cellulose sodium and silver nitrate. *Mater. Chem. Phys.*, **108**, 2-3, 421–424, 2008.

84. Navaladian, S. *et al.*, Microwave-assisted rapid synthesis of anisotropic Ag nanoparticles by solid state transformation. *Nanotechnology*, **19**, 4, 045603, 2008.
85. Luo, B., Liu, S., Zhi, L., Chemical approaches toward graphene-based nanomaterials and their applications in energy-related areas. *Small*, **8**, 5, 630–646, 2012.
86. Ellabban, O., Abu-Rub, H., Blaabjerg, F., Renewable energy resources: Current status, future prospects and their enabling technology. *Renew. Sustain. Energy Rev.*, **39**, 748–764, 2014.
87. Zhou, J.-H. and Zhang, Y.-W., Metal-based heterogeneous electrocatalysts for reduction of carbon dioxide and nitrogen: Mechanisms, recent advances and perspective. *Reaction Chem. Eng.*, 3, 5, 591–625, 2018.
88. Gawande, M.B. *et al.*, Core–shell nanoparticles: synthesis and applications in catalysis and electrocatalysis. *Chem. Soc. Rev.*, **44**, 21, 7540–7590, 2015.
89. Lee, S.M. *et al.*, All-water-based solution processed Ag nanofilms for highly efficient electrocatalytic reduction of CO2 to CO. *Appl. Catalysis B: Environ.*, **259**, 118045, 2019.
90. Dong, X.-Y. *et al.*, Nanosilver as a new generation of silver catalysts in organic transformations for efficient synthesis of fine chemicals. *Catalysis Sci. & Technol.*, **5**, 5, 2554–2574, 2015.
91. El Badawy, A.M. *et al.*, Surface charge-dependent toxicity of silver nanoparticles. *Environ. Sci. & Technol.*, **45**, 1, 283–287, 2011.
92. Wen, C., Yin, A., Dai, W.-L., Recent advances in silver-based heterogeneous catalysts for green chemistry processes. *Appl. Catalysis B: Environ.*, **160**, 730–741, 2014.
93. Xie, C. *et al.*, Surface and interface control in nanoparticle catalysis. *Chem. Rev.*, **120**, 2, 1184–1249, 2019.
94. Lamoth, M. *et al.*, Nanocatalysts unravel the selective state of Ag. *ChemCatChem*, **12**, 11, 2977–2988, 2020.
95. Campbell, F.W. *et al.*, The hydrogen evolution reaction at a silver nanoparticle array and a silver macroelectrode compared: Changed electrode kinetics between the macro-and nanoscales. *J. Phys. Chem. C*, **113**, 33, 14852–14857, 2009.
96. Yao, W. *et al.*, Engineering hollow porous platinum-silver double-shelled nanocages for efficient electro-oxidation of methanol. *Appl. Catalysis B: Environ.*, **282**, 119595, 2021.
97. Nazemi, M. and El-Sayed, M.A., The role of oxidation of silver in bimetallic gold–silver nanocages on electrocatalytic activity of nitrogen reduction reaction. *J. Phys. Chem. C*, **123**, 18, 11422–11427, 2019.
98. Fan, L. *et al.*, Strategies in catalysts and electrolyzer design for electrochemical CO_2 reduction toward C_{2+} products. *Sci. Adv.*, **6**, 8, 3111, 2020.
99. Bushuyev, O.S. *et al.*, What should we make with CO_2 and how can we make it? *Joule*, **2**, 5, 825–832, 2018.

100. Dahl, S. and Chorkendorff, I., Towards practical implementation. *Nat. Mater.*, **11**, 2, 100–101, 2012.
101. Mikkelsen, M., Jørgensen, M., Krebs, F.C., The teraton challenge. A review of fixation and transformation of carbon dioxide. *Energy & Environ. Sci.*, **3**, 1, 43–81, 2010.
102. Sun, D. *et al.*, Rational design of Ag-based catalysts for the electrochemical CO_2 reduction to CO: A review. *ChemSusChem*, **13**, 1, 39–58, 2020.
103. Wu, X. *et al.*, Fast operando spectroscopy tracking *in situ* generation of rich defects in silver nanocrystals for highly selective electrochemical CO_2 reduction. *Nat. Commun.*, **12**, 1, 1–11, 2021.
104. Hori, Y. *et al.*, Electrocatalytic process of CO selectivity in electrochemical reduction of CO_2 at metal electrodes in aqueous media. *Electrochim. Acta*, **39**, 11-12, 1833–1839, 1994.
105. Back, S., Yeom, M., Jung, Y., Catal., A.C.S., vol. 5, pp. 5089–5096, CrossRef CAS, 2015. Britt Hvolbæk, *et al.*, Catalytic activity of Au nanoparticles. *Nano Today*, 2, 4, 14–18, 2007.
106. Salehi-Khojin, A. *et al.*, Nanoparticle silver catalysts that show enhanced activity for carbon dioxide electrolysis. *J. Phys. Chem. C*, **117**, 4, 1627–1632, 2013.
107. Ma, S. *et al.*, Silver supported on titania as an active catalyst for electrochemical carbon dioxide reduction. *ChemSusChem*, 7, 3, 866–874, 2014.
108. Tornow, C.E. *et al.*, Nitrogen-based catalysts for the electrochemical reduction of CO_2 to CO. *J. Am. Chem. Soc.*, **134**, 48, 19520–19523, 2012.
109. Lu, Q. *et al.*, A selective and efficient electrocatalyst for carbon dioxide reduction. *Nat. Commun.*, 5, 1, 1–6, 2014.
110. Kim, C. *et al.*, Achieving selective and efficient electrocatalytic activity for CO_2 reduction using immobilized silver nanoparticles. *J. Am. Chem. Soc.*, **137**, 43, 13844–13850, 2015.
111. Hatsukade, T. *et al.*, Insights into the electrocatalytic reduction of CO_2 on metallic silver surfaces. *Phys. Chem. Chem. Phys.*, **16**, 27, 13814–13819, 2014.
112. Hsieh, Y.-C. *et al.*, Effect of chloride anions on the synthesis and enhanced catalytic activity of silver nanocoral electrodes for CO_2 electroreduction. *ACS Catalysis*, 5, 9, 5349–5356, 2015.
113. Xi, W. *et al.*, Ultrathin Ag nanowires electrode for electrochemical syngas production from carbon dioxide. *ACS Sustain. Chem. Eng.*, **6**, 6, 7687–7694, 2018.
114. Hu, H. *et al.*, Activation matters: Hysteresis effects during electrochemical looping of colloidal ag nanowire catalysts. *ACS Catalysis*, **10**, 15, 8503–8514, 2020.
115. Chang, Z. *et al.*, The tunable and highly selective reduction products on Ag@ Cu bimetallic catalysts toward CO_2 electrochemical reduction reaction. *J. Phys. Chem. C*, **121**, 21, 11368–11379, 2017.

116. Shan, C. *et al.*, Site-selective growth of AgPd nanodendrite-modified Au nanoprisms: High electrocatalytic performance for CO_2 reduction. *Chem. Mater.*, **29**, 14, 6030–6043, 2017.
117. Wang, J. *et al.*, Silver/copper interface for relay electroreduction of carbon dioxide to ethylene. *ACS Appl. Mater. Interfaces*, **11**, 3, 2763–2767, 2019.
118. Ting, L.R.L. *et al.*, Enhancing CO_2 electroreduction to ethanol on copper–silver composites by opening an alternative catalytic pathway. *ACS Catalysis*, **10**, 7, 4059–4069, 2020.
119. Park, H. *et al.*, AgIn dendrite catalysts for electrochemical reduction of CO_2 to CO. *Appl. Catalysis B: Environ.*, **219**, 123–131, 2017.
120. Lum, Y. and Ager, J.W., Sequential catalysis controls selectivity in electrochemical CO_2 reduction on Cu. *Energy Environ. Sci.*, **11**, 10, 2935–2944, 2018.
121. Zhang, Q. *et al.*, A cold plasma-activated *in situ* AgCo surface alloy for enhancing the electroreduction of CO_2 to ethanol. *J. Mater. Chem. A*, **8**, 17, 8410–8420, 2020.
122. Yang, Y. *et al.*, Insight into the Formation and Transfer Process of the First Intermediate of CO_2 Reduction over Ag-Decorated Dendritic Cu. *Chemistry–A Eur. J.*, **26**, 18, 4080–4089, 2020.
123. Dong, W.J. *et al.*, Tailoring electronic structure of bifunctional Cu/Ag layered electrocatalysts for selective CO_2 reduction to CO and CH4. *Nano Energy*, **78**, 105168, 2020.
124. Irtem, E. *et al.*, Ligand-mode directed selectivity in Cu–Ag core–shell based gas diffusion electrodes for CO_2 electroreduction. *ACS Catalysis*, **10**, 22, 13468–13478, 2020.
125. Martić, N. *et al.*, Ag 2 Cu 2 O 3–a catalyst template material for selective electroreduction of CO to C 2+ products. *Energy & Environ. Sci.*, **13**, 9, 2993–3006, 2020.
126. T., N., D., G., Y., Facile green synthesis and applications of silver nanoparticles: A state-of-the-art review. *RSC Adv.*, 9, 34926–34948, 2019, doi: 10.1039.C9RA04164H. [CrossRef][Google Scholar].
127. Xiong, Y. *et al.*, Cobalt single atom site catalysts with ultrahigh metal loading for enhanced aerobic oxidation of ethylbenzene. *Nano Res.*, **14**, 7, 2418–2423, 2021.
128. Gawande, M., Fornasiero, P., Zbořil, R., *ACS Catal.*, 10, 2231–2259, 2020. Wei, Z., Chen, Y., Wang, J., Su, D., Tang, M., Mao, S., Wang, Y., Carbon-based single-atom catalysts for advanced applications. *ACS Catal.*, **6**, 5816–5822, 2016.
129. Liu, N., Zhang, Q., Guan, J., A binuclear Co-based metal–organic framework towards efficient oxygen evolution reaction. *Chem. Commun.*, 57, 41, 5016–5019, 2021.
130. Huang, Z. *et al.*, Catalytically active single-atom sites fabricated from silver particles. *Angewandte Chemie*, **124**, 17, 4274–4279, 2012.

131. Huang, Z. *et al.*, Self-assembly of atomically dispersed Ag catalysts on polyhedral Co_3O_4 at elevated temperatures: A top-down nanofabrication of high-loading atomically dispersed catalysts. *ChemCatChem*, **12**, 2, 561–568, 2020.
132. Lv, K. *et al.*, Elastic Ag-anchored N-doped graphene/carbon foam for the selective electrochemical reduction of carbon dioxide to ethanol. *J. Mater. Chem. A*, **6**, 12, 5025–5031, 2018.
133. Wang, L. *et al.*, Selective reduction of CO to acetaldehyde with CuAg electrocatalysts. *Proc. Natl. Acad. Sci.*, **117**, 23, 12572–12575, 2020.
134. Wang, Y. *et al.*, CO_2 reduction to acetate in mixtures of ultrasmall $(Cu)_n$, $(Ag)_m$ bimetallic nanoparticles. *Proceedings of the National Academy of Sciences*, vol. **115**, pp. 278–283, 2018.
135. Fan, T. *et al.*, Electrochemically driven formation of sponge-like porous silver nanocubes toward efficient CO_2 electroreduction to CO. *ChemSusChem*, **13**, 10, 2677–2683, 2020.
136. Nam, D.-H. *et al.*, Intermediate binding control using metal–Organic frameworks enhances electrochemical CO_2 reduction. *J. Am. Chem. Soc.*, **142**, 51, 21513–21521, 2020.
137. Yuan, X. *et al.*, Interface engineering of silver-based heterostructures for CO_2 reduction reaction. *ACS Appl. Mater. & Interfaces*, **12**, 50, 56642–56649, 2020.
138. Zhang, X. *et al.*, Boosting CO_2 conversion with terminal alkynes by molecular architecture of graphene oxide-supported Ag nanoparticles. *Matter*, **3**, 2, 558–570, 2020.
139. Jiao, Y. *et al.*, Design of electrocatalysts for oxygen-and hydrogen-involving energy conversion reactions. *Chem. Soc. Rev.*, **44**, 8, 2060–2086, 2015.
140. Cao, R. *et al.*, Recent progress in non-precious catalysts for metal-air batteries. *Advanced Energy Mater.*, **2**, 7, 816–829, 2012.
141. Zuliani, A. *et al.*, Improving the electrocatalytic performance of sustainable Co/carbon materials for the oxygen evolution reaction by ultrasound and microwave assisted synthesis. *Sustain. Energy Fuels*, 5, 3, 720–731, 2021.
142. Chandrasekaran, S. *et al.*, Electronic structure engineering on two-dimensional (2D) electrocatalytic materials for oxygen reduction, oxygen evolution, and hydrogen evolution reactions. *Nano Energy*, 77, 105080, 2020.
143. Hou, Y. *et al.*, Ag@ Co x P core–shell heterogeneous nanoparticles as efficient oxygen evolution reaction catalysts. *ACS Catalysis*, 7, 10, 7038–7042, 2017.
144. Islam, D.A. *et al.*, Ag-nanoparticle-anchored rGO-coated MIL-88B (Fe) hybrids as robust electrocatalysts for the highly efficient oxygen evolution reaction at neutral pH. *ChemElectroChem*, **4**, 12, 3110–3118, 2017.
145. Dong, R. *et al.*, Selective reduction–oxidation strategy to the conductivity-enhancing Ag-decorated Co-based 2D hydroxides as efficient electrocatalyst in oxygen evolution reaction. *ACS Sustain. Chem. Eng.*, 6, 10, 13420–13426, 2018.

146. Yu, M. *et al.*, Dual role of silver moieties coupled with ordered mesoporous cobalt oxide towards electrocatalytic oxygen evolution reaction. *Angewandte Chemie*, **132**, 38, 16687–16695, 2020.
147. Zhang, Z. *et al.*, Spontaneous synthesis of silver-nanoparticle-decorated transition-metal hydroxides for enhanced oxygen evolution reaction. *Angewandte Chemie Int. Edition*, **59**, 18, 7245–7250, 2020.
148. Ashok, A. *et al.*, Synthesis of highly efficient bifunctional Ag/Co_3O_4 catalyst for oxygen reduction and oxygen evolution reactions in alkaline medium. *ACS Omega*, **3**, 7, 7745–7756, 2018.
149. Chala, S.A. *et al.*, Site activity and population engineering of NiRu-layered double hydroxide nanosheets decorated with silver nanoparticles for oxygen evolution and reduction reactions. *ACS Catalysis*, **9**, 1, 117–129, 2018.
150. Zhang, F.-F. *et al.*, Iridium oxide modified with silver single atom for boosting oxygen evolution reaction in acidic media. *ACS Energy Lett.*, **6**, 4, 1588–1595, 2021.
151. Dubouis, N. and Grimaud, A., The hydrogen evolution reaction: From material to interfacial descriptors. *Chem. Sci.*, **10**, 40, 9165–9181, 2019.
152. Asefa, T., Metal-free and noble metal-free heteroatom-doped nanostructured carbons as prospective sustainable electrocatalysts. *Accounts Chem. Res.*, **49**, 9, 1873–1883, 2016.
153. Zhang, L. *et al.*, Activity origins in nanocarbons for the electrocatalytic hydrogen evolution reaction. *Small*, **14**, 26, 1800235, 2018.
154. Ren, X. *et al.*, Current progress of Pt and Pt-based electrocatalysts used for fuel cells. *Sustain. Energy Fuels*, **4**, 1, 15–30, 2020.
155. Zheng, B. *et al.*, Genomic and phenotypic diversity of carbapenemase-producing enterobacteriaceae isolates from bacteremia in china: A multicenter epidemiological, microbiological, and genetic study. *Engineering*, 12, 90–100, 2020.
156. Lee, C.-L. and Tseng, C.-M., Ag– Pt nanoplates: Galvanic displacement preparation and their applications as electrocatalysts. *J. Phys. Chem. C*, **112**, 35, 13342–13345, 2008.
157. Xia, X. *et al.*, Operando Synthesis of a dendritic and well-crystallized molybdenum oxide/silver catalyst for enhanced activity in the hydrogen evolution reaction. *ChemCatChem*, 7, 16, 2517–2525, 2015.
158. Solomon, G. *et al.*, Ag2S/MoS2 nanocomposites anchored on reduced graphene oxide: Fast interfacial charge transfer for hydrogen evolution reaction. *ACS Appl. Mater. Interfaces*, **11**, 25, 22380–22389, 2019.
159. Yao, R.-Q. *et al.*, Nanoporous palladium–Silver surface alloys as efficient and pH-universal catalysts for the hydrogen evolution reaction. *ACS Energy Lett.*, **4**, 6, 1379–1386, 2019.
160. Lu, W. *et al.*, Nanoengineering 2D dendritic PdAgPt nanoalloys with edge-enrichedactivesitesforenhancedalcoholelectroxidationandelectrocatalytic hydrogen evolution. *ACS Appl. Mater. Interfaces*, **12**, 19, 21569–21578, 2020.

161. Du, N. *et al.*, Trimetallic tristar nanostructures: Tuning electronic and surface structures for enhanced electrocatalytic hydrogen evolution. *Advanced Mater.*, **28**, 10, 2077–2084, 2016.
162. Ge, J. *et al.*, Ultrathin palladium nanomesh for electrocatalysis. *Angewandte Chemie*, **130**, 13, 3493–3496, 2018.
163. Uchida, T. *et al.*, Hydrogen evolution reaction catalyzed by proton-coupled redox cycle of 4, 4′-bipyridine monolayer adsorbed on silver electrodes. *J. Am. Chem. Soc.*, **130**, 33, 10862–10863, 2008.
164. Nørskov, J.K. *et al.*, Towards the computational design of solid catalysts. *Nat. Chem.*, **1**, 1, 37–46, 2009.
165. Li, Z. *et al.*, A silver catalyst activated by stacking faults for the hydrogen evolution reaction. *Nat. Catalysis*, **2**, 12, 1107–1114, 2019.
166. Kang, W.J. *et al.*, Ultrafine Ag nanoparticles as active catalyst for electrocatalytic hydrogen production. *ChemCatChem*, **11**, 24, 5976–5981, 2019.
167. Lu, J. *et al.*, Preparation of Ag nanoparticles by spark ablation in gas as catalysts for electrocatalytic hydrogen production. *RSC Adv.*, **10**, 63, 38583–38587, 2020.
168. Ren, Y., Chen, Z., Yu, X., Ultrathin, porous and oxygen vacancies-enriched Ag/WO_{3-x} heterostructures for electrocatalytic hydrogen evolution. *Chemistry–An Asian J.*, **14**, 23, 4315–4321, 2019.
169. Wang, J. *et al.*, Different phases of few-layer MoS_2 and their silver/gold nanocomposites for efficient hydrogen evolution reaction. *Catalysis Sci. & Technol.*, **10**, 1, 154–163, 2020.
170. Majee, R. *et al.*, Tweaking nickel with minimal silver in a heterogeneous alloy of decahedral geometry to deliver platinum-like hydrogen evolution activity. *Angewandte Chemie Int. Edition*, **59**, 7, 2881–2889, 2020.
171. Rolly, G.S. *et al.*, New insights into HER catalysis: The effect of nano-silica support on catalysis by silver nanoparticles. *Phys. Chem. Chem. Phys.*, **22**, 11, 6401–6405, 2020.
172. Martín, A.J., Shinagawa, T., Pérez-Ramírez, J., Electrocatalytic reduction of nitrogen: From Haber-Bosch to ammonia artificial leaf. *Chem*, **5**, 2, 263–283, 2019.
173. Guo, X. *et al.*, Recent progress in electrocatalytic nitrogen reduction. *J. Mater. Chem. A*, 7, 8, 3531–3543, 2019.
174. Smil, V., Detonator of the population explosion. *Nature*, **400**, 6743, 415–415, 1999.
175. Ma, H., Chen, Z., Wang, Z., Electroreduction of nitrogen to ammonia on nanoporous gold. *Nanoscale*, **13**, 3, 1717–1722, 2021.
176. Garcia-Segura, S. *et al.*, Electrocatalytic reduction of nitrate: Fundamentals to full-scale water treatment applications. *Appl. Catalysis B: Environ.*, **236**, 546–568, 2018.
177. Xu, H. *et al.*, Electrochemical ammonia synthesis through N_2 and H_2O under ambient conditions: Theory, practices, and challenges for catalysts and electrolytes. *Nano Energy*, **69**, 104469, 2020.

178. Iwamoto, M. *et al.*, Ammonia synthesis on wool-like Au, Pt, Pd, Ag, or Cu electrode catalysts in nonthermal atmospheric-pressure plasma of N_2 and H_2. *ACS Catalysis*, 7, 10, 6924–6929, 2017.
179. Huang, H. *et al.*, Ag nanosheets for efficient electrocatalytic N_2 fixation to NH_3 under ambient conditions. *Chem. Commun.*, **54**, 81, 11427–11430, 2018.
180. Li, Y. *et al.*, Boron-doped silver nanosponges with enhanced performance towards electrocatalytic nitrogen reduction to ammonia. *Chem. Commun.*, **55**, 98, 14745–14748, 2019.

5

Noble Metal-Based Nanocatalysts Dispersed on Functionalized and Alternative Supports for Low-Temperature Fuel Cells and Electrolyzers

F.J. Rodríguez-Varela[1*], I.L. Alonso-Lemus[2], J.C. Martínez-Loyola[3], A. Torres-Núñez[1], R. Chávez-Alcázar[1], P.C. Meléndez-González[4] and M.E. Sánchez-Castro[1]

[1]*Sustentabilidad de los Recursos Naturales y Energía, Cinvestav Unidad Saltillo, Ramos Arizpe, Coahuila, México*

[2]*CONAHCyT-Cinvestav Saltillo. Sustentabilidad de los Recursos Naturales y Energía. Av. Industria Metalúrgica, Parque Industrial Ramos Arizpe. Ramos Arizpe, Coahuila, México*

[3]*Escuela de Ingeniería y Ciencias, Tecnológico de Monterrey, Monterrey, NL, México*

[4]*Facultad de Ciencias Químicas, Universidad Autónoma de Nuevo León (UANL), San Nicolás de los Garza, NL, México*

Abstract

Green electrochemical energy conversion and storage devices, such as acid proton exchange membrane fuel cells and electrolyzers (PEMFCs and PEMEs, respectively) and alkaline anion exchange membrane fuel cells and electrolyzers (AEMFCs and AEMEs, respectively), offer several advantages over their fossil fuel-based counterparts. From a sustainable point of view, these systems provide i) significantly lower greenhouse gas (GHG) emissions to the atmosphere, ii) two or even three times higher energy conversion efficiency, iii) adaptability to be interconnected in an energy power loop including renewable energy systems (i.e., solar, wind), and iv) versatility in the sense of their scalability. Noble metals, remarkably Pt, Pd, Ir, and Ru, are the most active and commonly used nanocatalysts in these electrochemical devices because of their high catalytic

**Corresponding author*: javier.varela@cinvestav.edu.mx

Sudheesh K. Shukla, Chaudhery Mustansar Hussain, Santanu Patra, Meenakshi Choudhary (eds.) Electrocatalytic Materials for Renewable Energy, (111–150) © 2024 Scrivener Publishing LLC

activity. It is acknowledged that noble metal nanoparticles must be anchored and dispersed on a carbon matrix support, aiming to increase their electrochemically active surface area and reduce the metal content of the nanocatalysts. This work discusses the recent advances on supported noble metal nanocatalysts for low-temperature fuel cells and electrolyzers. Studies on the high performance of nanocatalysts after successful covalent surface functionalization of carbon supports are discussed. Research on the development of alternative supports (e.g., biocarbons from cheap biomass) is reviewed. Proposals to reduce the noble metal contents on highly active nanocatalysts are introduced. The aims of these efforts are to produce high-performance and electrochemically stable nanocatalysts, having a lower cost, to compete with benchmark 20 wt.% Pt/C (and Vulcan therein) in PEMFC and AEMFC and Ir- and Ru-based nanocatalysts in PEME and AEME applications.

***Keywords*:** Noble metal nanocatalysts, carbon supports, covalent functionalization, surface functional groups, biocarbons from biomass, electrochemical energy conversion and storage, low-temperature fuel cells and electrolyzers

5.1 Introduction

Proton exchange membrane fuel cells and electrolyzers (PEMFCs and PEMEs) have positioned as advanced green energy technologies, operating at a high efficiency at low temperatures, typically 80°C or less [1]. The electrical efficiency of PEMFCs may reach up to 80%, while PEMEs can operate at efficiencies of ca. 60%–80% [1, 2]. One of the key aspects of their success is that both technologies rely on the commercial availability of Nafion® membranes, the polymer electrolyte from Chemours (Dupont in the past) for their operation [2]. Moreover, over the past 10 years, anion exchange membrane fuel cells (AEMFCs) have emerged as a viable technological alternative, supported by the development of different commercial anion exchange membranes (e.g., Fumasep® from Fumatech, Tokuyama A201 from Tokuyama, or Sustainion® 37–50 from Dioxide Materials) [3, 4]. Similarly, anion exchange membrane electrolyzers (AEMEs) have reached a high development, encouraged also by the performance of some of these anionic membranes [5], reaching an efficiency that may be higher than 80% [1].

Figure 5.1 shows the number of publications over the past 5 years related to a) PEMFCs and PEMEs and b) AEMFCs and AEMEs (including some of the different variations in names commonly used, see [6]). The search

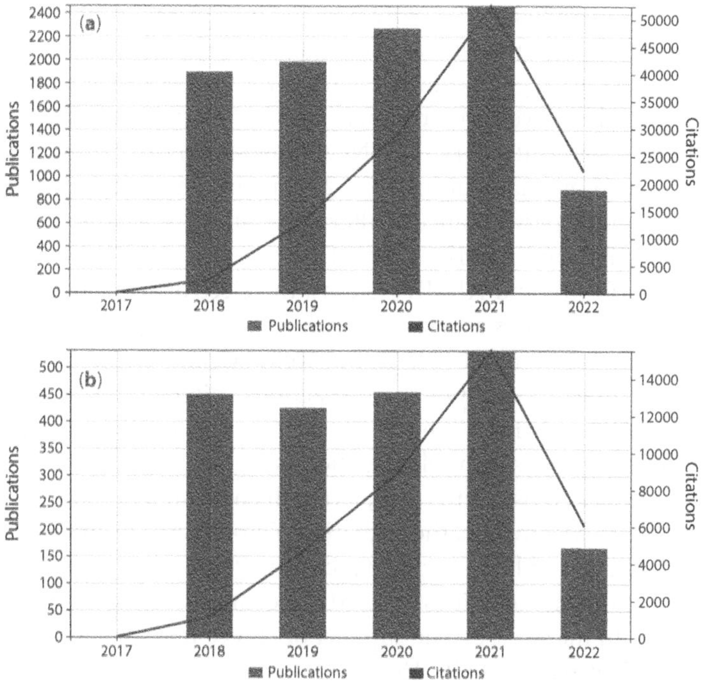

Figure 5.1 Number of publications and citations received on (a) PEMFC and PEME and (b) AEMFC and AEME. Source: Web of Science, as of June 10, 2022 (see [6]).

has been carried out on the Web of Science. During this period, ~9,500 works on PEMFC and PEME technologies have been published, with ca. 68,000 citations. Meanwhile, ~2,050 studies have been published on AEMFC and AEME technologies, receiving about 20,000 citations (without self-citations in both cases). Evidently, being a more mature technology, the systems based on PEM have been more deeply studied for a longer time, but those equipped with AEM are gaining attention rapidly.

The electrochemical reactions at the anodes and cathodes of PEMFCs and AEMFCs are the hydrogen oxidation reaction (HOR) and the oxygen reduction reaction (ORR), respectively. Studies show that the catalytic activity of the Pt nanocatalyst for the HOR is lower in alkaline than in acidic media [7]. Regarding the ORR, it has been reported that the high coverage of the Pt surface by OH^- groups in alkaline electrolyte causes an increase in overpotential (by ca. 25 mV) at 1 mA cm^{-2} kinetic current density, compared with its acidic counterpart. Interestingly, a higher percentage of intermediates, denoted by the more intense ring current, is formed at Pt/C in basic media, compared with acidic electrolyte [8]. Despite that, Pt-group metals (PGMs) reach good catalytic activity for the ORR in alkaline media. Moreover, the less severe environment of basic electrolytes allows for the use of non-PGM catalysts [9, 10]. Nevertheless, Pt/C (typically 20 wt.%, where C is commercial Vulcan XC-72) remains the benchmark for the comparison of the performance of alternative anode and cathode nanocatalysts of PEMFCs and AEMFCs [10, 11].

As for the PEME and AEME systems, the electrochemical reactions at the anodes and cathodes are oxygen evolution reaction (OER) and hydrogen evolution reaction (HER), respectively. Mainly noble metal nanocatalysts are used to catalyze these reactions, highlighting the catalytic activity of PGM [12–14]. Although PEME devices offer advantages over other water electrolysis technologies such as higher H_2 purity, lower ohmic losses, and less gas crossover [1, 13], its harsh environment requires the use of noble metals. Meanwhile, the operating pH of AEME technology allows to evaluate the use of high-performance non-PGM nanocatalysts, including non-precious metal oxides (e.g., Mn, Co, and perovskite-type, among others) [14]. In many cases, the support of choice for water-splitting electrolyzers is Vulcan as well [10, 15, 16].

On these issues, there is a worldwide tendency to develop nanocatalysts for low-temperature fuel cells and electrolyzers with novel compositions, covering several aspects aiming to improve their catalytic activity and stability while decreasing their cost. In this work, several of these topics are addressed, including i) the surface functionalization and evaluation of supports (mostly carbon-based), on this occasion focusing on covalent modification; ii) the research on novel carbon supports as an alternative to Vulcan; and iii) the synthesis and evaluation of nanocatalysts with low noble metal content (e.g., 5 wt.% Pt).

5.2 Electrochemical Reactions in Low-Temperature Fuel Cells and Electrolyzers

The HOR, ORR, and overall reaction in PEMFCs are shown in reactions (5.1–5.3):

$$\text{HOR (anode):}\quad 2H_2 \rightarrow 4H^+ + 4e^- \tag{5.1}$$

$$\text{ORR (cathode):} \quad O_2 + 4H^+ + 4e^- \rightarrow 2H_2O \tag{5.2}$$

$$\text{Overall:} \quad 2H_2 + O_2 \rightarrow 2H_2O \tag{5.3}$$

Meanwhile, those in AEMFCs are as follows:

$$\text{HOR (anode):} \quad 2H_2 + 4OH^- \rightarrow 4H_2O + 4e^- \tag{5.4}$$

$$\text{ORR (cathode):} \quad O_2 + 2H_2O + 4e^- \rightarrow 4OH^- \tag{5.5}$$

$$\text{Overall:} \quad 2H_2 + O_2 \rightarrow 2H_2O \tag{5.6}$$

It is well-known that the HOR is kinetically faster than the ORR. Interestingly, as already mentioned, it has been reported that both the HOR and the ORR proceed with slower kinetics in alkaline compared with acidic electrolytes [7]. More specifically, Pt-based nanocatalysts have a catalytic activity for the HOR about 2 orders of magnitude lower in alkaline media compared with acidic media [7]. Therefore, research has been oriented toward non-PGM catalysts for anodes and cathodes for AEMFCs that may compete with Pt/C, either supported on carbon structures alternative to Vulcan such as Ni/N-CNT, Ag/MWCNT, and NiMo/Ketjenblack 600J [7, 17, 18] or non-supported catalyst like Fe-N-C [19]. In the case of PEMFCs, several workers have proposed to implement metal-free catalysts based on heteroatom-doped carbon nanostructures such as N, S-doped graphene (NSG), N, P-doped graphene (NPG), and N, B, P-doped carbon, among others [20]. The reaction mechanism of the ORR in alkaline and acidic media has been extensively discussed previously [8, 20]. Similarly, the reaction mechanism of the HOR in both media has been reported elsewhere [21, 22].

Moreover, the HER, OER, and overall reaction in PEMEs and AEMEs are shown in reactions (5.7)–(5.9) and (5.10)–(5.12), respectively:

$$\text{HER (anode):} \quad 2H_2O \rightarrow O_2 + 4H^+ + 4e^- \tag{5.7}$$

$$\text{OER (cathode):} \quad 4H^+ + 4e^- \rightarrow 2H_2 \tag{5.8}$$

$$\text{Overall:} \quad 2H_2O \rightarrow 2H_2 + O_2 \tag{5.9}$$

And

$$\text{HER (anode):} \quad 4OH^- \rightarrow 2H_2O + O_2 + 4e^- \tag{5.10}$$

$$\text{OER (cathode):} \quad 4H_2O + 4e^- \rightarrow 2H_2 + 4OH^- \tag{5.11}$$

$$\text{Overall:} \quad 2H_2O \rightarrow 2H_2 + O_2 \tag{5.12}$$

The reaction mechanisms of the HER and OER in acidic and alkaline media have been discussed extensively elsewhere [23]. The HER is kinetically faster than the OER as well. Meanwhile, it has been reported that the overpotential of the HER at −10 mA cm^{-2} at an IrMo/CNT nanocatalyst is slightly smaller in 0.5 M H_2SO_4 compared with 1 M KOH (13 vs. 17 mV). The difference in performance in acidic and alkaline electrolyte is larger at commercial Pt/C (16 vs. 46 mV) [24]. Regarding the OER, it has been shown that the performance of IrO_2 nanoparticles is about the same in pH = 1 and pH = 13 over a wide potential interval [25].

Regardless of the pH of the electrolyte, the appropriate selection of nanocatalysts is relevant since some are highly active and stable for more than one electrochemical reaction. Such feature indicates that those nanocatalysts have bifunctional capabilities. Some nanocatalysts show high catalytic activity for the HER/OER, i.e., high performance for electrochemical water splitting [23, 26]. Moreover, nanocatalysts can demonstrate also a high catalytic activity for the ORR/OER, indicating excellent behavior in terms of the so-called oxygen potential [27, 28]. Furthermore, the bifunctionality of nanocatalysts regarding the HOR/HER has also been studied [29, 30].

5.3 Covalently Functionalized Supports for Fuel Cells and Electrolyzers

The most commonly used supports for low-temperature fuel cells and electrolyzers are amorphous carbon blacks [31]. Other supports are carbon nanofibers (CNFs), graphene, ordered mesoporous carbons (OMCs), and ordered mesoporous hollow carbon spheres (OMHCSs), among others, discussed extensively in ref. [32]. These supports have good textural and physicochemical properties but are prone to corrosion [31, 32], particularly at the highly positive potentials of the OER and to some extent the ORR. Even more, as pointed out earlier, defects on carbon surfaces promote a stronger bonding of nanoparticles preventing migration and coalescence, but their presence also decreases the corrosion resistance of the supports [33]. Other degradation mechanisms, observed particularly on Pt-based nanocatalysts, are particle migration and eventual agglomeration, particle detachment or dissolution, and Ostwald ripening [34].

Metal:support interactions influence the electrochemical performance and stability of supported nanocatalysts, and a practical way to promote a synergetic effect at the interface is to functionalize the carbon surface. Covalent functionalization has a beneficial effect promoting such interactions, since it produces hydroxyl (–OH), carbonyl (–C=O), and carboxyl (–COOH) surface chemical groups, among others, that result in a more polar and hydrophilic surface [35]. One of the most popular covalent functionalizations is the oxidative process (i.e., the use of concentrated aqua regia). The surface chemical groups formed thence act as anchoring sites of nanoparticles [33]. Some other oxidizing chemicals have been used, for example, H_2SO_4, H_2O_2, HF, and HNO_3, among others [36]. Nevertheless, the search for greener methods of synthesis has led to the successful implementation of softer chemical functionalization methodologies, e.g., using methanol and citric acid [37, 38]. In some cases, N-atoms have been considered as functionalizing species [33] although heteroatoms incorporated into the structure of carbon materials are known as doping agents.

Reports by several groups indicate that attention should be paid to the content of O-functionalities on the surface of supports. A relatively large concentration may be the cause of carbon corrosion as well as dissolution and agglomeration of nanoparticles, attributed in part to the formation of Pt-oxides [39, 40]. In the case of CNF, it may provoke structural damage [37]. Alternative functionalization of supports like Vulcan, OMHCS, or reduced graphene oxide (rGO), such as the use of organometallic compounds, develops metal and metal oxide sites (e.g., Ru, RuO_2, CrO_2, Cr_2O_3, and CrO_3) along with surface functionalities. Under this approach, species such as Ru^0 and Ru^{4+} act as nucleation sites to grow Pt nanoparticles, leading to the development of alloyed Pt–Ru phases once the Pt precursor is reduced at the surface [41–45]. In another procedure, Vulcan has been functionalized with 4-aminomethylpyridine, developing surface N-pyridinic which binds with Pt nanoparticles, showing high durability [46].

The surface chemistry of carbon supports after functionalization plays a key role in the adsorption of precursors, which act as seeds of metal nanoparticles. Xie *et al.* took advantage of the positively charged groups covalently grafted on Vulcan after functionalization with PBI, aiming to drive the adsorption of $PtCl_6^{2-}$ ions on them enhancing the dispersion of nanoparticles [47]. The PtNiN/XC72-PBI nanocatalyst had nitrided PtNi nanoparticles, while the support was N-doped (Figure 5.2a). It showed high catalytic activity for the ORR in acidic media and high performance in an H_2/air fuel cell (Figures 5.2b–e), delivering a rated power density of 1,059 mW cm^{-2} at 0.6 V and mass activity of 440 mA mg_{Pt}^{-1}. The methodology was effective in producing nanocatalysts with a small particle size, exceeding the 2022 target by the US Department of Energy (DOE). It was proposed that doping the support with N-atoms improves the ionomer/catalyst interface, enhancing the catalyst utilization.

Surface functionalization of the support may also have a positive effect promoting a high dispersion and small size of nanoparticles. Kwon *et al.* evaluated the citric acid-functionalized Vinatech and Vulcan porous supports for the dispersion of Pt nanoparticles [48]. The former showed a higher relative concentration of oxygen functionalities and a larger number of mesopores (Figure 5.3a). The functionalization provided nucleation sites for the growing of nanoparticles (Figure 5.3b). Despite having more oxygen groups, the anchoring of smaller nanoparticles was observed at Pt/Vinatech, which showed a higher catalytic activity for the ORR than Pt/Vulcan in 0.1 M $HClO_4$. It demonstrated an enhanced performance and high stability in PEMFC tests (Figures 5.3c, d), increasing the current density by 1.7% after 3k cycles. The catalytic activity for the ORR was improved at Pt/Vinatech due to a faster charge transfer over the highly dispersed nanoparticles. Meanwhile, its good behavior in PEMFC was ascribed to a higher electrochemical surface area (ECSA).

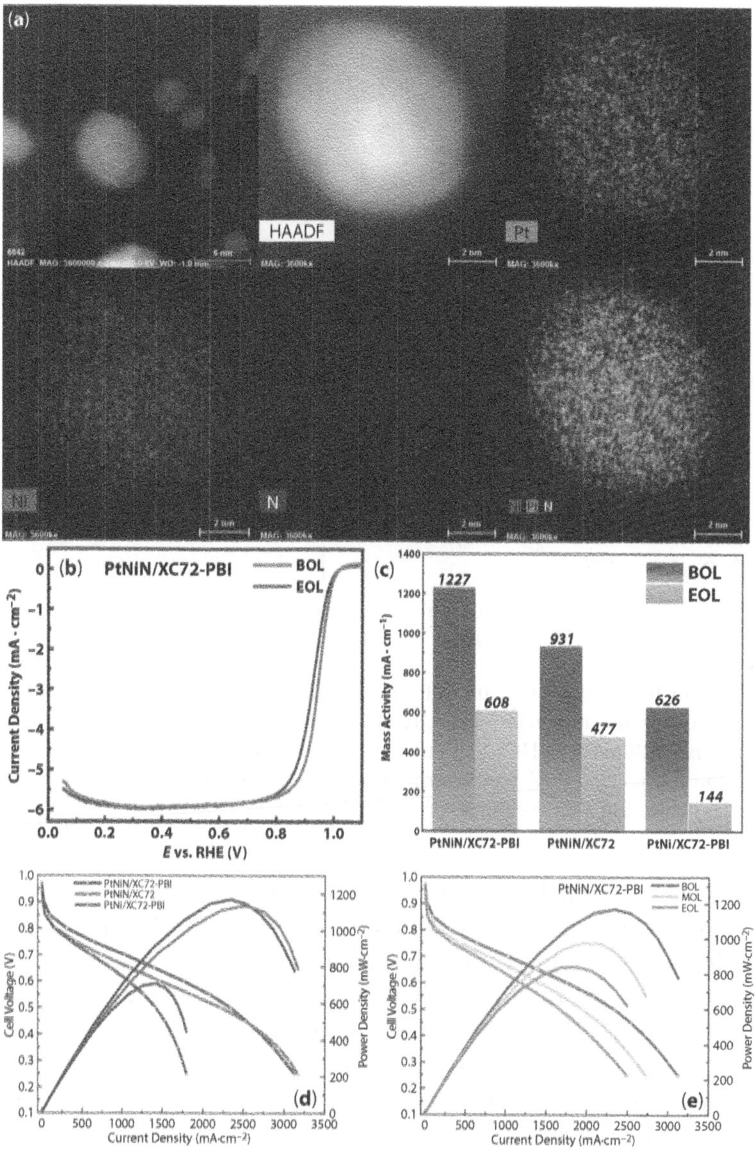

Figure 5.2 (a) Chemical mapping of PtNiN/XC72-PBI. (b) Polarization curves of the ORR in 0.1 M $HClO_4$. (c) Mass activity of PtNiN/XC72-PBI. (d) Polarization and power density curves of MEAs containing PtNiN/XC72-PBI and other nanocatalysts. (e) Polarization and power density curves of MEAs containing PtNiN/XC72-PBI at BOL, MOL, and EOL (30,000 AST cycling). Reprinted with permission from ref. [47]. Copyright 2022, American Chemical Society.

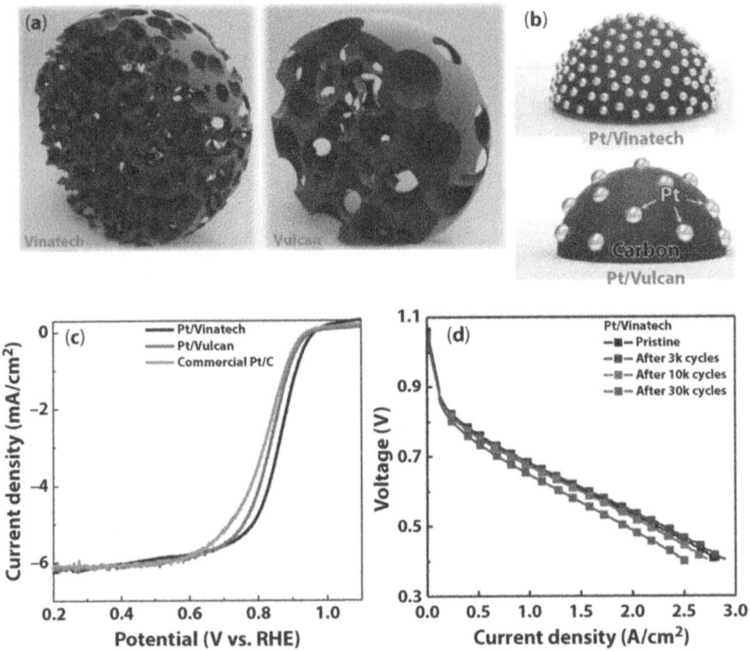

Figure 5.3 (a, b) Schemes of the porosity and dispersion of Pt nanoparticles on citric acid-functionalized commercial Vinatech and Vulcan porous carbons. (c) Polarization curves of the ORR in 0.1 M $HClO_4$ of Pt/Vinatech, surpassing the catalytic activity of Pt/Vulcan and commercial Pt/C. (d) Performance of MEA containing Pt/Vinatech at both electrodes before and after cycling. Reprinted from ref. [48].

Covalent functionalization of carbon supports can also be achieved via the use of organic molecules, which are capable of stabilizing nanoparticles. Jeong *et al.* utilized the 1,5-diaminonaphthalene (DAN) molecule to functionalize electrochemically reduced graphene oxide (ERGO) [49]. The NH_2 groups at DAN formed C–N bonds from the reaction with acidic groups, stabilizing the dispersion of Pd nanoparticles (Figures 5.4a, b). The relative concentration of oxygen functionalities at graphene oxide was reduced using an electroreduction method. The ERGO-DAN-Pd nanocatalyst showed improved catalytic activity for the ORR, promoting the reaction via a 4 e^- transfer mechanism with a low production of H_2O_2 (less than 5% at –0.6 vs. Ag/AgCl) (Figures 5.4c, d). ERGO-DAN-Pd showed a high electrochemical stability after 500 cycles in 0.1 M KOH (Figure 5.4e). It was reported that the surface chemistry of ERGO-DAN-Pd improved its corrosion resistance compared with Pt/C.

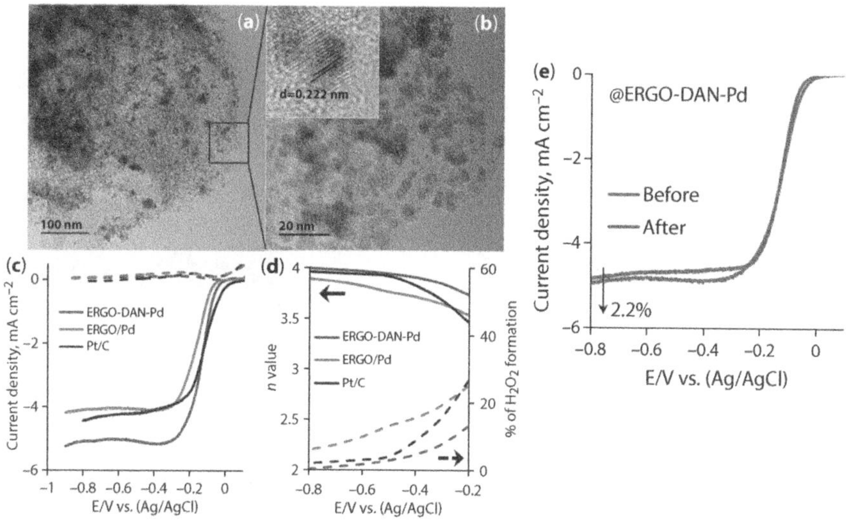

Figure 5.4 (a, b) TEM micrographs of the ERGO-DAN-Pd nanocatalyst. (c, d) Polarization curves of the ORR, electron transfer number, and % of H_2O_2 production at ERGO-DAN-Pd and other nanocatalysts in 0.1 M KOH. (e) Polarization curves of the ORR at ERGO-DAN-Pd before and after cycling. Reprinted with permission from ref. [49]. Copyright 2017, Elsevier.

Meanwhile, a series of supported noble metals (i.e., Ru, Pd, Rh, Pt) have been identified as highly active for the HER. For their high dispersion on a carbon matrix, the functionalization of the former brings advantageous surface chemistry characteristics. Besides such functionalities, the preformation of surface metal oxides can foster nucleation, followed by the high dispersion of metal nanoparticles. Fan *et al.* proposed the formation of metal oxides and the subsequent anchoring of metal nanoparticles on oxygen-rich Vulcan, OMC, and Ketjenblack supports [50]. A high dispersion of Rh, Ru, and Ir nanoparticles was achieved (Figures 5.5a–c). The Rh/C-300A-350H nanocatalyst showed enhanced catalytic activity for the HER in 1.0 M KOH, with an overpotential of 29 mV at 10 mA cm^{-2}, surpassing the performance of Pt/C, Ru, and Ir nanocatalysts (Figure 5.5d). The OH^--rich surface had a positive effect on weakening the adsorption energy of hydrogen, promoting the cleavage of O–H bonds and a high utilization of metal nanoparticles, thus enhancing the performance of the nanocatalyst (Figures 5.5d–f).

Therefore, the advantageous conditioning of the carbon surface should not be limited to developing functional groups, but it may be extended to the formation of metal or metal-oxide sites for further nucleation and

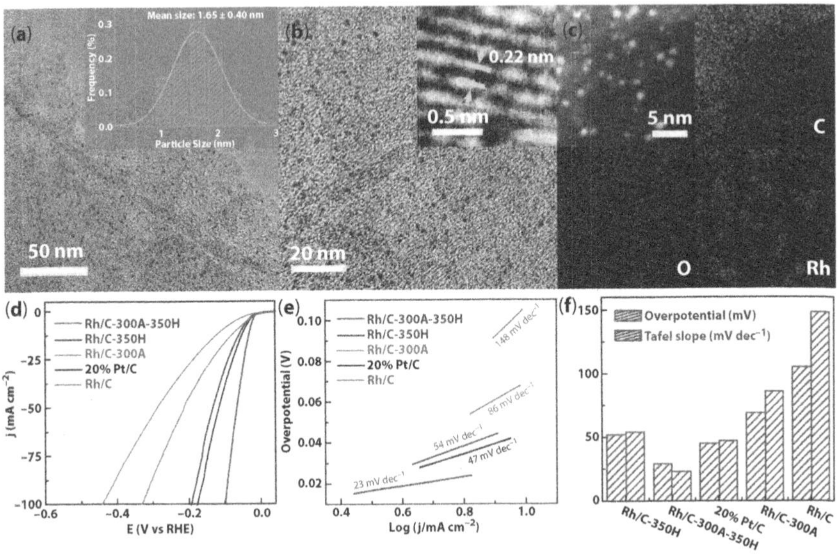

Figure 5.5 (a–c) TEM analysis and chemical mapping of the Rh/C-300A-350H nanocatalyst. (d–f) Polarization curves of the HER in 1.0 M KOH, Tafel plots, and overpotentials and Tafel slopes, comparing the performance of Rh/C-300A-350H with those of benchmark Pt/C and other Rh-based nanocatalysts. Reprinted with permission from ref. [50]. Copyright 2022, Elsevier.

growth of metal nanoparticles. One recent proposal for obtaining such surface chemistry of carbon structures is that of organometallic functionalization [41–45]. Via such an innovative procedure, Rodríguez-Varela *et al.* incorporated several functional groups on OMHCS by using the $[(\eta^6\text{-}C_6H_5OCH_2CH_2OH)RuCl_2]_2$ (labeled as Ru-dim) and $[(\eta^6\text{-}C_6H_4CH(CH_3)_2CH_3)RuCl_2]_2$ (identified as Ru-cym) compounds. The functionalization developed Ru^0 and Ru^{4+} sites [42]. By using the microwave-assisted polyol method, the 5-wt.% $Pt/OMHCS_{Ru\text{-}dim}$ and $Pt/OMHCS_{Ru\text{-}cym}$ nanocatalysts were synthesized.

Chemical mapping demonstrated the presence of C, O, Pt, Ru, and Si (Figure 5.6a). The I_D/I_G Raman bands ratio was 1.44 and 1.22 for $OMHCS_{Ru\text{-}dim}$ and $OMHCS_{Ru\text{-}cym}$, respectively (1.21 in non-functionalized OMHCS [42], not shown here), as shown in Figures 5.6b, c. Thus, functionalization of OMHCS with Ru-dim proceeded via ionic bond interactions, increasing the degree of disorder of the spheres. Meanwhile, that with Ru-cym evidenced π-π interactions between its arene ligands and the graphitic domains of OMHCS, preserving their sp^2 domains (Figure 5.6d). $Pt/OMHCS_{Ru\text{-}cym}$ and $Pt/OMHCS_{Ru\text{-}dim}$ showed higher catalytic activity for the

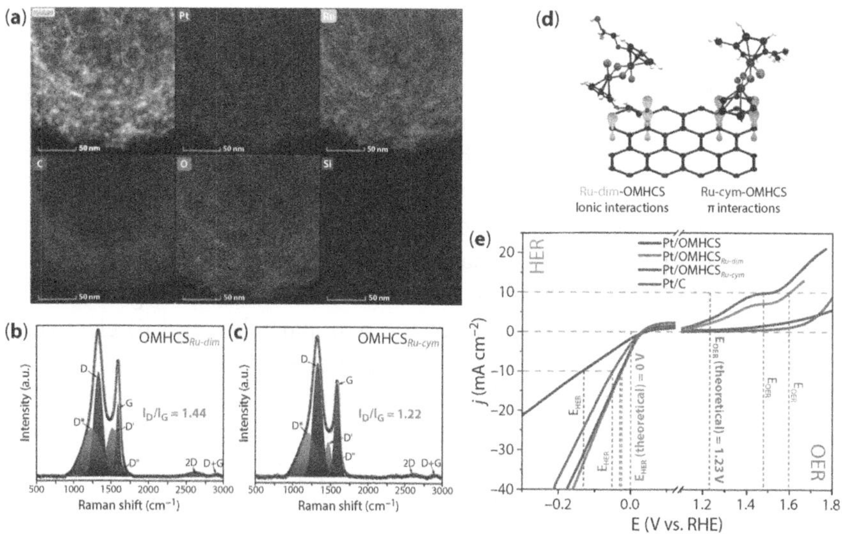

Figure 5.6 (a) HAADF image and chemical mapping of the Pt/OMHCS$_{Ru\text{-}cym}$ nanocatalyst (degree of Ru alloyed: ca. 35%). (b, c) Raman spectra of OMHCS$_{Ru\text{-}dim}$ and OMHCS$_{Ru\text{-}cym}$. The interactions are schematized in (d). (e) iR-compensated polarization curves of the HER and OER at Pt/OMHCS$_{Ru\text{-}cym}$, Pt/OMHCS$_{Ru\text{-}dim}$, Pt/C, and Pt/OMHCS. Electrolyte: 0.1 M KOH; scan rate: 5 mV s^{-1}.

HER and OER in 0.1 M KOH than 20 wt.% Pt/C (Figure 5.6e). The overpotentials at −10 mA cm^{-2} (HER) and 10 mA cm^{-2} (OER) were −0.029 and 1.52 V vs. RHE, respectively, on Pt/OMHCS$_{Ru\text{-}cym}$. Thus, its water-splitting potential was significantly smaller than that of Pt/C. The high performance of Pt/OMHCS$_{Ru\text{-}cym}$ and Pt/OMHCS$_{Ru\text{-}dim}$ was ascribed to the presence of Ru^0 and RuO_2 sites, active species for the reactions, and to the modification of the *d*-valence band of Pt because of the alloying with Ru atoms from the functionalization.

5.4 Alternative Carbon Supports for Fuel Cells and Electrolyzers

The membrane electrode assembly (MEA) is the core component of PEMFCs and AEMFCs. Their performance largely depends on their behavior under a severe environment involving oxidation and reduction reactions. The MEA is also the most expensive component of these electrochemical devices. Figure 5.7(a) shows the components of an MEA, such as the gas diffusion layer (GDL), the catalyst layer (CL), and the ion exchange membrane (i.e., PEM or AEM). In this sense, the CL is one of the most complex parts, since it contains the catalysts, commonly Pt/C, where oxidation and reduction reactions of the species (H_2 or O_2) take place. In addition, the CL must facilitate the transport of ions and electrons and the diffusion of species at the triple-phase boundary [51].

In this regard, the support used for the dispersion of nanoparticles plays a crucial role in improving the nanocatalysts' performance (and, thus, that of the fuel cells) because it interacts with the active sites causing two effects: i) geometric, changing the shape of the nanoparticles, and ii) electronic, modifying their structural density. These effects influence the amount of available active sites affecting the catalytic activity of the nanocatalysts [52]. As already mentioned, carbon blacks have been widely used as supports of metal nanoparticles. These structures are a by-product of the pyrolysis of hydrocarbons (e.g., oil, natural gas, etc.), which makes them inexpensive and broadly available. Vulcan XC-72 is a carbon black possessing meso-porosity (20–40 nm average pore size), small particle size (~30 nm), and fair specific surface area (SSA_{BET} of ~250 $m^2\ g^{-1}$), making it very suitable as a support [53]. However, the main drawbacks of carbon blacks are their low electrochemical stability, fast corrosion rate, and loss of active sites in long-term tests [54].

To overcome these issues, several alternative carbon supports have been evaluated for PEMFC and AEMFC applications (Figure 5.7b), for example,

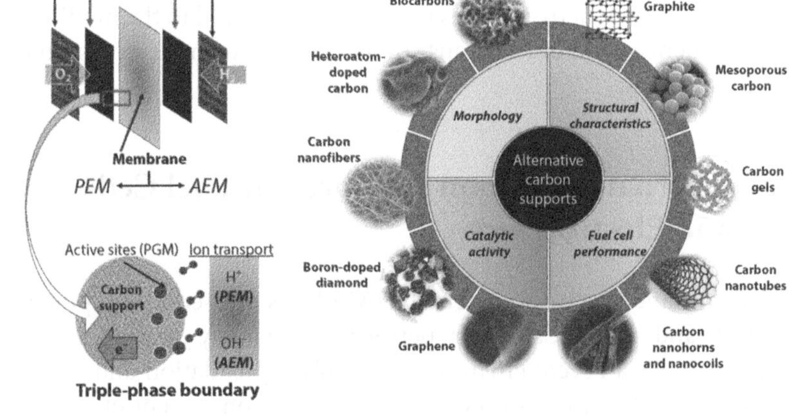

Figure 5.7 (a) Elements of the MEA and schematic representation of the triple-phase boundary. (b) Alternative carbon supports studied for PEMFC and AEMFC applications.

graphite [55–57], OMC [58–61], carbon gels [62, 63], carbon nanotubes (CNTs) [64–66], carbon nanohorns and nanocoils (CNHs and CNCs, respectively) [67, 68], graphene [57, 69], boron-doped diamond [70], CNF [32], heteroatom-doped carbons [71], and more recently, biocarbons [35]. Most reports on alternative carbon supports focus on cathode catalysts, due mainly to the fact that the kinetics of the ORR is about 6 orders of magnitude slower than the HOR [72]. In addition, as mentioned previously, oxidation of carbon supports predominantly occurs at the cathode, where harsh conditions such as high potentials (0.6–1.2 V vs. RHE), high oxygen concentration, and extreme pH are normally found [73].

Recently, Chourashiya *et al.* [57] screened a series of Pt-based cathode catalysts using graphite powder (SG20 series) and graphene (IG300 series) as supports. All graphite-supported catalysts showed higher stability than two commercial Pt-based catalysts after 7,500 stress cycles. The performance of the PtSG20-trt23 nanocatalysts stood out in terms of electrochemical surface area (ECSA = 48.3 m^2 gPt^{-1}), ECSA retention (ECSA%) after 7,500 cycles (~70%), and low-Pt loading (~5 wt.%). Overall, the graphite-based supports had improved stability because of their good corrosion resistance at potentials more positive than 1.1 V vs. RHE. This behavior can be attributed to the most stable sp^2 hybridization formed by the C=C bonds. Graphite supports also have higher electrical conductivity than carbon blacks [55, 56]. Furthermore, graphite is inexpensive and abundant. However, it is expected that the demand for graphite will show a three-fold increase by 2030 [74] because it is intensively used in the battery industry, which could be a drawback for its application in PEMFCs and AEMFCs.

Meanwhile, OMCs are very interesting alternative supports, offering major advantages when used as support in cathode electrodes. Because of their textural and morphological features, OMCs enhance the transport of O_2 molecules. Also, OMCs anchor nanoparticles within their pores preventing poisoning of active sites by sulfonates contained in the ionomer, which is reflected in enhanced catalytic activity, performance, and durability [52, 58]. Yarlagadda *et al.* [58] showed that the local oxygen transport resistance ($R(O_2)$-local) of 30%Pt/OMC is 3 s cm^{-1}, i.e., lower than that of Pt supported on carbon blacks (8.3 to 15.9 s cm^{-1}). Moreover, 30%Pt/OMC had an ECSA close to 65 m^2 gPt^{-1} and ECSA% = 24% after 30,000 cycles. The commonly used OMC supports, such as CMK3, are obtained by hard-templating methods using silicon oxide (e.g., MCM-41 or SBA-15), which allows the tailoring of porosity with high surface area accessible to O_2 [59, 60].

Meléndez-González *et al.* [61] reported the synthesis of low-Pt loading nanocatalyst (~5 wt.%) supported on OMHCS, synthesized by the polyol and the BAE methods (labeled as Pt/HCS-P and Pt/HCS-B, respectively). The high surface area of the OMHCS (SSA_{BET} = 1,084 $m^2\ g^{-1}$) inhibited particle size growth (d < 3nm), improving their high dispersion. It also enhanced the transport of oxygen and other species to the active sites. Pt/HCS-P showed a six-fold increase in mass current density (i_m) compared with benchmark 20 wt.% Pt/C in 0.5 M KOH (Figure 5.8a). As can be seen, the OMHCS showed characteristics that make them promising supports for the ORR in AEMFC applications. However, attention should be paid to the serious restrictions, which may hamper their wider use in fuel cells, such as time-consuming and expensive synthesis procedures, limited scalability, somehow low conductivity (0.3 × 10^{-2} S cm^{-1}), and poor ionomer distribution [52, 58].

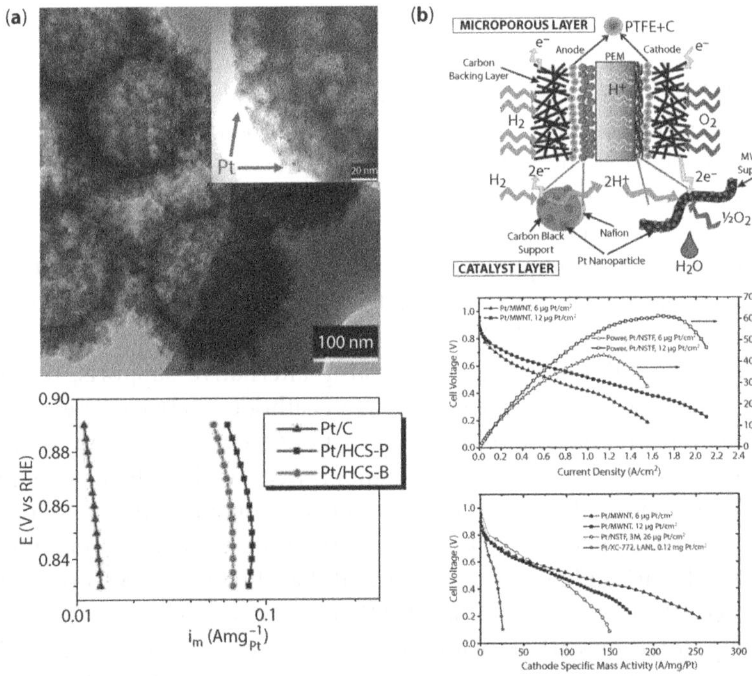

Figure 5.8 (a) HRTEM micrograph of Pt/HCS-B (top) and comparison of mass catalytic activity of Pt/HCS-B, Pt/HCS-P, and Pt/C (bottom). Reprinted from ref. [61], with permission from Springer Nature. (b) Scheme of the MEA of a PEMFC containing Pt/MWCNT nanocatalysts at cathode and anode (top), polarization and power density curves of the PEMFC with cathodes having 6 and 12 μg Pt cm^{-2} loadings (middle), and cathode-specific mass activity (bottom). Reprinted from ref. [76]. Copyright 2007, American Chemical Society.

Nanostructured carbon supports such as CNT, CNH, CNC, and lately graphene have shown attractive features to improve fuel cell performance in terms of power density and stability, even with a reduced Pt loading [57, 64–69]. According to the DOE, the target for the total loading of PGM on both electrodes was 0.125 mg cm^{-2} (electrode area) by 2020 [75]. In this sense, CNTs have been widely studied as alternative support since their discovery in the 1990s, with promising results. For example, Tang *et al.* [76] reported a novel MEA design with an ultra-low Pt loading (6 and 12 μg Pt cm^{-2}) on multiwall CNT (Pt/MWCNT) at the cathode and conventional 20 wt.% Pt/C at the anode (loading: 0.2 mg Pt cm^{-2}). The MEA sustained a high performance in a H_2/O_2 PEMFC (Figure 5.8b). The power density was 613 mW cm^{-2} with a mass activity (on the basis of cathode loading) of 250 A mg Pt^{-1}. The findings of this research were interesting from the point of view of the ultra-low cathode Pt loading, aiming to eventually achieve the DOE target.

Another type of carbon structures is generally known as gels. Carbon xerogels [62] and aerogels [63] have been evaluated as supports of Pt nanoparticles for cathode applications in PEMFC. The former are easier to synthesize than the latter. Typically, carbon aerogels have been prepared by supercritical drying and pyrolysis of a 3D polymeric gel (e.g., resorcinol-formaldehyde), while carbon xerogels can be obtained by simple evaporative drying and pyrolysis methods. For instance, Job *et al.* [62] obtained carbon xerogel-supported Pt nanocatalysts, with the carbon showing SSA_{BET} up to 716 m^2 g^{-1}. Meanwhile, the highest ECSA among the Pt nanocatalysts was of 42 m^2 g_{Pt}^{-1} of the X1-Pt sample, measured by CO chemisorption (27 m^2 g_{Pt}^{-1} measured by cyclic voltammetry in the MEA). The authors showed that xerogels can have a similar performance to aerogels in PEMFC tests. However, it should be noted that the active sites had poor dispersion because a large percentage of Pt nanoparticles (50% to 60%) formed agglomerated particles. Nevertheless, because of the complexities of their synthesis and the somehow poor metal:support interactions, carbon gels have not been explored more thoroughly.

Many research groups have agreed that supporting Pt alloys on CNT is an effective strategy to increase the performance of fuel cell nanocatalysts. It has been reported that CNTs improve the long-term stability, corrosion resistance, wettability, and dispersion of active sites [64–66]. Currently, the large-scale production of high-purity CNT, one of the main limitations of their use in the past two decades, is feasible. For instance, currently, the price of MWCNTs has dropped to less than USD 100 kg^{-1}, while that of single-wall carbon nanotubes (SWCNTs) is around USD 2,000 kg^{-1} [77]. CNH and CNC are derivatives of the synthesis of CNT, the three structures having a similar performance. However, to date, no synthesis procedures have been developed for the large-scale production of CNH and CNC, which is one of the main limiting factors for their widespread application in PEMFCs and AEMFCs.

Graphene is one of the newest nanostructured carbons that have been evaluated as a support for fuel cell nanocatalysts. Overall, the performance of graphene-supported PGM nanocatalysts toward the HOR and the ORR is similar to that of carbon black-supported Pt nanocatalysts [69]. It has been observed that graphene having a 2D sheet structure blocks the diffusion of gaseous species, opposite to the high diffusion rate reported for the porous semispherical particles of carbon blacks and mesoporous carbons [78]. However, some Pt/graphene nanocatalysts have shown outstanding performance for the oxidation of liquid organic molecules because this support has demonstrated to improve the tolerance of the metal nanoparticles to CO, which is not always the case for other carbon supports.

CNFs are perhaps one of the most promising supports for fuel cell applications. A comprehensive review of these supports has recently been reported by Peera *et al.* [32]. The two most widely reported methods for obtaining CNFs are chemical vapor deposition (CVD) and electrospinning. Highly graphitic nanofibers (GNFs) are obtained by the former method. GNFs have high corrosion resistance and good electrical conductivity, promote a high dispersion of active sites, and are of low cost. For example, Peera *et al.* [79] reported the superior performance of Pt/GNF-DNA as a cathode catalyst (0.1 mg Pt cm^{-2}). A PEMFC equipped with such a cathode generated a power density of 675 mW cm^{-2}, achieving a current density of 1.3 A cm^{-2}. Pt/GNF-DNA showed high stability, with ECSA losses of only ~40% after 10,000 cycles (Figure 5.9a).

On the other hand, the wide group of alternative carbon supports includes, for example, heteroatom-doped CNT, graphene, and OMC, among others. Commonly, heteroatoms (e.g., N, S, P, B, O, etc.) can substitute an atom from the carbon lattice, modifying the electronic density of adjacent carbon atoms. This modification or defect in the carbon lattice enhances the catalytic activity, particularly for the ORR in alkaline electrolytes [71]. The role of the heteroatoms on the reaction mechanisms is not yet completely clear; however, it has been shown that low Pt-content nanocatalysts with high catalytic activity and stability can be obtained using such supports [56, 80].

Moreover, heteroatom self-doped biocarbons are emerging supports recently used for electrochemical energy applications. Biocarbons are obtained from biomass by hydrothermal carbonization and/or pyrolysis. One important advantage of using biomass as a raw material is its

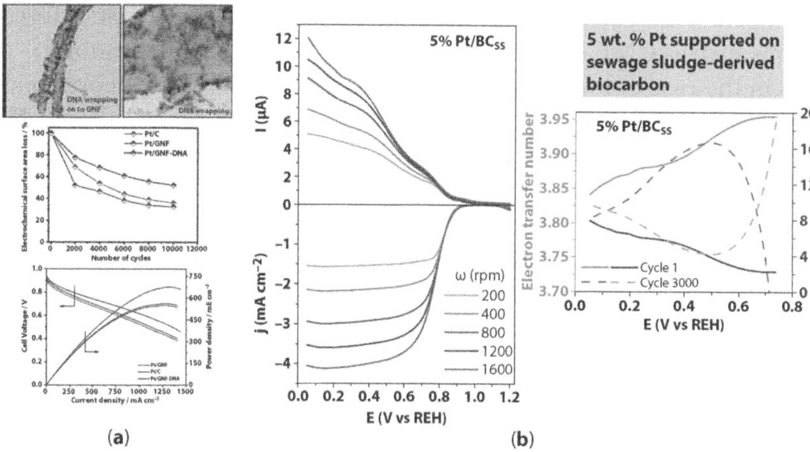

Figure 5.9 (a) Morphology and polarization curves of the Pt nanocatalyst supported on graphitic carbon nanofibers (Pt/GNF-DNA) with high performance and stability. Reprinted from ref. [79]. Copyright 2015, Elsevier. (b) Disk–ring polarization curves of the ORR at 5 wt.% Pt/BC_{SS} (supported on a sewage sludge-derived biocarbon, BC_{SS}) labeled as Pt/in 0.1 M KOH, collected at 5 mV s^{-1} (left), electron transfer number and %HO_2^- before and after an ADT of 3,000 cycles (right).

abundance and low cost. There are few reports of biocarbons used as supports of PGM for PEMFC or AEMFC applications [81].

Our group has developed biocarbons from several sources of biomass, with the purpose of using them as supports of noble metal nanoparticles or as metal-free catalysts. Figure 5.9(b) (left) shows the disk and ring polarization curves of the ORR of a 5 wt.% Pt/BC_{SS} nanocatalyst collected at 5 mV s^{-1}. Here, BC_{SS} (surface area of 412 m^2 g^{-1}, average pore diameter of 4.3 nm) was a biocarbon obtained from sewage sludge from the wastewater treatment plant in the northern city of Saltillo, Coahuila, Mexico. The sewage sludge was hydrothermally treated at 200°C for 12 h, then mixed with $ZnCl_2$ and pyrolyzed at 800°C for 2 h. The Pt/BC_{SS} nanocatalyst was synthesized by the BAE method.

It was demonstrated that BC_{SS} has excellent behavior during the ORR as support of the low Pt-content nanocatalyst in 0.1 M KOH. The electron transfer number was ca. 4 before and after an accelerated degradation test (ADT, 3,000 cycles between 0.6 and 1 V vs. RHE at 50 mV s^{-1}). The percentage of hydroperoxide ion produced (%HO_2^-) remained below 8% after ADT (Figure 5.9b, right). The study showed that biocarbons from biomass should be developed and evaluated as supports of nanoparticles, aiming to compete with Vulcan and replace the more expensive nanostructured carbons such as CNT, MWCNT, and OMHCS in fuel cell applications.

As in the case of low-temperature fuel cells, the MEA is the core component of PEMEs and AEMEs. Within the MEA, supported noble metal nanocatalysts promote the HER and OER. Xie *et al.* [82] used carbon cloth as support of PtNi-Ni nanoarrays for the HER in alkaline media, with a low Pt loading of 7.7 wt.%. The nanocatalyst labeled as PtNi-Ni/CC showed an overpotential of 38 mV at −10 mA cm^{-2}, a high stability, and a faradaic efficiency close to 100% after 90 h of operation. Its performance was higher than that of Pt/C (overpotential of 51 mV at −10 mA cm^{-2}). The authors attributed the high performance of the PtNi-Ni/CC nanocatalyst to its 3D nanoarray architecture.

For PEME applications, our group has also developed biocarbons as supports. A 5 wt.% Pt/BC_{GP} nanocatalyst was synthesized by the BAE method, where BC_{GP} is a self-doped biocarbon derived from garlic peel (surface area of 1,329 m^2 g^{-1}, average pore diameter of 2.5 nm). The BC_{GP} support was obtained in the same fashion as the BC_{SS} biocarbon described above. The Pt/BC_{GP} nanocatalyst was evaluated as a cathode for the HER in 0.1 M $HClO_4$ and compared with the performance of a 5% Pt/C nanocatalyst. Pt/BC_{GP} showed an overpotential of 37 mV at −10 mA cm^{-2} before ADT, which compared well with 23 mV of Pt/C (Figure 5.10a). Pt/BC_{GP} demonstrated an extraordinarily high stability after 3,000 cycles, with a practically unchanged overpotential of 38 mV at −10 mA cm^{-2}. Thus, it was demonstrated that biocarbons can act as supports of PGM-based nanocatalysts for the HER.

The atomic layer deposition technique (ADL) has shown to be a technique that allows the precise control of particle size and dispersion, with the potential of dramatically reducing the Pt loading. Figure 5.10(b) (top) shows a scheme of atom-size Pt or Pt clusters obtained by this method [83]. Employing it, Cheng *et al.* [84] supported single atom-size Pt on graphene and N-doped graphene (labeled as ALDPt/NGNs), evaluating their performance in acidic media toward the HER, as shown in Figure 5.10b) (bottom). The Pt loading of the ALDPt/NGN nanocatalysts was 2.1% and 7.6%. Moreover, the mass activity of the ALD50Pt/NGN catalyst was 37.4 times higher than that of Pt/C at an overpotential of 0.05 V (10.1 vs. 0.27 A mg^{-1}), with a Pt loading of 2.1%, showing a Tafel slope of 29 mV dec^{-1}.

Mahmood *et al.* [85] screened a variety of noble (Pt, Pd, and Ru) and non-noble metal (Co and Ni) catalysts supported on nitrogenated 2D carbon structure (C_2N). The catalysts were evaluated in acidic and alkaline media, and the best performance was shown by the Ru@C_2N material. This universal pH catalyst showed overpotentials of 13 and 17 mV, at -10 mA

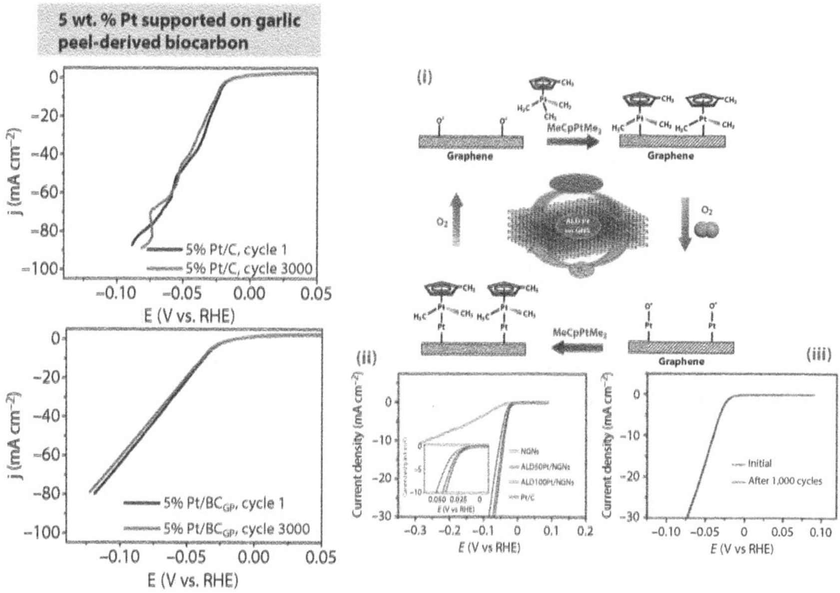

(a) (b)

Figure 5.10 (a) iR-compensated polarization curves of the HER at Pt/C (C = Vulcan XC-72) and Pt/BC_{GP} (BC_{GP} = garlic peel-derived biocarbon) acquired at 5 mV s^{-1} in 0.1 M $HClO_4$ after 1 and 3,000 cycles. (b) (i) Scheme of the ALD of atomic size Pt clusters supported on graphene supports. Reprinted from ref. [83] with permission of Springer Nature; (ii) polarization curves of the HER at Pt/C and the ALDPt/NGNs in 0.5 M H_2SO_4 electrolyte and a scan rate of 2 mV s^{-1}; and (iii) polarization curves of the HER at cycles 1 and 1,000 on ALD50Pt/NGNs. Reprinted from ref. [84] with permission of Springer Nature.

cm^{-2}, in 0.5 M H_2SO_4 and 1.0 M KOH, respectively. The authors attributed the high catalytic activity of Ru@C_2N to the Ru–H bonding mechanism.

In general, the literature on alternative carbon supports is not as extensive when it comes to applications in electrolyzers as compared with fuel cells. As an alternative, molecular catalysis using carbon supports (or substrates) to promote the HER is a promising alternative for PEME and AEME applications [12]. The main advantages of such an approach are the precision in the structural design, a more efficient use of the active sites, and a better understanding of the reaction mechanisms. The preparation of this type of catalysts consists of the immobilization of the molecule with the active site (e.g., Pt, Ru, Ir, etc.) on a support, in this case carbon-based (e.g., CNT, CNF, and graphene, among many others).

Regarding the OER, attempts to develop corrosion resistance supports have been made, in view of the low stability of most carbons at high positive potentials. For PEME applications, metal oxides such as SnO_2, TiO_2, or doped Sb-SnO_2, with improved conductivity, and metal carbides like TaC and TiC have been proposed, as reviewed recently [86]. Maillard *et al.* reported that supporting IrO_x nanoparticles increases their catalytic activity for the OER but reduces their stability [87]. In their study, the authors found that tin oxides doped with tantalum (TaTO) supporting IrO_x nanoparticles have a higher stability than tin oxides doped with antimony (ATO) and Vulcan. They concluded that the mixed oxidation states of Ir and the high density of active sites were responsible for the high catalytic activity of the nanocatalysts for the OER.

Similarly, alternative stable supports must be tested for the OER in AEME devices, which allow for the use of abundant materials [88]. In alkaline media, carbon black-supported non-noble metals have shown a high catalytic activity for the OER. Li *et al.* [89] developed a $Ni_{0.69}Fe_{0.31}O_x$/C nanocatalyst (where C = Vulcan) with higher catalytic activity than commercial Ir/C. The faradaic efficiency of $Ni_{0.69}Fe_{0.31}O_x$/C was 97%, which suggested that the corrosion of Vulcan and the oxidation of the catalyst were negligible. Moreover, the $Ni_{0.69}Fe_{0.31}O_x$/C nanocatalyst showed a lower overpotential of 280 mV at 10 mA cm^{-2} compared with 305 mV at Ir/C, along with a high stability in 1.0 M KOH. $Ni_{0.69}Fe_{0.31}O_x$/C also showed a lower Tafel slope, compared with Ir/C as shown in Figure 5.11. Thus, the less aggressive environment in AEMEs offers some advantages regarding the use of materials, compared with PEMEs.

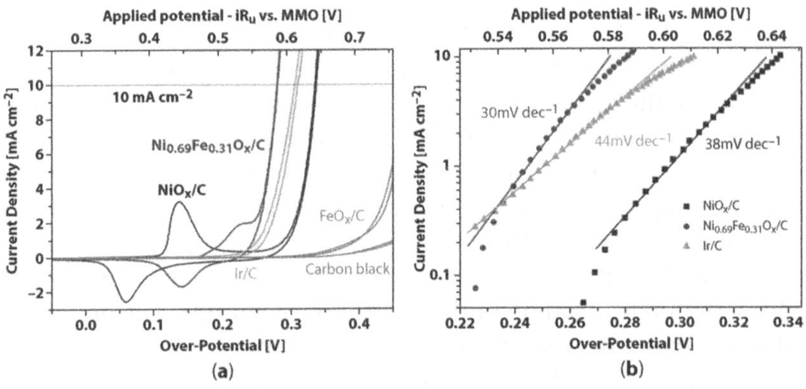

Figure 5.11 (a, b) OER curves at $Ni_{0.69}Fe_{0.31}O_x$/C, Ir/C, and other nanocatalysts acquired at 5 mV s^{-1} and 1,600 rpm in 1.0 M KOH. (b) Tafel slopes at the nanocatalysts. Reprinted from ref. [89]. Copyright 2014, American Chemical Society.

5.5 Comparison of the Performance of Nanocatalysts for Fuel Cell and Electrolyzer Reactions

Table 5.1 shows a comparison of electrochemical parameters from the ORR of some of the nanocatalysts discussed in the previous sections and other studies from the literature. In terms of the half-cell potential ($E_{1/2}$), most of the Pt-based nanocatalysts perform well in 0.1 M $HClO_4$, reaching values close to 0.9 V vs. RHE or higher, e.g., the PtNiN/XC72-PBI [47], Pt/Vinatech [48], Pd_2Cu_1@Pt/C [90], and Mo–PtFe/C–H [91] nanocatalysts, which also deliver a high limiting current density (j_L). It is interesting, however, that the Pd_3/Co-nanofilm/C nanocatalyst in [92] shows a good positive $E_{1/2}$ value in alkaline electrolyte compared with the acidic media. When reported by the authors, the values indicate a mechanism

Table 5.1 Electrochemical parameters of the ORR at PGM-based nanocatalysts.

Nanocatalyst	Electrolyte	$E_{1/2}$ (V vs. RHE)	N	j_L at 1,600 rpm (mA cm²)	Ref.
PtNiN/XC72-PBI	0.1 M $HClO_4$	~0.95	–	–6.0	[47]
Pt/Vinatech	0.1 M $HClO_4$	~0.90	4.0	–6.0	[48]
Pt/BC_{SS}	0.1 M KOH	0.74	3.8	–4.0	Figure 5.9b
Pd_2Cu_1@Pt/C	0.1 M $HClO_4$	0.90	–	–5.5	[90]
Mo–PtFe/C–H	0.1 M $HClO_4$	0.92	–	–5.9	[91]
Pd_3/ Co-nanofilms/C	0.5 M H_2SO_4/0.1 M NaOH	0.79 / 0.91	–	–	[92]
Pt/GO-C	0.5 M H_2SO_4	0.79	3.9	–3.3	[93]
Pt/C-TKK	0.1 M $HClO_4$	0.90	–	–5.1	[94]
NCM ⊂ Pt/C	0.1 M $HClO_4$	0.89	–	–5.6	[95]
Pd/CeO_2-NR/C	0.5 M KOH	0.87	4.0	–3.7	[96]

–: not found/not reported.

Table 5.2 Electrochemical parameters of the HER at PGM-based nanocatalysts.

HER				
Nanocatalyst	**Electrolyte**	**Overpotential at −10 mA cm^2 (mV)**	**Tafel slope (mV·dec^{-1})**	**Ref.**
Pt/OMHCS$_{Ru\text{-}dim}$	0.1 M KOH	28	–	Figure 5.6e
Rh/C-300A-350H	1 M KOH	29	–	[50]
Pt/BC$_{GP}$	0.1 M $HClO_4$	37	–	Figure 5.10a
Ru-1.0	1 M KOH	13	25.3	[98]
Pt/Co-N-C	1 M KOH	33	36.8	[99]
Pt-316L-0.1	0.5 M H_2SO_4	44	55	[100]
$PtSe_2$/Pt	1 M KOH	42	53	[101]
OER				
Nanocatalyst	**Electrolyte**	**Overpotential at 10 mA cm^2 (mV)**	**Tafel slope (mV·dec^{-1})**	**Ref.**
$Sn_{0.1}$-RuO_2@NCP	0.5 M H_2SO_4	178	60.6	[102]
Pt/OMHCS$_{Ru\text{-}cym}$	0.1 M KOH	240	–	Figure 5.6e
IrO_2/GCN	0.5 M H_2SO_4	276	57	[103]
Ir/g-C_3N_4/NG (5.9 wt.%)	0.5 M H_2SO_4	287	72.8	[104]
IrO_2-BN-rGO	0.5 M H_2SO_4	300	72.1	[105]
IrO_2/CNT	0.1 H_2SO_4	270	–	[106]
Ru^0/CeO_2	0.5 M KOH	420	122	[107]
MWCNT-bpyRutpy	1 M KOH	436	77.2	[108]

–: not found/not reported.

involving a 4 e^- transfer, including the good performance of 5 wt.% Pt/BC_{SS} with biocarbon from sewage sludge as support, with $E_{1/2}$ = 0.74 V vs RHE. Regarding the stability of nanocatalysts in acidic and alkaline media, a recent study shows that O–C=O surface functional groups play a significant role in the degradation of catalytic activity of Pt in acidic media, which is less severe in alkaline electrolyte [97].

Table 5.2 shows the performance of some PGM nanocatalysts discussed earlier in this chapter for the HER, along with several others from reports elsewhere. The overpotential at −10 mA cm^{-2} falls below 90 mV in acidic and alkaline media. It calls the attention that Rh [50] and Ru [98] nanocatalysts demonstrate a catalytic activity comparable to that of Pt-based materials. The low Tafel slope obtained from the Ru nanocatalyst is 25.3 mV dec^{-1}, having a low overpotential of only 13 mV [98]. It is to be highlighted that the 5 wt.% Pt/$OMHCS_{Ru\text{-}dim}$ nanocatalysts, i.e., the support functionalized with an organometallic compound, shows a high catalytic activity with an overpotential of only 28 mV. Moreover, the 5 wt.% Pt/BC_{GP} nanocatalyst, i.e., a biocarbon derived from garlic peel as support, has a high performance for the HER in 0.1 M $HClO_4$ electrolyte, with an overpotential of 37 mV.

Table 5.2 also shows the performance of nanocatalysts for the OER. It has been reported that the $Sn_{0.1}$-RuO_2@NCP nanocatalyst has an overpotential at 10 mA cm^{-2} of only 178 mV in acidic 0.5 M H_2SO_4 [102]. Meanwhile, the Pt/$OMHCS_{Ru\text{-}cym}$ nanocatalyst, i.e., the support functionalized with an organometallic support, shows an overpotential of 240 mV in 0.1 M KOH. In most cases, the Tafel slope has a value below 80 mV dec^{-1}. Only in the case of Ru^0/CeO_2, a slope of 122 mV dec^{-1} has been reported [106].

Outlook and Future Scope

Among the PGMs, or the noble metals in general, Pt, Pd, Ir, and Ru are considered the most active catalysts for the electrochemical devices discussed here. However, these are costly nanocatalysts. Carbon (and other) supports help decrease the PGM content in nanocatalysts, which is very important since their price has increased drastically. Figure 5.12 shows the price of Pt, Pd, Ir, and Ru as of June 21, 2022 [109]. Ir is the most expensive PGM, reaching ca. 176.3 USD/g at the time. The price of Pd shows a two-fold increase compared with that of Pt (about 33.1 USD/g of the latter). Meanwhile, Ru remains the least expensive among these four metals, even though it is not among the cheapest metals.

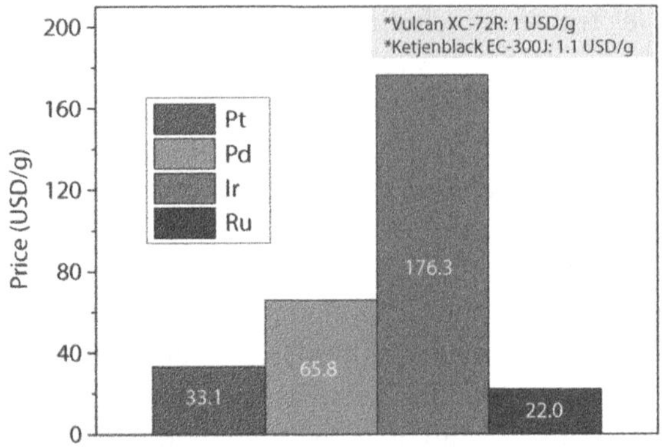

Figure 5.12 Price of some PGMs. Inset: Price of Vulcan XC-72R and Ketjenblack EC-300J as of June 21, 2022.

By analyzing the data in Figure 5.12, the use of supports such as Vulcan XC-72R and Ketjenblack EC-300J becomes very relevant since their commercial price for electrochemical applications is ca. 1 and 1.1 USD g^{-1}, respectively, also on June 21, 2022 [110]. Evidently, some other providers may have a lower price, and other carbon blacks may be even cheaper. On this matter, the application of biocarbons should also keep the price of supported PGM-based nanocatalysts in the low spectrum, adding the positive sustainable aspect of revalorizing waste biomass.

Therefore, the tendency among industrial and academic research groups is that of decreasing the noble metal content in the electrodes of these electrochemical devices, following the targets set by the DOE. For practical applications such as passenger vehicles, the total Pt content in PEMFCs is estimated to be in the 25–35-g range, with a total cost of $750–1,050 USD (at ~30 USD/g) [111]. Considering the current Pt price in Figure 5.12, the total cost would be ca. $828–1,160 USD. Thus, the importance of achieving the DOE target for 2025 of ≤0.1 total PGM content (rated in g kW^{-1}) [112], down from 0.125 g kW^{-1} currently (which is 0.125 mg cm^{-2} total PGM loading on both electrodes) [75, 112], is even greater. Attention should be paid to the fact that 0.125 mg cm^{-2} total loading refers to 0.1 at the cathode and 0.025 at the anode [111]. If the earth's crust abundance of Pt and Pd of 0.001 and 0.00064 is considered (as reported in [113], even though several other references inform different values), reaching the 2025 DOE's goal is of paramount importance in terms of costs and viability of PEMFCs as well as AEMFCs.

Similarly, the noble metal content in the electrodes of PEMEs and AEMEs devices should be reduced. As highlighted elsewhere, the AEME technology is in the earlier stages of development and more barriers must be overcome if the status of PEME systems is to be reached [5]. At the latter, the current target of Ir loading for the OER is 0.33 g kW^{-1}, while that for 2035 is 0.05 g kW^{-1}. This means 1.0 and 0.4 mg cm^{-2}, respectively [114]. Decreasing the Ir content is also highly relevant because of its five times higher price (Figure 5.12) and its lower earth's crust abundance (0.000003 ppm, [114]), compared with Pt. It has been reported that the price of Ir is volatile, showing a drastic increase in the 2020–2021 period, risking the extensive deployment of PEME systems [115]. An alternative to decrease the Ir content and thus the cost of anode electrodes of PEMEs is to develop IrRu alloys for the OER. Such an approach increases the catalytic activity while improving the stability due to the presence of RuO_2 [116].

Even more, Ru is active for the HER in AEME applications. Ru-based nanocatalysts have shown a higher intrinsic catalytic activity for the HER than Pt/C, considering the turnover frequency [85]. Although a less costly noble metal than Pt and Ir, the earth's crust abundance of Ru is reported as 0.001, i.e., it is a scarcely available noble metal [80]. Nevertheless, the annual production of Ru is higher than that of Ir but lower than Pt (25–36, 8–10, and 187–200 tons/year, respectively) [115].

Figure 5.13 shows that, under several metal loadings of Pt, Ir, and Ru, the amount of catalyst required to install electrolyzers and meet the H_2 production demand varies significantly [115]. According to the authors, it would take ca. 4 years to cover the ~400 GW of H_2 global annual demand if 10% of the annual production of Pt is used to deploy PEMEs (Figure 5.13a, solid pink line), i.e., the capacity rate would be limited to 100 GW/year. In the case of Ir, if 10% of its annual production is compromised by this technology, it would take 200 years to reach the 2-GW/year deployment rate that would be the limit if using this metal (Figure 5.13b, solid cyan line). The figure for 10% annual production of Ru used in electrolyzers is 50 years at a limit capacity rate of 8 GW/year (Figure 5.13c, solid green line). Figure 5.13d shows the tendency of the average prices of Pt, Ir, and Ru over the past years, which overall is in agreement with the data in Figure 5.12.

Therefore, the use of Pt, Pd, Ir, and Ru, or their combination, in the electrochemical energy conversion and storage systems discussed here seems to be unavoidable. In some cases, non-noble metals show promising performances although their catalytic activity and stability must equal those of the aforementioned noble metals. It is acknowledged that the cost of electrodes must be reduced, and this is achievable by lowering the noble metal loading. To do so, dispersing nanoparticles on a suitable support and paying attention to the design of nanocatalysts is highly recommended. A novel approach to maintain the low costs of electrodes is

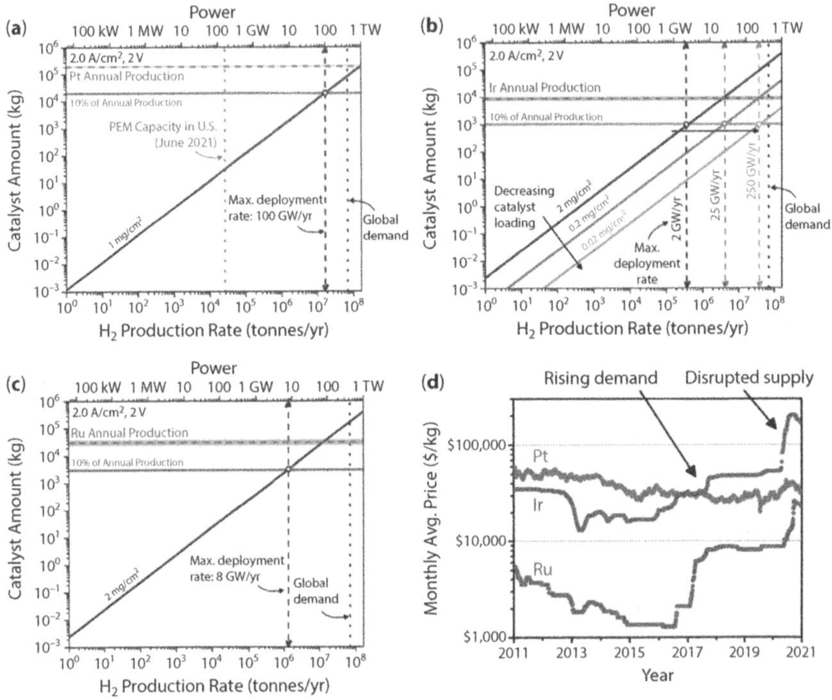

Figure 5.13 The amount of (a) Pt, (b) Ir, and (c) Ru required to implement PEME at different loading (mg/cm^2_{geo}), considering the scale (blue solid lines). (d) Tendency of the average prices of Pt, Ir, and Ru in the 2011–2021 period. Reprinted with permission from ref. [115]. Copyright 2022, American Chemical Society.

producing value-added biocarbons from widely available and very cheap biomass, which demonstrated that they can be used as supports. Under this approach, sustainable issues are added beyond the fact that fuel cells and electrolyzers are green energy systems.

The future scope for the broad application of PEMFC, AEMFC, PEME, and AEME systems should be put in terms of the recent reports of the Hydrogen Council, where [117–119]:

i) Roadmaps have been developed in 30 countries.
ii) Proposals for the development of 680 large-scale projects (worth: USD 240 billion) have been submitted.
iii) By 2030, companies foresee a 16-fold increase in their investments.

iv) Leaders in electrolyzer deployment and the manufacturing of fuel cells are China and Japan and South Korea, respectively.
v) In a net-zero scenario by 2050, the demand for clean hydrogen is projected to be around 660 million metric tons (MT).
vi) Hydrogen is considered a highly relevant piece of the decarbonization puzzle.

In conclusion, a large participation of these technologies should be expected in the following years, aiming to have a positive impact on the decarbonization of the economy globally.

References

1. Rodríguez-Varela, J., Alonso-Lemus, I.L., Savadogo, O., Palaniswamy, K., Overview: Current trends in green electrochemical energy conversion and storage. *J. Mater. Res.*, 36, 4071, 2021.
2. Wang, Y., Pang, Y., Xu, H., Martinez, A., Chen, K.S., PEM fuel cell and electrolysis cell technologies and hydrogen infrastructure development–A review. *Energy Environ. Sci.*, 15, 2288, 2022.
3. Dekel, D.R., Review of cell performance in anion exchange membrane fuel cells. *J. Power Sources*, 375, 158, 2018.
4. Das, G., Choi, J.-H., Nguyen, P.K.T., Kim, D.-J., Yoon, Y.S., Anion exchange membranes for fuel cell application: A review. *Polymers*, 14, 1197, 2022.
5. Du, N., Roy, C., Peach, R., Turnbull, M., Thiele, S., Bock, C., Anion-exchange membrane water electrolyzers. *Chem. Rev.*, 122, 11830, 2022.
6. The search included: Proton Exchange Membrane Fuel Cell, Polymer Electrolyte Membrane Fuel Cell, Proton Exchange Membrane Electrolyzer, Polymer Electrolyte Membrane Electrolyzer, Polymer Electrolyte Water Electrolyzer, Proton Exchange Water Electrolyzer; and Anion Exchange Membrane Fuel Cell, Alkaline Polymer Electrolyte Fuel Cell, Anion Exchange Membrane Electrolyzer, Anion Exchange Water Electrolyzer, Alkaline Polymer Electrolyte Electrolyzer, Anion Exchange Membrane Electrolysis, Anion Exchange Membrane water Electrolysis.
7. Davydova, E.S., Mukerjee, S., Jaouen, F., Dekel, D.R., Electrocatalysts for hydrogen oxidation reaction in alkaline electrolytes. *ACS Catal.*, 8, 6665, 2018.
8. Ramaswamy, N. and Mukerjee, S., Fundamental mechanistic understanding of electrocatalysis of oxygen reduction on Pt and non-Pt surfaces: Acid versus alkaline media. *Adv. Phys. Chem.*, 2012, 491604, 2012.

9. Sgarbi, R., Kumar, K., Jaouen, F., Zitolo, A., Ticianelli, E., Maillard, F., Oxygen reduction reaction mechanism and kinetics on M-N_xC_y and M@N- C active sites present in model M-N-C catalysts under alkaline and acidic conditions. *J. Solid State Electrochem.*, 25, 45, 2021.
10. Zhang, J., Zhu, W., Huang, T., Zheng, C., Pei, Y., Shen, G., Nie, Z., Xiao, D., Yin, Y., Guiver, M.D., Recent insights on catalyst layers for anion exchange membrane fuel cells. *Adv. Sci.*, 8, 2100284, 2021.
11. Chourashiya, M., Sharma, R., Gyergyek, S., Andersen, S.M., Gram-size Pt/C catalyst synthesized using Pt compound directly recovered from an end-of-life PEM fuel cell stack. *Mater. Chem. Phys.*, 276, 125439, 2022.
12. Zhang, B., Fan, L., Ambre, R.B., Liu, T., Meng, Q., Timmer, B.J.J., Sun, L., Advancing proton exchange membrane electrolyzers with molecular catalysts. *Joule*, 4, 1408, 2020.
13. Chen, Z., Guo, L., Pan, L., Yan, T., He, Z., Li, Y., Shi, C., Huang, Z.-F., Zhang, X., Zou, J.-J., Advances in oxygen evolution electrocatalysts for proton exchange membrane water electrolyzers. *Adv. Energy Mater.*, 12, 2103670, 2022.
14. Wan, L., Xu, Z., Xu, Q., Wang, P., Wang, B., Overall design of novel 3D-ordered MEA with drastically enhanced mass transport for alkaline electrolyzers. *Energy Environ. Sci.*, 15, 1882, 2022.
15. Guo, H., Fang, Z., Li, H., Fernandez, D., Henkelman, G., Humphrey, S.M., Yu, G., Rational design of Rhodium–Iridium alloy nanoparticles as highly active catalysts for acidic oxygen evolution. *ACS Nano*, 13, 13225, 2019.
16. He, C., Yang-Neyerlin, A.C., Pivovar, B.S., Probing anion exchange membrane fuel cell cathodes by varying electrocatalysts and electrode processing. *J. Electrochem. Soc.*, 169, 024507, 2022.
17. Truong, V.M., Duong, N.B., Yang, H., Comparison of carbon supports in anion exchange membrane fuel cells. *Materials*, 13, 5370, 2020.
18. Kabir, S., Lemire, K., Artyushkova, K., Roy, A., Odgaard, M., Schlueter, D., Oshchepkov, A., Bonnefont, A., Savinova, E., Sabarirajan, D.C., Mandal, P., Crumlin, E.J., Zenyuk, I.V., Atanassov, P., Serov, A., Platinum group metal-free NiMo hydrogen oxidation catalysts: High performance and durability in alkaline exchange membrane fuel cells. *J. Mater. Chem. A*, 5, 24433, 2017.
19. Adabi, H., Shakouri, A., Hassan, N.U., Varcoe, J.R., Zulevi, B., Serov, A., Regalbuto, J.R., Mustain, W.E., High-performing commercial Fe–N–C cathode electrocatalyst for anion-exchange membrane fuel cells. *Nat. Energy*, 6, 834, 2021.
20. Ma, R., Lin, G., Zhou, Y., Liu, Q., Zhang, T., Shan, G., Yang, M., Wang, J., A review of oxygen reduction mechanisms for metal-free carbon-based electrocatalysts. *NPJ Comput. Mater.*, 5, 78, 2019.
21. Cong, Y., Yi, B., Song, Y., Hydrogen oxidation reaction in alkaline media: From mechanism to recent electrocatalysts. *Nano Energy*, 44, 288, 2018.

22. Wang, J.X., Hydrogen oxidation and evolution on platinum in acids, in: *Encyclopedia of Applied Electrochemistry*, G. Kreysa, K. Ota, R.F. Savinell (Eds.), pp. 1045–1049, Springer, New York, NY, 2014.
23. Al-Naggar, A.H., Shinde, N.M., Kim, J.-S., Mane, R.S., Water splitting performance of metal and non-metal-doped transition metal oxide electrocatalysts. *Coord. Chem. Rev.*, 474, 214864, 2023.
24. Zhang, J., Zhang, L., Liu, J., Zhong, C., Tu, Y., Li, P., Du, L., Chen, S., Cui, Z., OH spectator at IrMo intermetallic narrowing activity gap between alkaline and acidic hydrogen evolution reaction. *Nat. Commun.*, 13, 5497, 2022.
25. Fornaciari, J.C., Weng, L.-C.-., Alia, S.M., Zhan, C., Pham, T.A., Bell, A.T., Ogitsu, T., Danilovic, N., Weber, A.Z., Mechanistic understanding of pH effects on the oxygen evolution reaction. *Electrochim. Acta*, 405, 139810, 2022.
26. Debata, S., Patra, S., Banerjee, S., Madhuri, R., Sharma, P.K., Controlled hydrothermal synthesis of graphene supported $NiCo_2O_4$ coral-like nanostructures: An efficient electrocatalyst for overall water splitting. *Appl. Surf. Sci.*, 449, 203, 2018.
27. Hou, S., Kluge, R.M., Haid, R.W., Gubanova, E.L., Watzele, S.A., Bandarenka, A.S., Garlyyev, B., A review on experimental identification of active sites in model bifunctional electrocatalytic systems for oxygen reduction and evolution reactions. *ChemElectroChem*, 8, 3433, 2021.
28. Mamtani, K., Jain, D., Dogu, D., Gustin, V., Gunduz, S., Co, A.C., Ozkan, U.S., Insights into oxygen reduction reaction (ORR) and oxygen evolution reaction (OER) active sites for nitrogen-doped carbon nanostructures (CNx) in acidic media. *Appl. Catal. B: Environ.*, 220, 88, 2018.
29. Panigrahy, S., Samanta, R., Panda, P., Mishra, R., Barman, S., RuO_2 as promoter in Pt-RuO2-nanostructures/carbon composite, a pH-universal catalyst for hydrogen evolution/oxidation reactions. *Int. J. Energy Res.*, 46, 6406, 2022.
30. Samanta, R., Mishra, R., Barman, S., Interface-engineered porous Pt–PdO nanostructures for highly efficient hydrogen evolution and oxidation reactions in base and acid. *ACS Sustain. Chem. Eng.*, 10, 3704, 2022.
31. Bharti, A. and Cheruvally, G., Application of novel carbonaceous materials as support for fuel cell electrocatalysts, in: *Advanced Electrocatalysts for Low- Temperature Fuel Cells*, F.J. Rodríguez-Varela and T.W. Napporn (Eds.), pp. 175–213, Springer Nature, Switzerland AG, 2018.
32. Peera, S.G., Koutavarapu, R., Akula, S., Asokan, A., Moni, P., Selvaraj, M., Balamurugan, J., Kim, S.O., Liu, C., Sahu, A.K., Carbon nanofibers as potential catalyst support for fuel cell cathodes: A review. *Energy Fuels*, 35, 11761, 2021.
33. Fang, Z., Lee, M.S., Kim, J.Y., Kim, J.H., Fuller, T.F., The effect of carbon support surface functionalization on PEM fuel cell performance, durability, and ionomer coverage in the catalyst layer. *J. Electrochem. Soc.*, 167, 064506, 2020.

34. Karczmarska, A., Adamek, M., El Houbbadi, S., Kowalczyk, P., Laskowska, M., Carbon-supported noble-metal nanoparticles for catalytic applications—A review. *Crystals*, 12, 584, 2022.
35. Rodríguez-Varela, J. and Alonso-Lemus, I.L., Waste-derived biocarbons as nonnoble metal catalysts and supports for nanocatalysts for fuel cells reactions in alkaline media, in: *Encyclopedia of Electrochemistry*, Wiley-VCH GmbH, Germany, 2021.
36. Shi, X., Pérez-Salcedo, K., Hanif, S., Anwar, R., Cindrella, L., Iqbal, N., Jose, S., Kannan, A.M., Progress on the functionalization of carbon nanostructures for fuel cell electrocatalysts, in: *Advanced Electrocatalysts for Low-Temperature Fuel Cells*, F.J. Rodríguez-Varela and T.W. Napporn (Eds.), pp. 215–234, Springer Nature, Switzerland AG, 2018.
37. Pech-Rodríguez, W.J., Rodríguez-Varela, F.J., Rocha-Rangel, E., González-Quijano, D., Martínez-Loyola, J.C., Vargas-Gutiérrez, G., Suarez-Velazquez, G., Morais, C., Napporn, T.W., Cost-effective functionalization of Vulcan XC-72 by using the intermittent microwave heating process as nanocatalyst support for ethanol electrooxidation in acid media. *Mex. J. Chem. Eng.*, 20, 955, 2021.
38. Pech-Rodríguez, W.J., Rodríguez-Varela, F.J., Suarez-Velázquez, G.G., Meléndez-González, P.C., Armendariz-Mireles, E.N., Insight into the effect of green methanol functionalization of Vulcan on the catalytic activity Pt1Sn1 alloy for the Ethanol Oxidation Reaction in acid and alkaline media. *J. Mater. Res.*, 36, 4182, 2021.
39. Bai, J., Ke, S., Song, J., Wang, K., Sun, C., Zhang, J., Dou, M., Surface engineering of carbon-supported platinum as a route to electrocatalysts with superior durability and activity for PEMFC cathodes. *ACS Appl. Mater. Interfaces*, 14, 4, 5287, 2022.
40. Kim, J.H., Cheon, J.Y., Shin, T.J., Park, J.Y., Joo, S.H., Effect of surface oxygen functionalization of carbon support on the activity and durability of Pt/C catalysts for the oxygen reduction reaction. *Carbon*, 101, 449, 2016.
41. Hernández-Ramírez, A., Rodríguez-Varela, F.J., Meléndez-González, P.C., Sánchez-Castro, M.E., Catalytic activity of Pt–$CoTiO_3$ nanocatalysts supported on reduced graphene oxide functionalized with Cr organometallic compounds for the oxygen reduction reaction. *J. Mater. Res.*, 36, 4192, 2021.
42. Martínez-Loyola, J.C., Alonso-Lemus, I.L., Sánchez-Castro, M.E., Escobar-Morales, B., Torres-Lubián, J.R., Rodríguez-Varela, F.J., Surface functionalization of ordered mesoporous hollow carbon spheres with ru organometallic compounds as supports of low-Pt content nanocatalysts for alkaline hydrogen and oxygen evolution reactions. *MRS Adv.*, 5, 2973, 2020.
43. Martínez-Loyola, J.C., Siller-Ceniceros, A.A., Sánchez-Castro, M.E., Sánchez, M., Torres-Lubián, J.R., Escobar-Morales, B., Ornelas, C., Alonso-Lemus, I.L., Rodríguez-Varela, F.J., High performance pt nanocatalysts for the oxidation of methanol and ethanol in acid media by effect of functionalizing

carbon supports with ru organometallic compounds. *J. Electrochem. Soc.*, 167, 164502, 2020.

44. Siller-Ceniceros, A.A., Sánchez-Castro, E., Morales-Acosta, D., Torres-Lubián, J.R., Martínez-Guerra, E., Rodríguez-Varela, J., Functionalizing reduced graphene oxide with ru-organometallic compounds as an effective strategy to produce high-performance Pt nanocatalysts for the methanol oxidation reaction. *ChemElectroChem*, 6, 4902, 2019.
45. Siller-Ceniceros, A.A., Sánchez-Castro, M.E., Morales-Acosta, D., Torres-Lubian, J.R., Martínez G., E., Rodríguez-Varela, F.J., Innovative functionalization of vulcan xc-72 with ru organometallic complex: Significant enhancement in catalytic activity of Pt/C electrocatalyst for the Methanol Oxidation Reaction (MOR). *Appl. Catal. B: Environ.*, 209, 455, 2017.
46. Rich, S.S., Burk, J.J., Kong, C.S., Cooper, C.D., Morse, D.E., Buratto, S.K., Nitrogen functionalized carbon black: A support for Pt nanoparticle catalysts with narrow size dispersion and high surface area. *Carbon*, 81, 115, 2015.
47. Li, C., Song, L., Zhao, X., Sasaki, K., Xie, J., Nitrogen-doped PtNi catalysts on polybenzimidazole-functionalized carbon support for the oxygen reduction reaction in polymer electrolyte membrane fuel cells. *ACS Appl. Mater. Interfaces*, 14, 26814, 2022.
48. Lee, W.J., Bera, S., Kim, C.M., Koh, E.-K., Hong, W.-P., Oh, S.-J., Cho, E., Kwon, S.-H., Synthesis of highly dispersed Pt nanoparticles into carbon supports by fluidized bed reactor atomic layer deposition to boost PEMFC performance. *NPG Asia Mater.*, 12, 40, 2020.
49. Park, D., Ahmed, M.S., Jeon, S., Covalent functionalization of graphene with 1,5-diaminonaphthalene and ultrasmall palladium nanoparticles for electrocatalytic oxygen reduction. *Int. J. Hydrogen Energy*, 42, 206, 2017.
50. Peng, Y., Zhang, Y., Guo, A., Mao, M., Wang, Y., Long, Y., Fan, G., Universal low-temperature oxidative thermal redispersion strategy for green and sustainable fabrication of oxygen-rich carbons anchored metal nanoparticles for hydrogen evolution reactions. *Chem. Eng. J.*, 433, 133648, 2022.
51. Xing, L., Shi, W., Su, H., Xu, Q., Das, P.K., Mao, B., Scott, K., Membrane electrode assemblies for PEM fuel cells: A review of functional graded design and optimization. *Energy*, 177, 445, 2019.
52. Antolini, E., Carbon supports for low-temperature fuel cell catalysts. *Appl. Catal. B: Environ.*, 88, 1, 2009.
53. Uchida, M., Aoyama, Y., Tanabe, M., Yanagihara, N., Eda, N., Ohta, A., Influences of both carbon supports and heat-treatment of supported catalyst on electrochemical oxidation of methanol. *J. Electrochem. Soc.*, 142, 2572, 1995.
54. Soo, L.T., Loh, K.S., Mohamad, A.B., Daud, W.R.W., Wong, W.Y., An overview of the electrochemical performance of modified graphene used as an electrocatalyst and as a catalyst support in fuel cells. *Appl. Catal. A: Gen.*, 497, 198, 2015.

55. Zhang, S., Shao, Y., Li, X., Nie, Z., Wang, Y., Liu, J., Yin, G., Lin, Y., Low-cost and durable catalyst support for fuel cells: Graphite submicron particles. *J. Power Sources*, 195, 457, 2010.
56. Peera, S.G., Arunchander, A., Sahu, A., Platinum nanoparticles supported on nitrogen and fluorine co-doped graphite nanofibers as an excellent and durable oxygen reduction catalyst for polymer electrolyte fuel cells. *Carbon*, 107, 667, 2016.
57. Chourashiya, M., Vindt, S.T., Velázquez-Palenzuela, A.A., Pedersen, C.M., Kallesøe, C., Andersen, S.M., Low-cost graphite as durable support for Pt-based cathode electrocatalysts for proton exchange membrane-based fuel cells. *Int. J. Hydrogen Energy*, 43, 23275, 2018.
58. Yarlagadda, V., Ramaswamy, N., Kukreja, R.S., Kumaraguru, S., Ordered mesoporous carbon supported fuel cell cathode catalyst for improved oxygen transport. *J. Power Sources*, 532, 231349, 2022.
59. Park, K.W., Yang, H.N., Lee, W.H., Choi, B.S., Kim, W.J., Effect of hybridization of Pt supported mesoporous-CMK-3 into Pt-CB as cathode catalyst on cell performance and durability in proton exchange membrane fuel cell. *Microporous Mesoporous Mater.*, 220, 282, 2016.
60. Ambrosio, E.P., Dumitrescu, M.A., Francia, C., Gerbaldi, C., Spinelli, P., Ordered mesoporous carbons as catalyst support for PEM fuel cells. *Fuel Cells*, 9, 197, 2009.
61. Meléndez-González, P.C., Garza-Duran, E., Martínez-Loyola, J.C., Quintana-Owen, P., Alonso-Lemus, I.L., Rodríguez-Varela, F.J., Enhanced catalytic activity of low-Pt content nanocatalysts supported on hollow carbon spheres for the ORR in alkaline media. *MRS Adv.*, 5, 2961, 2020.
62. Job, N., Marie, J., Lambert, S., Berthon-Fabry, S., Achard, P., Carbon xerogels as catalyst supports for PEM fuel cell cathode. *Energy Convers. Manage.*, 49, 2461, 2008.
63. Marie, J., Berthon-Fabry, S., Chatenet, M., Chainet, E., Pirard, R., Cornet, N., Platinum supported on resorcinol–formaldehyde-based carbon aerogels for PEMFC electrodes: Influence of the carbon support on electrocatalytic properties. *J. Appl. Electrochem.*, 37, 147, 2007.
64. Luo, C., Xie, H., Wang, Q., Luo, G., Liu, C., Review of the application and performance of carbon nanotubes in fuel cells. *J. Nanomater.*, 560392, 10, 2015.
65. Wang, C., Waje, M., Wang, X., Tang, J.M., Haddon, R.C., Yan, Proton exchange membrane fuel cells with carbon nanotube based electrodes. *Nano Lett.*, 4, 345, 2004.
66. Lina, J.F., Kamavaram, V., Kannan, A.M., Synthesis and characterization of carbon nanotubes supported platinum nanocatalyst for proton exchange membrane fuel cells. *J. Power Sources*, 195, 466, 2010.
67. Sano, N. and Ukita, S., One-step synthesis of Pt-supported carbon nanohorns for fuel cell electrode by arc plasma in liquid nitrogen. *Meter. Chem. Phys.*, 99, 447, 2006.

68. Sevilla, M., Lota, G., Fuertes, A.B., Saccharide-based graphitic carbon nanocoils as supports for PtRu nanoparticles for methanol electrooxidation. *J. Power Sources*, 171, 546, 2007.
69. Dwivedi, S., Graphene based electrodes for hydrogen fuel cells: A comprehensive review. *Int. J. Hydrogen Energy*, 47, 41848, 2022.
70. Salazar-Banda, G.R., Eguiluz, K.I.B., Avaca, L.A., Boron-doped diamond powder as catalyst support for fuel cell applications. *Electrochem. Commun.*, 9, 59, 2007.
71. Peera, S.G., Kwon, H.-J., Lee, T.G., Hussain, A.M., Heteroatom- and metalloid-doped carbon catalysts for oxygen reduction reaction: A mini-review. *Ionics*, 26, 1563, 2020.
72. Gasteiger, H.A., Kocha, S.S., Sompalli, B., Wagner, F.T., Activity benchmarks and requirements for Pt, Pt-alloy, and non-Pt oxygen reduction catalysts for PEMFCs. *Appl. Catal. B: Environ.*, 56, 9, 2005.
73. Cheng, N., Banis, M.N., Liu, J., Riese, A., Li, X., Li, R., Ye, S., Knights, S., Sun, X., Extremely stable platinum nanoparticles encapsulated in a zirconia nanocage by area-selective atomic layer deposition for the oxygen reduction reaction. *Adv. Mater.*, 27, 277, 2015.
74. S&P Global Commodity Insights. Feature: Graphite supply a concern in meeting growing battery demand, https://www.spglobal.com/commodity insights/ en/market-insights/latest-news/energy-transition/021622-feature-graphite- supply-a-concern-in-meeting-growing-battery-demand, consulted in June of 2022.
75. DOE Technical Targets for Polymer Electrolyte Membrane Fuel Cell Components. Hydrogen and Fuel Cell Technologies Office, https:// www.energy.gov/eere/fuelcells/doe-technical-targets-polymer-electrolyte-membrane-fuel-cell-components, consulted in June of 2022.
76. Tang, J.M., Jensen, K., Waje, M., Li, W., Larsen, P., Pauley, K., Chen, Z., Ramesh, P., Itkis, M.E., Yan, Y., Haddon, R.C., High performance hydrogen fuel cells with ultralow Pt loading carbon nanotube thin film catalysts. *J. Phys. Chem. C*, 111, 17901, 2007.
77. Jia, X. and Wei, F., Advances in production and applications of carbon nanotubes, in: *Single-Walled Carbon Nanotubes Preparation, Properties and Applications*, M. Olivucci and W.-Y. Wong (Eds.), pp. 299–333, Springer Nature, Switzerland AG, 2019.
78. Xin, Y., Liu, J.-G., Zhou, Y., Liu, W., Gao, J., Xie, Y., Yin, Y., Zou, Z., Preparation and characterization of Pt supported on graphene with enhanced electrocatalytic activity in fuel cell. *J. Power Sources*, 196, 1012, 2011.
79. Peera, S.G., Sahua, A.K., Arunchander, A., Nath, K., Bhat, S.D., Deoxyribonucleic acid directed metallization of platinum nanoparticles on graphite nanofibers as a durable oxygen reduction catalyst for polymer electrolyte fuel cells. *J. Power Sources*, 297, 379, 2015.
80. Duo, W. and Xu, D., Active sites derived from heteroatom doping in carbon materials for oxygen reduction reaction, in: *Electrocatalysts for Fuel Cells and*

Hydrogen Evolution, A. Ray, I. Mukhopadhyay, R.K. Pati (Eds.), p. 61784, Intechopen, United Kingdom, 2017.

81. Gao, J., Mao, M., Li, P., Liu, R., Song, H., Sun, K., Zhang, S., Segmentation and re-encapsulation of porous PtCu nanoparticles by generated carbon shell for enhanced ethylene glycol oxidation and oxygen-reduction reaction. *ACS Appl. Mater. Interfaces*, 12, 6298, 2020.
82. Xie, L., Liu, Q., Shi, X., Asiri, A.M., Luo, Y., Sun, X., Superior alkaline hydrogen evolution electrocatalysis enabled by an ultrafine PtNi nanoparticle-decorated Ni nanoarray with ultralow Pt loading. *Inorg. Chem. Front.*, 5, 1365, 2018.
83. Sun, S., Zhang, G., Gauquelin, N., Chen, N., Zhou, J., Yang, S., Chen, W., Meng, X., Geng, D., Banis, M.N., Li, R., Ye, S., Knights, S., Botton, G.A., Sham, T.-K., Sun, X., Single-atom catalysis using Pt/Graphene achieved through atomic layer deposition. *Sci. Rep.*, 3, 1775, 2013.
84. Cheng, N., Stambula, S., Wang, D., Banis, M.N., Liu, J., Riese, A., Xiao, B., Li, R., Sham, T.-K., Liu, L.-M., Botton, G.A., Sun, X., Platinum single-atom and cluster catalysis of the hydrogen evolution reaction. *Nat. Commun.*, 7, 13638, 2016.
85. Mahmood, J., Li, F., Jung, S.-M., Okyay, M.S., Ahmad, I., Kim, S.-J., Park, N., Jeong, H.Y., Baek, J.-B., An efficient and pH-universal ruthenium-based catalyst for the hydrogen evolution reaction. *Nat. Nanotech.*, 12, 441, 2017.
86. Qu, H.-Y., He, X., Wang, Y., Hou, S., Electrocatalysis for the oxygen evolution reaction in acidic media: Progress and challenges. *Appl. Sci.*, 11, 4320, 2021.
87. Ferreira da Silva, C.D., Claudel, F., Martin, V., Chattot, R., Abbou, S., Kumar, K., Jiménez-Morales, I., Cavaliere, S., Jones, D., Rozière, J., Solà-Hernandez, L., Beauger, C., Faustini, M., Peron, J., Gilles, B., Encinas, T., Piccolo, L., Barros de Lima, F.H., Dubau, L., Maillard, F., Oxygen evolution reaction activity and stability benchmarks for supported and unsupported IrOx electrocatalysts. *ACS Catal.*, 11, 4107, 2021.
88. Plevová, M., Hnát, J., Bouzek, K., Electrocatalysts for the oxygen evolution reaction in alkaline and neutral media. A comparative review. *J. Power Sources*, 507, 230072, 2021.
89. Qiu, Y., Xin, L., Li, W., Electrocatalytic oxygen evolution over supportedsmall amorphous Ni–Fe nanoparticles in alkaline electrolyte. *Langmuir*, 30, 7893, 2014.
90. Shi, W., Park, A.-H., Kwon, Y.-W., Scalable synthesis of (Pd,Cu)@Pt core-shell catalyst with high ORR activity and durability. *J. Electroanal. Chem.*, 918, 116451, 2022.
91. He, C., Ma, Z., Wu, Q., Cai, Y., Huang, Y., Liu, K., Fan, Y., Wang, H., Li, Q., Qi, J., Li, Q., Wu, X., Promoting the ORR catalysis of Pt-Fe intermetallic catalysts by increasing atomic utilization and electronic regulation. *Electrochim. Acta*, 330, 135119, 2020.

92. An, L., Chen, Y., Shi, J., Cao, J., Liu, B., Yang, J., Oxygen reduction activity and stability of Composite Pdx/Co-Nanofilms/C electrocatalysts in acid and alkaline media. *Front. Chem.*, 6, 596, 2018.
93. Ruiz-Camacho, B., Palafox-Segoviano, J.A., Pérez-Díaz, P.J., Medina-Ramírez, A., Synthesis of supported Pt nanoparticles by sonication for ORR: Effect of the graphene oxide-carbon composite. *Int. J. Hydrogen Energy*, 46, 26027, 2021.
94. Chung, S., Ham, K., Kang, S., Ju, H.K., Lee, J., Enhanced corrosion tolerance and highly durable ORR activity by low Pt electrocatalyst on unique pore structured CNF in PEM fuel cell. *Electrochim. Acta*, 348, 136346, 2020.
95. Chen, T., Xu, Y., Meng, D., Guo, X., Zhu, Y., Peng, L., Hu, J., Ding, W., Nest-type NCM ⊂ Pt/C with oxygen capture character as advanced electrocatalyst for oxygen reduction reaction. *J. Energy Chem.*, 71, 304, 2022.
96. Meléndez-González, P.C., Sánchez-Castro, E., Alonso-Lemus, I.L., Pérez-Hernández, R., Escobar-Morales, B., Garay-Tapia, A.M., Pech-Rodríguez, W.J., Rodríguez-Varela, J., Bifunctional Pd-CeO_2 Nanorods/C nanocatalyst with high electrochemical stability and catalytic activity for the ORR and EOR in alkaline media. *ChemistrySelect*, 5, 14032, 2020.
97. Labata, M.F., Li, G., Ocon, J., Abel Chuang, P.-Y., Insights on platinum-carbon catalyst degradation mechanism for oxygen reduction reaction in acidic and alkaline media. *J. Power Sources*, 487, 229356, 2021.
98. Hu, Q., Gao, K., Wang, X., Zheng, H., Cao, J., Mi, L., Huo, Q., Yang, H., Liu, J., He, C., Subnanometric Ru clusters with upshifted D band center improve performance for alkaline hydrogen evolution reaction. *Nat. Commun.*, 13, 3958, 2022.
99. Hong, S., Kim, J.H., Shin, D., Bak, G., Jang, D., Kim, W.B., Hwang, Y.J., Synergistic effect of Pt-loaded Co-N-C electrocatalysts for hydrogen evolution reaction in alkaline conditions. *Appl. Surf. Sci.*, 610, 155523, 2023.
100. Ren, B., Zhang, H., Lu, G., Ge, P., Jia, X., Synthesis of self-supported metal fiber felt electrode for electrocatalytic hydrogen evolution. *Mater. Lett.*, 330, 133260, 2023.
101. Wang, Z., Xiao, B., Lin, Z., Xu, Y., Lin, Y., Meng, F., Zhang, Q., Gu, L., Fang, B., Guoand, S., Zhong, W., PtSe2/Pt heterointerface with reduced coordination for boosted hydrogen evolution reaction. *Angew. Chem. Int. Ed.*, 60, 23388, 2021.
102. Qiu, L., Zheng, G., He, Y., Lei, L., Zhang, X., Ultra-small Sn-RuO_2 nanoparticles supported on N-doped carbon polyhedra for highly active and durable oxygen evolution reaction in acidic media. *Chem. Eng. J.*, 409, 128155, 2021.
103. Chen, J., Cui, P., Zhao, G., Rui, K., Lao, M., Chen, Y., Zheng, X., Jiang, Y., Pan, H., Dou, S.X., Sun, W., Low-coordinate iridium oxide confined on graphitic carbon nitride for highly efficient oxygen evolution. *Angew. Chem., Int. Ed.*, 58, 12540, 2019.

104. Jiang, B., Wang, T., Cheng, Y., Liao, F., Wu, K., Shao, M., Ir/g-C_3N_4/Nitrogen-doped graphene nanocomposites as bifunctional electrocatalysts for overall water splitting in acidic electrolytes. *ACS Appl. Mater. Interfaces*, 10, 39161, 2018.
105. Joshi, P., Yadav, R., Hara, M., Inoue, T., Motoyama, Y., Yoshimura, M., Contribution of B,N-co-doped reduced graphene oxide as a catalyst support to the activity of iridium oxide for oxygen evolution reaction. *J. Mater. Chem. A*, 9, 9066, 2021.
106. Badam, R., Hara, M., Huang, H.-H., Yoshimuram, M., Synthesis and electrochemical analysis of novel IrO2 nanoparticle catalysts supported on carbon nanotube for oxygen evolution reaction. *Int. J. Hydrogen Energy*, 43, 18095, 2018.
107. Demir, E., Akbayrak, S., Önal, A.M., Özkar, S., Ceria supported ruthenium(0) nanoparticles: Highly efficient catalysts in oxygen evolution reaction. *J. Colloid Interface Sci.*, 534, 704, 2019.
108. Wu, Z.-Y., Zhang, Q.-X., Huang, L.-J., Xu, Y.-J., Tang, D.-L., Covalent immobilization of ruthenium polypyridyl complex on multi-walled carbon nanotube supports for oxygen evolution reaction in an alkaline solution. *J. Power Sources*, 488, 229448, 2021.
109. https://www.metalsdaily.com/live-prices/pgms/, consulted on June 21, 2022.
110. https://www.fuelcellstore.com/vulcan-xc-72r, consulted on June 21, 2022.
111. Banham, D., Zou, J., Mukerjee, S., Liu, Z., Yang, D., Zhang, Y., Peng, Y., Dong, A., Ultralow platinum loading proton exchange membrane fuel cells: Performance losses and solutions. *J. Power Sources*, 490, 229515, 2021.
112. Wang, Y., Seo, B., Wang, B., Zamel, N., Jiao, K., Cordobes Adroherd, X., Fundamentals, materials, and machine learning of polymer electrolyte membrane fuel cell technology. *Energy AI*, 1, 100014, 2020.
113. https://www.angelo.edu/faculty/kboudrea/periodic/physical_abundances.htm
114. Minke, C., Suermann, M., Bensmann, B., Hanke-Rauschenbach, R., Is iridium demand a potential bottleneck in the realization of large-scale PEM water electrolysis? *Int. J. Hydrogen Energy*, 46, 23581, 2021.
115. Hubert, M.A., King, L.A., Jaramillo, T.F., Evaluating the case for reduced precious metal catalysts in proton exchange membrane electrolyzers. *ACS Energy Lett.*, 7, 17, 2022.
116. Hrbek, T., Kúš, P., Košutová, T., Veltruská, K., Dinhová, T.N., Dopita, M., Matolín, V., Matolínová, I., Sputtered Ir–Ru based catalysts for oxygen evolution reaction: Study of iridium effect on stability. *Int. J. Hydrogen Energy*, 47, 21033, 2022.
117. https://hydrogencouncil.com/en/hydrogen-insights-2021 consulted on November 7, 2022.

118. https://hydrogencouncil.com/en/hydrogen-insights-2022 consulted on November 7, 2022.
119. https://hydrogencouncil.com/en/hydrogen-for-net-zero/ consulted on November 7, 2022.

6

Metal Oxide-Based Electrocatalytic Materials for Hydrogen Evolution and Hydrogen Oxidation Reaction

Amit Mall[1], Akshaya K. Palai[2], Pratap Chandra Padhi[1], Sudheesh K. Shukla[3], Rashmiprava Sahoo[4], Trupti R. Das[1*], Santanu Patra[5] and Deepak Kumar[6]

[1]CIPET: Institute of Petrochemicals Technology (IPT) - Bhubaneswar, Bhubaneswar, India
[2]CIPET: SARP-Laboratory for Advanced Research in Polymeric Materials (LARPM), Bhubaneswar, India
[3]School of Chemical Engineering and Physical Sciences, Lovely Professional University, Jalandhar, Punjab, India
[4]Central Institute of Mining and Fuel Research, Dhanbad, India
[5]IDUN, Department of Health Technology, Technical University of Denmark, Kgs. Lyngby, Denmark
[6]Atotech Development Centre, IMT Manesar, Gurugram, Haryana, India

Abstract
The widespread environmental damage caused by the combustion of fossil fuels has posed significant global challenges. To achieve better electrocatalytic behavior with less expansion, oxygen evolution reaction (OER), oxygen reduction reaction (ORR) and hydrogen evolution reaction (HER) are very essential for the extensive commercialization of cells, water electrolyzers, and metal–air batteries. Hydrogen production from electrochemical water splitting is a highly promising technology for sustainable energy storage. On the other hand, metal oxides with enormous compositions and structures paved the way for HER electrocatalysis. In this chapter, we have summarized recent advances in the design and development of different metal oxides for hydrogen evolution and oxidation reactions. A brief description of some key fundamental concepts of HER and HOR has been discussed. This discussion is followed by a review of various metal oxides, such as

**Corresponding author*: trdas1893@gmail.com

Sudheesh K. Shukla, Chaudhery Mustansar Hussain, Santanu Patra, Meenakshi Choudhary (eds.) Electrocatalytic Materials for Renewable Energy, (151–164) © 2024 Scrivener Publishing LLC

nickel oxide, iron oxide, iridium oxide, and copper oxide, used for HER electrocatalysts. Lastly, some concluding remarks and perspectives on the future scope of this existing field are provided.

Keywords: Electrocatalyst, metal oxide, hydrogen evolution reaction (HER), hydrogen oxidation reaction (HOR), water splitting

6.1 Introduction

Nowadays, the energy crisis is the biggest issue across the world due to the threat of global warming; hence, clean energy sources are required in order to overcome the burden of shrinking energy storage. The water-splitting method is considered to be one of the most important net zero energy generation processes among hydrogen production technologies [1]. The two important chemical reactions for water splitting follow the anodic oxygen evolution reaction (OER) and the cathodic hydrogen evolution reaction. Generally, in the case of the fuel cell, the oxygen evolution reaction (OER) and the hydrogen evolution reaction (HER) are reversed, and the total amount of chemical energy reserved in the form of bonds in fuels is converted to electrical energy. The electrochemical reactions involved in the fuel cell are the hydrogen–oxygen evolution reaction and the oxygen reduction reaction [2]. Hydrogen oxidation reaction (HOR) is the reverse process of HER when considered alongside the OER, involving the conversion of hydrogen gas into protons and electrons.

Metal oxides are well-known catalysts used for water splitting. The influence of the controlled morphology of various metal oxide nanocatalysts has shown excellent behavior in electrolysis [3].

Figure 6.1 shows the schematic representation of water splitting. The evolution of oxygen near the anode is governed by the reaction mechanism $4OH^- \rightarrow 2H_2O + O_2 + 4e^-$ and the evolution of hydrogen at the cathode is governed by $2H_2O + 2e- \rightarrow 2OH^- + H_2$. The parameters that affect electrochemical water splitting are absorption of light, charge separation, and surface reactions. Many materials have been used to enhance the water-splitting efficiency such as nitrides, semiconductors, and chalcogenides, but these materials lead to low stability and less photocurrent generation. In order to improve water splitting, metal oxides are conveniently used as catalysts for electrochemical water splitting. The electrocatalytic properties of various metal oxides are used as promising electrode

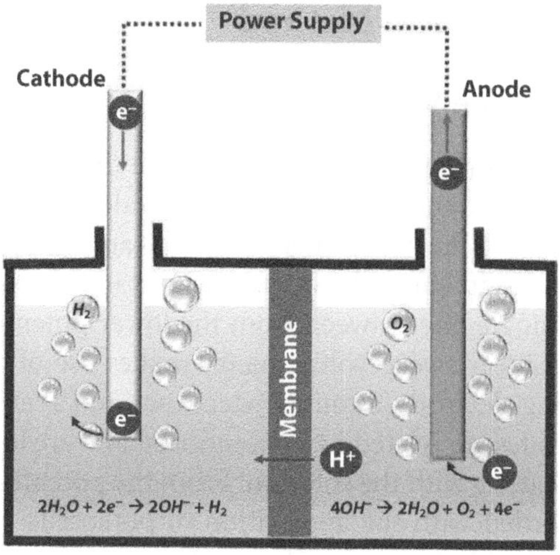

Figure 6.1 Schematic representation showing electrochemical water splitting.

materials for the fuel cell. Under ambient conditions, metals are prone to oxides, resulting in the formation of a 1–2-nm-thick layer composed of metal oxide and metal hydroxide. The presence of oxides on the metal surface may greatly influence its catalytic properties, especially low-temperature electrocatalysis [4–6].

A catalyst in electrochemical water splitting plays a vital role in lowering the potential barrier. The performance parameters calculated during electrolysis are activity, stability, and efficiency. The activity includes the calculation of overpotential, Tafel slope, and exchange current density. From the polarization curve in the linear sweep voltammetry (LSV) technique, the activity can be calculated. To design a catalyst without consuming extra energy, it must be there to offer small overpotential. The catalyst would initiate proton reduction at a very low overpotential (η) and produce hydrogen and oxygen with high faradaic efficiency [7].

Since the last decades, the global importance of non-noble metal oxide as a better electrocatalyst for HER and hydrogen oxidation reaction (HOR) has attracted a lot of attention. This chapter discusses the development of a metal oxide-based electrocatalyst in HER and HOR. Here, a brief discussion of the basic mechanism of the catalyst, the linear sweep voltammetry technique, and the mechanism behind HER and HOR has been discussed. Moreover various metal oxides like, nickel oxide-, iron oxide-, iridium oxide-, and copper oxide-based nanocomposites have been studied.

6.2 Electrochemical Method

6.2.1 Linear Sweep Voltammetry for Catalysis

Linear sweep voltammetry is used to determine the peak current, current density, and half-peak current potential. Generally, it is simple and effective compared with cyclic voltammetry which is an irreversible system. It is equivalent to one-segment cyclic voltammetry. Linear sweep voltammetry involves only a single linear sweep from the lower potential to the higher potential limit. Linear sweep voltammetry makes use of three electrodes, namely, a working electrode, a counter electrode, and a reference electrode. The three electrodes are related to a potentiostat and are located in an electrochemical cell along with the solution. From the potentiostat, the current response is recorded with the application of potentials. This suggests an electrochemical (oxidation and reduction) reaction. An electrochemical spectra will appear based on the corresponding oxidation and reduction reaction [8, 9]. Figure 6.2 shows the different parameters studied during the linear sweep voltammetry plot of FeCoNi sulfide composites. These parameters are calculated and compared with existing techniques. Figure 6.2(a) suggests the current density with a change in applied potential, from which the onset and overpotentials are calculated. Figures 6.2(b) and (c) represent the plots for calculating tafel's slope and stability of the electrode with respect to no. of cycles, respectively. Figure 6.2(d) represents the time dependence of the current density at a constant potential [10].

Most metallic oxides exhibit superior electrochemical activity, energy storage, and fuel cell reactions. By using linear sweep voltammetry, important parameters can be observed that can used to determine the thermodynamic reversibility of the material. These parameters include the peak current and the potential at half the peak current. LSV measurements were made on the oxygen reduction reaction material [11].

6.3 Electrocatalysis

6.3.1 Hydrogen Evolution Reaction (HER)

HER is an electrochemical reaction where the hydrogen ions from a solution (from water in case of water splitting) are reduced to form hydrogen gas (H2) at an electrode surface (2 H+ + 2 e− → H_2). The HER is a crucial reaction in numerous applications, including hydrogen production through fuel cells, and electrochemical water splitting, and it plays a

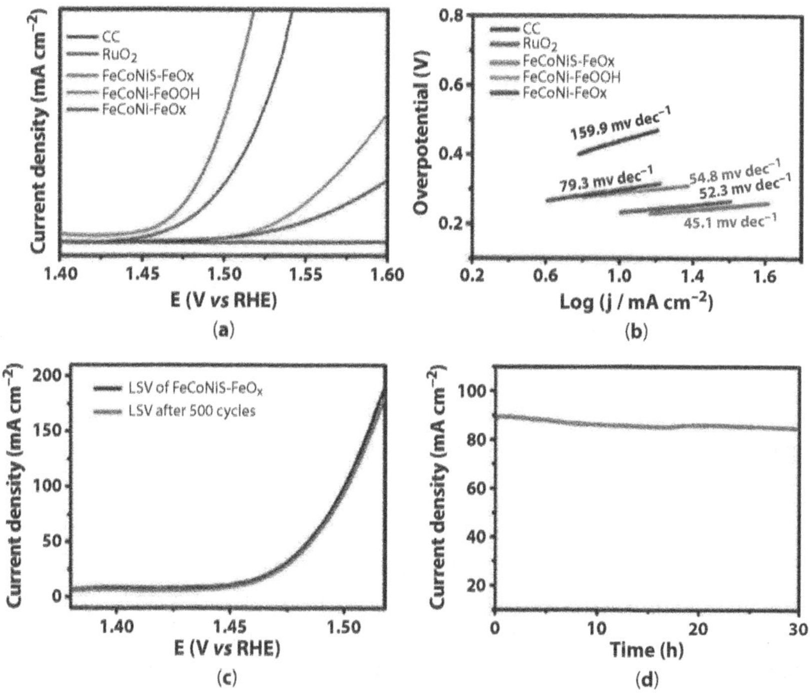

Figure 6.2 (a) LSV curves of different OER catalysts. (b) Tafel curves of different OER catalysts. (c) LSV curve of FeCoNiS-FeO_X catalysts. (d) Time-dependent current density curve of FeCoNiS-FeO_X catalysts under constant potential. This figure has been adopted from ref. [10].

critical role in the development of clean and sustainable energy technologies. An awesome manner to drive the HER with renewable resources of power can result in a sustainable supply of hydrogen fuel, which could be reserved, transported, and used in a zero-emission combustion engine. Accomplishing excessive energetic overall performance for water splitting calls for the use of a catalyst to decrease the overpotential which is important for HER activity. Platinum is know as the state-of-the-art electrocatalyst for HER having a very small overpotentials even at high response rates in acidic media. However, the excessive cost of Pt limits its real-time uses. The development of a novel electrocatalyst with low overpotential, low cost, higher current density, and greater stability is both necessary and urgent. It also plays a key role in the production of hydrogen [12].

A great deal of attention has been given to the HER in electrochemistry because of its great significance to many electrochemical technologies. Additionally, basically, it involves one of the simplest electrode reactions,

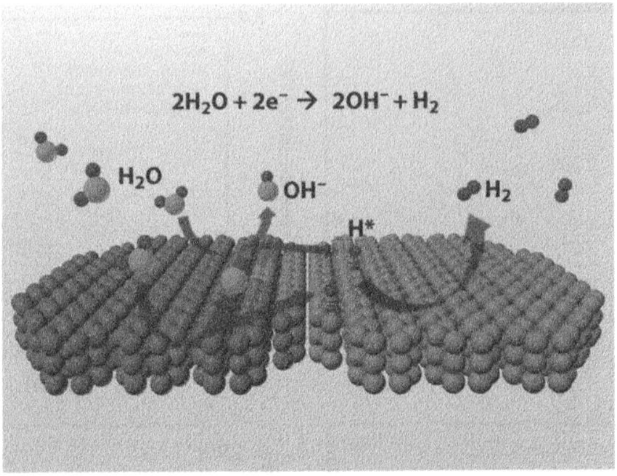

Figure 6.3 Hydrogen evolution reaction. This figure has been adopted from ref. [14].

and critically, this reaction is given consideration as it provides hydrogen as a gas that has the potential to update fossil-based transportation fuels. HER has become crucial in fuel cells research after the development of electrolysis, wherein hydrogen is immediately produced on the cathode using either an inorganic catalyst or an electroactive area; hence, much less energy is used than that in the traditional water electrolysis due to the favorable anode potential [13]. Figure 6.3 shows the schematic representation of the hydrogen evolution process at the cathode where in the presence of a catalyst, the water molecule accepts two-electrons and releases two hydroxide ions along with hydrogen gas [14].

6.3.2 Hydrogen Oxidation Reaction (HOR)

To gain green energy, hydrogen can be provided to proton exchange membrane fuel cells (PEMFCs), which might be environmentally friendly for energy conversion without being restricted via the Carnot cycle. PEMFCs perform the direct conversion of chemical energy to electricity through the use of HOR and ORR on the anode and cathode factors, respectively [15]. The triple-phase boundary (TPB) is a place in a gasoline cell where three stages are in contact necessary for an electrochemical reaction at the anode. These three phases are called ion-carrying out phase, electron-carrying out section, and gas section [16].

6.4 Metal Oxide-Based Catalyst

Numerous metal oxide-based electrocatalysts used in HER and HOR are shown in Table 6.1. Heterostructure-based nanocomposites having nickel foam have shown excellent behavior toward HER. The microporous structure of a catalyst has influenced its catalytic activity. The presence of a porous structure along with nanowires has established low overpotential and Tafel slope. Bimetal oxide-based electrocatalysts have also shown tremendous behavior toward hydrogen production.

Table 6.1 Different metal oxide-based catalysts.

Serial no.	Metal oxide-based oxide	Developed material	Type	References
1.	Nickel oxide	NiO	HER	[12]
		Manganese-doped nickel oxide/nickel hetero-established arrays at the skeleton of nickel foam (Mn-NiO-Ni/Ni-F)	HER	[17]
2.	Iron oxide	Iron oxide and phosphide encapsulated within N,P-doped microporous carbon nanofibers	HER/HOR	[18]
		Hierarchically porous Mo-doped Ni–Fe oxide nanowires	HER	[19]
		Nickel–cobalt and nickel–iron oxide catalysts	HER	[20]
		Fe		[21]
3.	Iridium oxide	Iridium oxide thin films were electrochemically grown on Ir(111) substrate		[22]
4.	Copper oxide	CuO		[23]

6.4.1 Nickel Oxide

Nickel-based metal oxide catalysts have received popularity because of their stability in robust alkaline environments, coupled with advanced kinetics over acidic media and reduced cost. The electrocatalytic interest of NiO was tested in each hydroxide and carbonate alkaline media. Concentrations of KOH (0.5 M) and Na_2CO_3 (0.1 M) electrolytes have been decided on to normalize the pH at around 11. Figure 6.4 suggests the CVs for both reflux-NiO and RT NaOH-NiO in 0.5 M KOH (Figure 6.4a) and 0.5 M KOH/0.1 M CH_3OH (Figure 6.4b) [12].

The concept behind the green electrocatalytic activity of nickel oxide nanoparticles is investigated by sweeping the RDE capability from 1.0 to 2.0 V. X. Lu *et al.* have suggested a properly ordered manganese-doped nickel oxide/nickel hetero-established arrays at the skeleton of nickel foam (Mn-NiO-Ni/Ni-F) in order to generate hydrogen. The material is

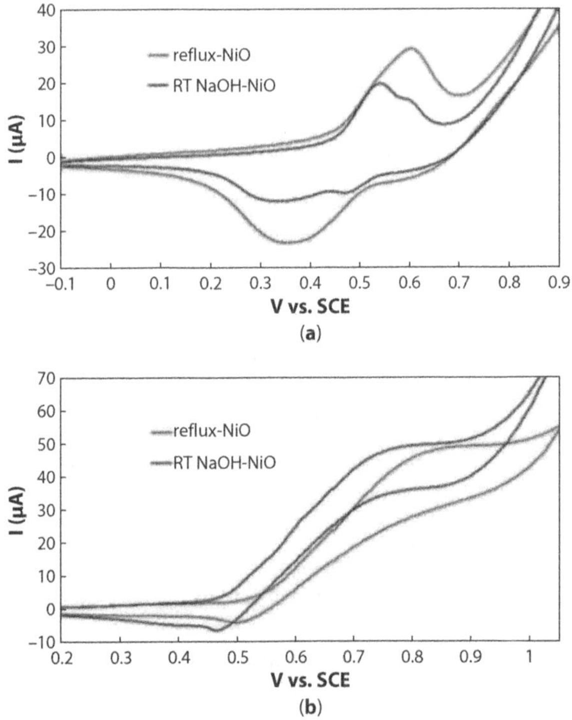

Figure 6.4 CVs from −0.1 V to 1.05 V vs. SCE in (a) 0.005 M KOH and (b) 0.005 M KOH/0.1 M CH_3OH for reflux-NiO and RT NaOH-NiO. Scan number 20 is shown for each CV, and the scan rate was 20 mV/s. This figure has been adopted from ref. [17].

particularly much effective under impartial electrolyte and herbal seawater. The well-known electrocatalyst shows negligible onset overpotential and brilliant long-time period stability (>500 h in impartial electrolytes), outperforming the Pt/C benchmark [17].

6.4.2 Iron Oxide-Based Catalyst

Iron oxide based electrocatalyts have also been utilized for HOR, where density functional theory (DFT) based calculations have been performed

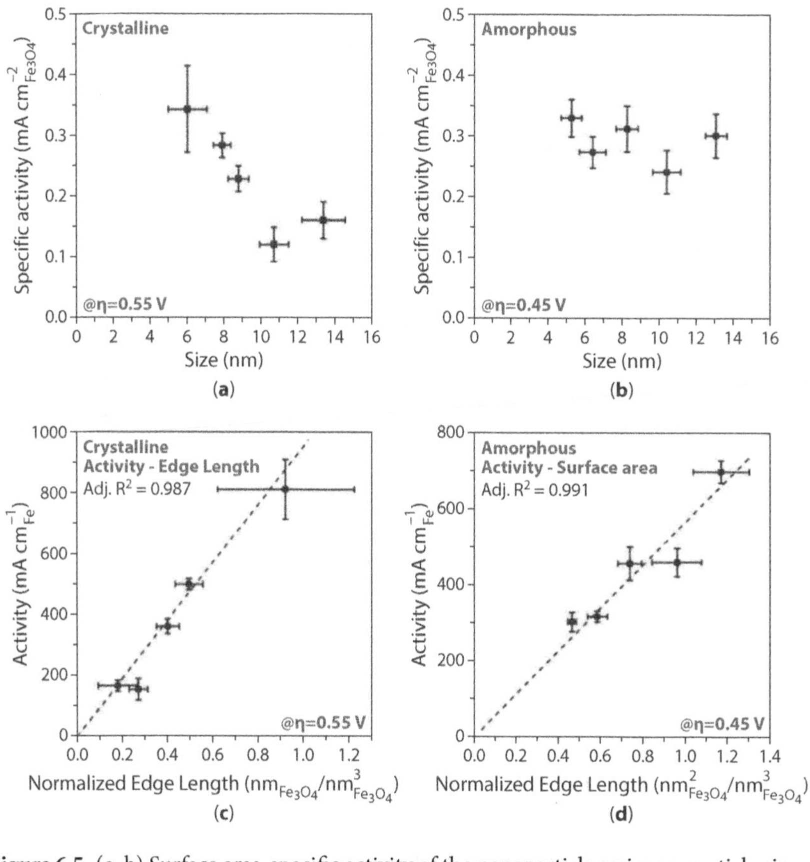

Figure 6.5 (a, b) Surface area-specific activity of the nanoparticle series vs. particle sizes. (a) C-Fe_3O_4 nanoparticles. (b) A-FeO_x nanoparticles. (c) The plot of mass activity at 0.55 V overpotential vs. volume-normalized edge length for C-Fe_3O_4 nanoparticles. (d) The plot of mass activity at 0.45 V overpotential vs. volume-normalized surface area for A-FeO_x nanoparticles. The overpotential is marked at the lower right corner. This figure has been adopted from ref. [21] (Zhongbin Zhuang *et al.* 2021 J. Electrochem. Soc. 168 034518).

to investigate the OER activity of the materials and to study the OER active sites. These calculations elucidated the mechanism of the localized linear correlation between the electrocatalytic behavior and a wide range of defect regions/sites on crystalline/amorphous Fe oxide nanoparticles. The calculations show that increasing the amount of side/defect sites by reducing the size/crystallinity of Fe oxide nanoparticles is a feasible method to enhance the activity of fully OER Fe oxide-based electrocatalysts (Figure 6.5) [21].

6.4.3 Iridium Oxide-Based Catalyst

Iridium oxide based electrocatalysts have gained a significant research attention towards hydrogen generation. Here the cathodic reaction ($2H + 2e^{-} \rightarrow H_2$) can basically be described using the *H bond adsorption energy descriptor. The effect of electrolyte composition was also investigated for OER on model Ir-oxide electrodes. Iridium oxide thin films were grown electrochemically on Ir(111) substrate. Anodically patterned

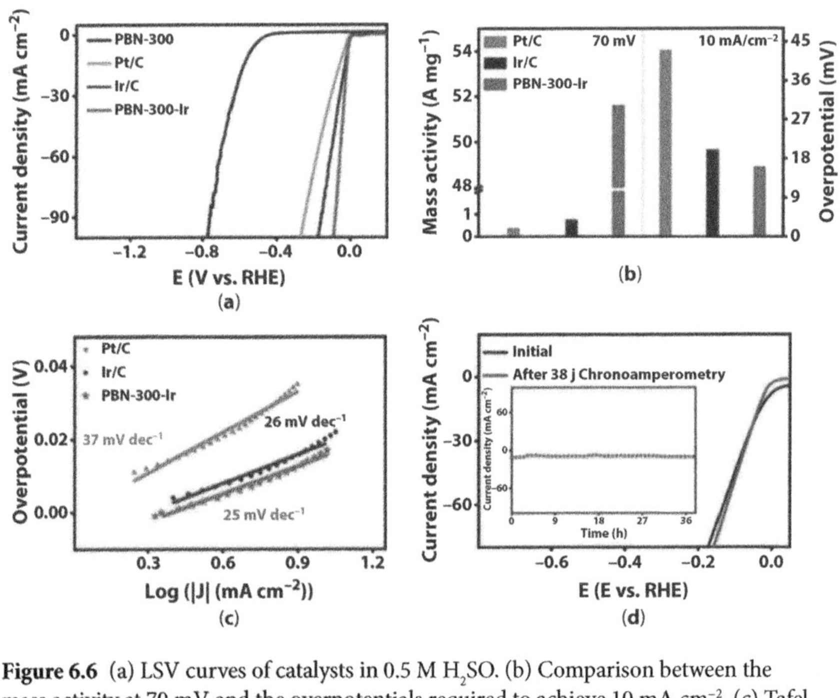

Figure 6.6 (a) LSV curves of catalysts in 0.5 M H_2SO. (b) Comparison between the mass activity at 70 mV and the overpotentials required to achieve 10 mA cm^{-2}. (c) Tafel plots. (d) Comparison of LSV curves (black) and after (red) the chronoamperometry test. The inset is the current density vs. time (*I–T*) curves of PBN-300-Ir recorded for 38 h at –0.019 V vs. RHE. This figure has been adopted from the Ir catalyst (https://doi.org/10.3390/nano12152618).

iridium oxide films (AIROFs) were grown via voltammetry at 1.575 V and 0.7 V vs. RHE until constant CVs were obtained [22].

Figure 6.6 shows the use of a holey carbon skeleton (PBN), produced from the polyhexaphenylbenzene network (PHN), as an electrocatalyst for water splitting. The iridium element with an approximate concentration of 0.74% has been incorporated to support the PBN (PBN-300-Ir). An overpotential of just 17 mV has been observed at 10 mA cm^{-2} in 0.5M H_2SO_4 (Figure 6.6a). The study suggests a very high utilization efficiency of the Ir atom (Figures 6.6b, c). After 38 h of the HER reaction in a highly acidic electrolyte, the current density of PBN-300-Ir remained almost unchanged (Figure 6.6d), demonstrating that PBN-300-Ir has high stability [24].

6.4.4 Copper Oxide-Based Catalyst

Similar to the other metal oxides, the LSV experiments on the Cu_2O-Co_3O_4/carbon (CN) composite electrodes were studied in the 1-M KOH

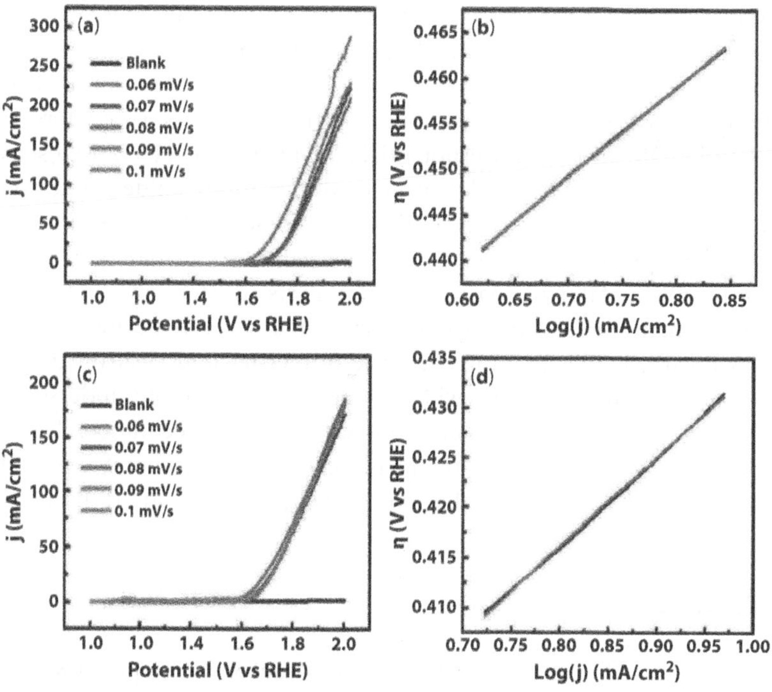

Figure 6.7 LSV curves of Cu_2O-Co_3O_4/CN composite synthesized at (a) 400°C and (c) 500°C and the corresponding Tafel slope (b, d) at different scan rates, in 1 M KOH. This figure has been adopted from ref. [25].

solution, considering the anodic reaction during HER. The effect of different scan rates on the LSV response of the Cu_2O-Co_3O_4/carbon (CN) composite synthesized at different temperatures of 400°C (Figure 6.7a) and 500°C (Figure 6.7c) is shown in Figure 6.7. The Tafel slope for the product synthesized at 400°C is 281 mV/dec, and the current density is 0.599 mA/cm^2, as shown in Figure 6.7b. At 500°C, the overall catalytic performance is low, the Tafel slope is 261 mV/dec, and the current density is 0.771 mA/cm^2, as shown in Figure 6.7d [25].

The catalytic activity of the material is determined by the long-term electrolysis at 10 mA/cm^2 in 1.0 M NaOH which was no longer found when catalytic controls were completed at reduced current densities as previously found in a single species of CuO catalysts. For example, the catalytic activity remained strong for more than a hundred hours in 1.0 M NaOH when electrolysis was performed at 1 mA/cm^2 [23].

Conclusion and Future Scope

In this chapter, a brief introduction to the current energy problems has been addressed. The basics of hydrogen evolution reactions and hydrogen reduction reactions have been discussed. Linear sweep voltammetry is the technique most commonly used to determine the process parameters of the HER and HOR study, i.e., overpotential and Tafel slope. Here, we discussed various reported nickel oxide-, iron oxide-, iridium oxide-, and copper oxide-based electrocatalysts used in the electrochemical water splitting. The growing energy demand suggests that we need to work more on the cheapest non-noble metal oxide-based electrocatalysts. The ternary combinations, porous structure-based metal oxide nanocatalysts, can be more effective in this regard. Understanding the basic properties and electron transport phenomena at the nanoscale will provide a more accurate understanding of the large-scale production of hydrogen using this technique. This field has a broader vision for the future to address the world's energy crisis. More research needs to be conducted, encompassing materials and techniques.

References

1. Zahra, T., Ahmad, K.S., Thomas, A.G., Zequine, C., Gupta, R.K., Malik, M.A., Sohail, M., Phyto-inspired and scalable approach for thesynthesis of PdO–2Mn2O3: A nanomaterial for application in water splitting electrocatalysis. *RSC Adv.*, 10, 29961, 2020.

2. Ariful, A. *et al.*, Tissue paper-derived porous carbon encapsulated transition metal nanoparticles as advanced non-precious catalysts: Carbon- shell influence on the electrocatalytic behaviour. *J. Colloid Interface Sci.*, 581 Pt B, 905–918, 2020.
3. Jayaseelan, S.S., Bhuvanendran, N., Xu, Q., Su, H., Co3O4 nanoparticles decorated Polypyrrole/carbon nanocomposite as efficient bi-functional electrocatalyst for electrochemical water splitting. Hydrogen Energy Publications LLC. *RSC Adv.*, 10, 49, 29326–29335, 2020, https://doi.org/10.1016/j.ijhydene.2019.12.0850360-3199/© 2019All rights reserved.
4. Yang, M.-Q., Wang, J., Wu, H., Ho, G.W., Noble metal-free nanocatalysts with vacancies for electrochemical water splitting. *Small*, 14, 15, 1703323, 2018.
5. Gu, X.-K., Camayang, J.C.A., Samira, S., Nikolla, E., Oxygen evolution electrocatalysis using mixed metal oxides under acidic conditions: Challenges and opportunities. *J. Catal.*, 388, 130–140, 2020.
6. Wang, J., Cherevan, A.S., Hannecart, C., Naghdi, S., Nandan, S.P., Gupta, T., Eder, D., Ti-based MOFs: New insights on the impact of ligand composition and hole scavengers on stability, charge separation and photocatalytic hydrogen evolution. *Appl. Catal. B: Environ.*, 282, 119626, 2021.
7. Wang, S., Lu, A., Zhong, C.J., Hydrogen production from water electrolysis: Role of catalysts. *Nano Converg.*, 8, 1, 4, 2021. doi:10.1186/s40580-021-00254-x.
8. Bird, A.J. *et al.*, *Electrochemical methods : Fundamentals and applications*, 2nd edn, John Willey and Son, Austin, 2001.
9. Gueshi, T., Tokuda, K., Matsuda, H., Voltammetry at partially covered electrodes: Part II. Linear potential sweep and cyclic voltammetry. *J. Electroanal. Chem. Interfacial Electrochem.*, 101, 29–38, 1979.
10. Teng, W., Huo, M., Sun, Z., Yang, W., FeCoNi sulfides derived from *In situ* sulfurization of precursor oxides as oxygen evolution reaction catalyst. *Front. Chem.*, 8, 334, 2020. doi: 10.3389/fchem.2020.00334.
11. Li, Q., Song, L., Liang, Z., Sun, M., Wu, T., Huang, B., Luo, F., Du, Y., Yan, C.-H., A review on Ceo2-based electrocatalyst and photocatalyst in energy conversion. *Adv. Energy Sustainability Res.*, 2, 2000063, 2021, https://doi.org/10.1002/aesr.202000063.
12. Dubouis, N. and Grimaud, A., The hydrogen evolution reaction: From material to interfacial descriptors. *Chem. Sci.*, 10, 9165–9181, 2019.
13. Strmcnik, D., Lopes, P.P., Genorio, B., Stamenkovic, V.R., Markovic, N.M., Design principles for hydrogen evolution reaction catalyst materials. *Nano Energy*, 51, 29–36, 2016.
14. Wei, J., Zhou, M., Long, A., Xue, Y., Liao, H., Wei, C., Xu, Z.J., Heterostructured electrocatalysts for hydrogen evolution reaction under alkaline conditions. *Nano-Micro Lett.*, 10, 75, 2018, https://doi.org/10.1007/s40820-018-0229-x.

15. Chen, M., Xie, X., Guo, J., Chena, D., Xu, Q., Space charge layer effect at the platinum anode/BaZr0.9Y0.1O3d electrolyte interface in proton ceramic fuel cells. *J. Mater. Chem. A*, 8, 12566–12575, 2020. doi: 10.1039/d0ta03339a.
16. Karuppanan, K.K., Panthalingal, M.K., Biji, P., Nanoscale, catalyst support materials for proton-exchange membrane fuel cells. *Front. Energy Res.*, 11, 1091105, 2023. doi: 10.3389/fenrg.2023.1091105.
17. Lu, X., Pan, J., Lovell, E., Tan, T.H., Ng, Y.H., Amal, R., A sea-change: Manganese doped nickel/nickel oxide electrocatalysts for hydrogen generation from seawater. *Energy Environ. Sci.*, 11, 1898–1910, 2018. doi: 10.1039/C8EE00976G.
18. Wang, M., Zhang, C., Meng, T., Pu, Z., Jin, H., He, D., Zhang, J., Mua, S., Iron oxide and phosphide encapsulated within N,P-doped microporous carbon nanofibers as advanced tri-functional electrocatalyst toward oxygen reduction/evolution and hydrogen evolution reactions and zinc-air batteries. *J. Power Sources*, 413, 367–375, 2019.
19. Chen, Y., Dong, C., Zhang, J., Zhang, C., Zhang, Z., Hierarchically porous Mo-doped Ni-Fe oxide nanowires efficiently catalysing oxygen/hydrogen evolution reactions. *J. Mater. Chem. A*, 6, 8430–8440, 2018, doi: 10.1039/C8TA00447A.
20. Faid, A.Y., Barnett, A.O., Seland, F., Sunde, S., Optimized Nickel-Cobalt and Nickel-Iron oxide catalysts for the hydrogen evolution reaction in alkaline water electrolysis. *J. Electrochem. Soc.*, 166, 8, 519–533, 2019.
21. Zhuang, Z. *et al.*, Oxygen evolution on iron oxide nanoparticles: The impact of crystallinity and size on the overpotential. *J. Electrochem. Soc.*, 168, 034518, 2021.
22. Ganassin, A., Colic, V., Tymoczko, J., Bandarenkaa, A.S., Schuhmann, W., Non-covalent interactions in water electrolysis: Influence on the activity of Pt(111) and iridium oxide catalysts in acidic media. *Phys. Chem. Chem. Phys.*, 17, 8349–8355, 2015. doi: 10.1039/x0xx00000x.
23. Ngoc, H.T., Rousse, M., Zanna, S., Lucas, I.T., Xu, X., Menguy, N., Mougel, M., Fontecave, M., A dendritic nanostructured copper oxide electrocatalyst for the oxygen evolution reaction. *Angew Chem. Int. Ed. Engl.*, 18, 56, 17, 4792–4796. doi: 10.1002/anie.201700388. Epub 2017 Apr 4.
24. Peng, Y., Liao, Y., Ye, D., Meng, Z., Wang, R., Zhao, S., Tian, T., Tang, H., Recent advances regarding precious metal-based electro catalysts for acidic water splitting. *Nanomaterials, 12,* 2618, 2022, https://doi.org/10.3390/nano12152618.
25. Shaikh, Z.A., Moiseev, N., Mikhaylov, A., Yüksel, S., Facile synthesis of Copper Oxide-Cobalt Oxide/Nitrogen-Doped Carbon (Cu2O-Co3O4/CN) composite for efficient water splitting. *Appl. Sci., 11,* 9974, 2021, https://doi.org/10.3390/app11219974.

7

Metal–Organic Framework-Based Electrocatalytic Materials

Athira Krishnan[1]*, Rijith S.[2], Sumi V. S.[3] and Bhagya T. C.[4]

[1]*Department of Chemistry, Govt. College, Chittur, Palakkad, Kerala, India*
[2]*Post-Graduate and Research Department of Chemistry, DST-FIST Supported Department, Sree Narayana College, Kollam, Affiliated to University of Kerala, Kerala, India*
[3]*Department of Polymer Chemistry, Government College, Attingal, Thiruvananthapuram, Kerala, India*
[4]*Department of Chemistry, University of Kerala, Kariavattom Campus, Thiruvananthapuram, Kerala, India*

Abstract

Metal–organic framework (MOF)-based electrocatalysis is pivotal in many chemical transformations. MOF, constituted with the merits of both organic and inorganic parts, outweighs the demerits of noble/transition metal-based conventional electrocatalysts. The importance of MOF in various electrochemical transformations arises due to its unique structure and excellent physicochemical properties. Pristine MOF-based electrocatalysis is widely employed in various applications. However, its efficiency is mainly limited by its low conductivity. As a remedy, researchers are trying to synthesize conductive MOFs. The present study explores the state-of-the-art findings on the development of conductive MOFs (including their synthetic strategy) and their catalytic application in various electrochemical reactions (HER, HOR, OER, ORR, and CO_2 reduction). Furthermore, the conductive mechanism in MOFs is discussed. Additionally, challenges and future perspectives in this emerging field are also addressed.

***Keywords*:** MOF, Electrocatalyst, Conduction mechanism, HER, OER, ORR, HOR, CO_2 reduction

**Corresponding author*: athikrishnan91@gmail.com

Sudheesh K. Shukla, Chaudhery Mustansar Hussain, Santanu Patra, Meenakshi Choudhary (eds.) Electrocatalytic Materials for Renewable Energy, (165–194) © 2024 Scrivener Publishing LLC

7.1 Introduction

Electrocatalyst-assisted electrochemical transformation is an energy-efficient and cost-effective green technology. It is beneficial for many electrochemical reactions such as hydrogen evolution reaction (HER), hydrogen oxidation reaction (HOR), oxygen evolution reaction (OER), oxygen reduction reaction (ORR), and CO_2 reduction reaction [1–7]. Noble metals such as platinum, ruthenium (Ru), and palladium (Pd) are well-recognized conventional electrocatalysts. Although noble metals are promising in their activity toward electrochemical transformations, their high cost and low availability make them less relevant for industrial application. Therefore, researchers strive to find a cost-effective alternative with activity and electrochemical stability, and there aroused an era of transition metal/oxide/sulfide/phosphide-based electrocatalysis [8–11]. Still, it is a hot area of research. Lanthanide-based materials are also known for their activity in electrocatalysis. Even after years of active research in the field, it is a fact that the majority of earth-abundant metals/their compounds are still unable to attain the efficiency of noble metal-based catalytic systems. The urgency for low-cost and readily available catalytic systems with activity and stability, even comparable to or better than noble metal-based ones, arrived at metal–organic frameworks (MOFs). MOF-based catalysts hold the advantage of organic and inorganic parts within the system. The high specific surface area and highly porous structure, with pores of definite size and shape, expose more active sites to specific reactants and allow easy mass transport at the electrode/electrolyte interface [12]. Pristine MOF-based materials synthesized by various synthetic strategies are reported for their electrocatalytic application in many energy conversions [13, 14].

In 2015, J-S Qin *et al.* reported the fabrication of a polymolybdate-based MOF as a catalyst for active HER from water [15]. In the same year, N. Kornienko and coworkers introduced thin films of cobalt-porphyrin MOF to reduce carbon dioxide to carbon monoxide in an aqueous medium efficiently and selectively [16]. Kung C.-W *et al.* proposed the fabrication of Co-modified MOF thin films for water oxidation [17]. In 2016, a research group reported the fabrication and application of an active and stable ultrathin MOF nanosheet for OER. After that, intensive studies have been carried out on MOF-based electrocatalysts [18]. The efforts resulted in the development of MOF-based systems with activity superior to noble metal-based systems [19]. Despite these attractive features, MOF-based systems suffer from poor conductivity and pore/active center blockage by organic ligands [20]. So, the current science tries to improve the activity of MOFs by methods such as size tailoring, defect engineering, and composite formation.

On account of the importance of improving the conductivity of MOFs for electrocatalytic applications, in the present work, the authors focused on possible conductive mechanisms in MOFs, the state-of-the-art studies on their synthesis, and their catalytic activity in various electrochemical reactions, viz., HER, HOR, OER, ORR, and CO_2 reduction. Furthermore, the challenges and future perspectives in this emerging field are also addressed.

7.2 Mechanism of Conduction in MOFs

MOFs having high conductivity can be widely applied as auspicious candidates for electrocatalysis, chemiresistive sensing, energy storage, and several other applications. In general, electron conduction is a result of the directional movement of readily available electrons, whereas ionic conduction is a conductive process due to the directional movement of positive and negatively charged ions influenced by the presence of an electric field. In general, the electronic conductivity of MOFs relies upon the charge carriers (electrons), charge density, and mobility, whereas ionic conductivity is determined by the mobility and concentration of the charge carriers (protons) [21]. The parameters such as highly functional pore surface, high surface area, and rich structural tunability of conductive MOFs offer boundless potencies to enhance both electronic as well as proton conductivity [22].

Facilitating the charge transfer pathway of electrons and protons is considered as the foremost objective for designing conductive MOFs. Based on the pathway of charge carriers and different characteristics of MOFs, electron and proton conductions in MOFs have different mechanisms.

7.2.1 Mechanism of Electron Transport in MOFs

The electrical conductivity (σ) [23] of MOFs relies on charge density (n) and mobility (μ) of electrons and holes as represented by the following equation:

$$\sigma = e(\mu_e n_e + \mu_h n_h) \tag{7.1}$$

Charge density (n) can be calculated using the Arrhenius equation as follows:

$$n = n_0 \exp(-\frac{E_a}{kT}) \tag{7.2}$$

where n_0 is the prefactor, E_a is the activation energy of electronic transport, T is the absolute temperature, and k is the Boltzmann constant. Accordingly, for a higher conductivity, both charge density and mobility of charge carriers should be maximum. A higher concentration of lightly bound charge carriers is essential for attaining a high charge density ($>10^{15}$ cm^{-3}). The source of charge carriers in MOFs is generally from metal ions and redox-active organic ligands. Moreover, a higher charge density reduces the activation energy at a particular temperature [24]. The mobility of charge carriers in comparison reveals the efficiency of the charge transport mechanism.

In MOFs, the electron conduction can have two types of charge transport mechanisms: hopping transport and band transport. For the former, the charge carriers, i.e., electrons and holes, are at certain sites with distinct energy levels and hop in between the adjacent sites owing to thermal activation. This type of mechanism is usually associated with the MOFs, comprising node-to-node or ligand redox hopping. The band transport mechanism is applicable when the charge carriers are typically delocalized.

The mobility of charge carriers depends on the charge scattering frequency and the effective mass (m^*) of charge carriers as represented by the following equation:

$$\mu = \frac{e\tau}{m^*} \tag{7.3}$$

Where e is the elemental charge, and τ is the mean time between two charge scattering events.

From a design point of view, energetic and spatial overlap of appropriate symmetric orbitals is mandatory for both hopping and band transport mechanisms, so that with intensive orbital overlapping, the mobility of charge carriers and the charge transport path length also increase. Furthermore, based on the above-discussed charge transport mechanisms,

MOFs can have three electron transfer pathways, specifically "through-space," "through-bond," and "through-guest" conduction approaches.

7.2.1.1 "Through-Space"

"Through-space" or otherwise the conduction via overlapping and delocalization of orbitals is considered a non-covalent approach, where additional charge transfer pathways in MOFs are generated by means of non-covalent interactions for establishing enhanced electrical conductivity. This phenomenon can evade short chemical bonds between electroactive building blocks [25]. The "through-space" electron transport approach in conductive MOFs can be enabled typically by means of pi–pi stacking, the extent of which depends on the overlapping of adjacent ligand orbitals [26–28].

7.2.1.2 "Through-Bond"

The "through-bond" charge transport depends on the nature of metal–ligand pairs, and it occurs during the overlapping of metal–ligand orbitals, thereby leading to covalent bond formation for electron transport. Hence, MOFs should possess a metal–organic coordination structure, where metal ions are covalently bonded with conjugated organic ligands and loosely bound electrons. The effective overlapping of metal ion orbitals with ligands leads to high conductivity [29]. In redox-active MOFs and 2D graphene-like MOFs, the charge transfer during through-bond (through-linker) conduction takes place via consecutive chains of coordination bonds [30], and the mechanism is often hopping transport, which is generally a temperature-dependent mechanism, i.e., the conductivity increases with the rise in temperature. Nonetheless, those MOFs have a band gap value lower than 0.3 eV and are intrinsically semiconducting [31].

7.2.1.3 "Through-Guest"

The "through-guest" charge transport pathway is exhibited by MOFs having permanent porosity. For instance, the introduction of proton carriers in ionically conductive MOFs can induce their proton conductivity [32]. The proton-conducting nature of porous MOFs can be enhanced by impregnating them with strong acids of non-volatile nature (e.g., H_2SO_4 and H_3PO_4) [33]. Furthermore, guest molecules, which are probably redox active in

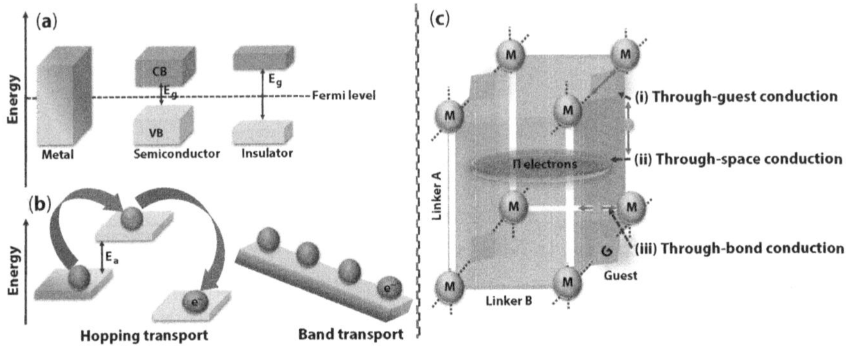

Figure 7.1 (a) Schematic representation of the electronic structure of metal, semiconductor, and insulator (E_g—band gap). (b) Hopping and band-like charge transport mechanism. (c) Three types of electron transport mechanisms in MOFs (electrons are transferred from the region at high electric potential energy to the region at lowest potential energy) (E_a—activation energy barrier) [31].

nature, can encourage the generation of free charge carriers in skeletons of MOF via the interaction of charge carriers within the framework [34]. The effectual application of such a conduction mode can enhance the electronic conductivity significantly; however, a porosity reduction is escorted along with [31]. The possible mechanisms of electron transport in MOFs are represented in Figure 7.1.

7.2.2 Mechanism of Ionic Conductivity in MOFs

The conductive performance of MOFs is mainly determined on the basis of ionic conductivity. Ionic conductivity (σ) is generally expressed as follows:

$$\sigma = \frac{n(Ze)^2 D}{kT} \exp(-\frac{E_a}{kT}) \tag{7.4}$$

Where n is the carrier concentration, D is the self-diffusion coefficient, Z_e is the carrier charge (for protons, $Z_e = 1$), E_a is the activation energy of ion transport, T is the absolute temperature, and k is the Boltzmann constant. Mobility (η) is proportional to the self-diffusion coefficient [$\eta = Z_e D/(kT)$]. Evidently, the larger the proton carriers, the larger the diffusion coefficient, and the activation energy would be smaller in MOFs, hence a higher proton conductivity [24, 35].

MOFs follow two major mechanisms for proton conduction: the Grotthuss mechanism and the vehicle mechanism [36–38]. The Grotthuss mechanism involves the proton transfer during the formation and breaking of hydrogen bonds via networks linked with hydrogen bonds [39–41]. Alternatively, the vehicle mechanism indicates proton transfer by solvent molecules and via concentration diffusion [42]. The term "vehicle" represents the fact that the proton in this case is not transferred but incorporated into "vehicle" molecules like H_2O, thereby resulting in the formation of complex ions, such as H_3O^+ [43, 44] (Figure 7.2). Hence, proton conduction is realized by the movement of these H_3O^+ complex ions, and such movements are dependent on concentration differences. Similar is the case with ammonia, where ammonia with its tetrahedral geometry is an auspicious proton-conducting medium having three proton donating sites and one acceptor site. Furthermore, it has a conjugate acid–base pair ($NH_4^+ + NH_3 \leftrightarrow NH_3 + NH_4^+$), and the porous MOFs would allow the conductivity of NH_3-mediated proton within a wide range of temperatures, by encapsulating NH_3 or NH_4^+ within its pores. Consequently, the Grotthuss mechanism in MOFs profoundly relies on the building of frameworks comprised of organic ligands or modified with acidic hydrophilic functionalities such as carboxylic and hydroxyl groups [45]. Instead, the vehicle mechanism mainly targets the selection of "vehicle" molecules inside the porous frameworks of MOFs, and the performance of MOFs is influenced by diffusion coefficients and the abilities of the "vehicle" molecules to bind protons. The Grotthuss mechanism is generally reported to be more efficient; however, based on the method of proton introduction, either mechanism can be dominant [31, 46]. Figure 7.2 illustrates the mechanisms of ionic conduction in MOFs.

All these mechanisms discussed above are the regularly applied theories during the designing of conductive MOFs. Ongoing research on conductive MOFs switches on practical applications, specifically for electrochemical energy conversion and storage applications. Accordingly, the beneficial characteristics and charge conduction mechanisms of MOFs are considered during the designing of conductive MOF, targeting to encourage the direct usage of pristine MOFs and thereby accomplish boosted performance, in comparison to non-conductive MOFs. Furthermore, the design strategies of different types of conductive MOFs with specific characteristics essential for various electrocatalytic applications are demonstrated in the below sections.

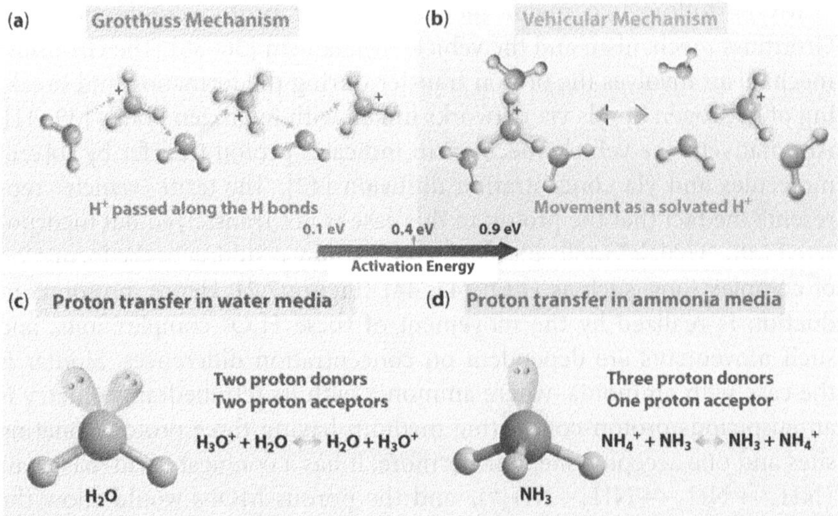

Figure 7.2 Schematic of (a, b) proton conduction mechanism and (c, d) proton transfer media. In the case of the Grotthuss mechanism, the transfer of protons is along the hydrogen bonds (reprinted with permission from ref. [31]).

7.3 Types of Conductive MOFs

Based on the design methods, conductive MOFs are classified as intrinsically conductive MOFs, guest-based conductive MOFs, and conductive MOF composites.

i) Intrinsically conductive MOFs: This type of MOFs possesses charge transport pathways such as redox-active metal ion ligands and conjugated organic ligands, and they can easily coordinate with the metal ion counterpart. These MOFs are conductive by the effective overlap between metal and ligand orbital and also the existence of non-covalent interactions. Intrinsically, conductive MOFs can further be classified into the following:

a) Mixed-valence metal ions and ligands: In this type, electron transfer happens through the metal cations inside the MOF [47, 48].
b) Infinite one-dimensional/two-dimensional metal–sulfur chains: In this case, 1D/2D metal–sulfur linkages are responsible for the conductivity of MOFs [25, 49].

c) π-interactions: In this type of conductive MOFs, efficient charge transport can be realized through non-covalent π-stacking interactions [25, 50].

ii) Guest-based conductive MOFs: The most frequently adopted strategy for enhancing the conductivity of MOF is introducing guest molecules to the pores of the framework. Guest molecules can be introduced either during the synthesis of MOFs or during the post-synthetic modification process. The production of redox-active sites inside the MOF can happen through the interaction between guest molecules and secondary building units (SBUs) of MOFs. The most widely used guest molecules include redox-active molecules, metal oxides, metal nanoclusters, cationic species, and conducting polymers.

a) Redox-active molecules: This type of guest molecules improves the conductivity of MOF by the charge transfer interaction between the molecule and the MOF. This may lead to the introduction of free charge carriers within the framework [51–53].
b) Metallic species: The assemblage of metal nanoclusters, metal oxides, or cationic molecules within the pores of MOFs improves the conductivity [54–56].
c) Conducting polymers: The introduction of large conducting polymers to the pores of MOF can also be used to increase conductivity [57–60].

iii) Conductive MOF composites: It is a very important method to facilitate the conductivity of MOF. A conductive MOF composite is synthesized by the hybridization of MOFs with carbon-based materials such as carbon nanotubes, graphene, graphene oxide, and quantum dots. Another pathway for conductive MOF composite preparation is the growth of MOF nanostructures on conductive substrates.

a) MOF–carbon composites: The conductivity of MOF can be enhanced by the incorporation of excellent electron acceptor carbonaceous materials such as carbon nanotubes, graphene, graphene oxide, C_{60}, and quantum dots [61–63].
b) MOF-conductive substrate composites: These composites can be synthesized by the *in-situ* synthesis of MOF on conductive substrates without using any binder materials. Conductive substrate enhances the electroactivity of

deposited MOFs, and therefore, MOF-conductive substrate composites are widely recognized in many of the electrochemical applications. Ni foam, Ni-Fe foam, etc. are some of the widely used conductive substrates [19, 64].

7.4 Conductive MOFs in Various Electrocatalytic Applications

Conductive MOFs as electrocatalysts are among the most interesting applications owing to their discrete characteristics of high porosity, boosted conductivity, and the presence of uniform distributed active sites. Moreover, the morphological features of MOFs play a major role in electrocatalytic applications. Architectural designing of MOFs with two-dimensional nanostructures could be beneficial for exposing their active sites and thereby an enhanced electrical conductivity with nanosized dimensions. All these open up wide horizons to overcome the limitations of MOFs for various electrocatalytic applications. Stability is a crucial factor to be considered for conductive MOFs during electrocatalysis, especially in different environments of highly acidic and alkaline media. Even though various conductive MOFs have been reported, their electrocatalytic applications are still in the progressing stage. In the following section, various concerns regarding the dominance of conductive MOFs over non-conductive MOFs with emphasis on conductive supports, long-term stability, and restrictions of conductive MOFs throughout electrocatalytic reactions are discussed. Furthermore, a recent approach of conductive MOFs for applications of hydrogen evolution reaction (HER), oxygen reduction reaction (ORR), and electrocatalytic carbon dioxide reduction reaction (CO_2RR) is tackled.

7.4.1 Some Concerns Related to the Electrocatalytic Applications of Conductive MOFs

MOFs have attractive characteristics of abundantly available active sites, highly accessible areas, predesigned frameworks, and precise catalytic environments. All these features make them promising electrocatalysts, and the electrons can be shifted to surface-active sites [65, 66]. Intrinsic conductive MOFs have more advantageous benefits when compared with the non-conductive MOFs [51, 67, 68]. In one way, the entire MOF framework is conductive, so that the transfer of electrons through the active sites is not hindered [69], while in the case of supported non-conductive MOFs, the electron transfer to the outer layer of MOF becomes more difficult [55, 69]. Additionally, in contrast to the upfront structural designing and porosity factors of intrinsic conductive MOFs, certain major factors are to be considered while selecting additional supports, mainly the appropriate matching and loading of MOFs onto conductive supports [55, 68, 70], the interfacial engineering of MOF and conductive support, and the negative influences on porosity during synthesis.

Stability factor is one of the major concerns to be considered for the performance evaluation of electrocatalysts. The stability of conductive MOFs varies with several materials and in different electrolyte media [71]. Several conductive MOFs in alkaline media are able to withstand their frameworks even after electrocatalysis, thereby exhibiting enhanced performances after the long-term stability analysis [72–74]. The stability of conductive MOFs throughout electrocatalysis was confirmed by several reports via multi-technique characterization methods [73], which evidenced the fact that, in conductive MOFs, the metal atoms are not engrossed and the valence energy levels remained the same. Moreover, their pore dimensions were also the same with no collapsing, hence revealing their superior stability. Solvent polarity is also a critical factor to be considered, which alters the structural characteristics of those conductive MOFs whose adjacent layers are linked via π–π stacking interaction. One of the conductive MOFs synthesized using a TTF monomer is conveyed to be highly delicate toward polar solvents [75]. If subjected to polar solvents like ethanol, the crystallinity of the conductive MOF decreases, and the rearrangement of the stacked structures occurs. Moreover, the conductivity increases instantaneously, owing to the increase in the rate of transfer of electrons and holes. Additionally, any change in organic ligands or their spatial orientations would oxidize or reduce the metal ions during electrocatalysis. Other than the abovementioned factors, higher temperature and incorporation of guest molecules could change the structural characteristics as well [76]. Few conductive MOFs are reported to be affected by water, resulting in the hydrolysis of secondary building units [77]. Furthermore, the conductivity of most of the MOFs could be optimized. The catalytic performance and conductivity would be affected by the structural changes in conductive MOFs. In addition, the effective mass transfer within the pores of conductive MOFs must be assured by the complete elimination of contaminants. Overall, there should be a balancing of structural stability, conductivity, and mass transfer for the conductive MOFs to exhibit maximum strength during electrocatalysis.

7.4.2 HER

Conductive MOFs of the dithiolene family have been widely reported as auspicious candidates for electrocatalytic HER. The conductive MOFs of Ni-THT, Co-THT, and Co-BHT are efficient electrocatalysts for HER [78, 79]. A thin film of Ni-THT was synthesized by the Langmuir–Blodgett method. Overpotential and Tafel slope were obtained as 80.5 mV dec^{-1} and 333 mV at 10 mA cm^{-2}. Remarkable stability in acidic media was obtained for Co-BHT and Co-THT, where Tafel slopes were obtained between 149 and 189 mV dec^{-1} (pH = 4.2; exchange current density = $10^{-5.3}$ A cm^{-2}). The corresponding overpotentials were 340 and 530 mV, respectively (pH = 1.3; current density = 10 mA cm^{-2}). Cu-BHT nanoparticles have been reported to exhibit good activity and stability under optimized conditions toward HER (Tafel slope: ~95 mV dec^{-1}, current density = 10 mA cm^{-2}) having an overpotential of 450 mV [80]. Several metal-dithiolene-based conductive MOFs with benzene-1,2,4,5-tetrathiolate (BTT) and benzene-1,2,4,5-tetraselenolate (BTSe) ligands have been reported by Downes and Marinescu as excellent HER electrocatalysts [81–83]. The same research group later reported the HER performance of Fe-BHT, Ni-BHT, and Co-BHT in acidic media at pH = 1.3, and the corresponding overpotentials followed the order of Fe- BHT, 405 mV > Ni-BHT, 331 mV > Co-BHT, 185 mV [84]. The augmented Co-BHT exhibited the lowest overpotential, which could be attributed to its enhanced charge transfer characteristics. Furthermore, the influence of film thickness on HER performance was also analyzed. Accordingly, overpotential was found to decrease initially with increased film thickness, owing to the fact that thicker films indicate bulk loadings of the catalyst realizing the accessibility of more active sites. However, the HER activity would be inhibited if the thickness is increased beyond a certain level, due to the poor diffusion of charge carriers within the thick film [85]. Huang *et al.* have reported the outstanding HER performance of a bimetallic sited conductive MOF ($Ni_3(Ni_3{\cdot}HAHATN)_2$) in an alkaline medium, with the lowest overpotential (115 mV at 10 mA cm^{-2}, Tafel slope at 45.6 mV dec^{-1}) and promising electrocatalytic stability [86]. Polyoxometalate-based conductive MOFs (POMOFs) have also been reported to exhibit enhanced electrocatalytic HER efficiency. Three POMOFs have been reported to exhibit a higher HER performance than Pt with ~95% hydrogen yield and 1.2×10^5 as the turnover number for 5 h [87]. Superior HER performance of another POMOF (NENU-500) has been reported by Qin *et al.* with an appreciable tolerance both in acidic and alkaline media [13].

7.4.3 OER

In general, for OER, when compared with monometallic MOFs, multimetallic MOFs exhibit higher conductivity and enhanced electrocatalytic activity. NiFe-MOF is a bimetallic MOF having 2,6-naphthalenedicarboxylic ligands and metal–oxygen (MO_6 = Fe or Ni) nodes within it [88]. As shown in Figure 7.3 (a, b), the NiFe-MOF/NF shows a smaller Tafel slope (34 mV dec^{-1}), when used for around 2,000 s with no considerable decay, whereas the bulk NiFe-MOF has a higher Tafel slope value (56 mV dec^{-1}). The NiFe-MOF (2D) nanosheets reveal electrical conductivity ($1 \pm 0.2 \times 10^{-5}$ S cm^{-1}) superior to the bulk NiFe-MOF ($1 \pm 0.5 \times 10^{-8}$ S cm^{-1}). This enhanced the charge transport mechanism and resulted in a higher electrocatalytic OER activity. Li *et al.* have reported the pronounced electrocatalytic OER of a well-organized MOF composite designed by incorporating ultrathin nanosheets of NiFe MOF into three-dimensionally ordered microporous hydroxides [89]. Later, Shen *et al.* reported the synthesis of $[\{Fe_3(\mu_3\text{-}O)\ (bdc)_3\}_4\{Co_2(na)_4(L^T)_2\}_3]$, ($Fe_3$-$Co_2$, H_2bdc = 1,4-benzenedicaboxylic acid, Hna = nicotinic acid, L^T = terminal ligand) hybrid MOF as an efficient OER electrocatalyst (pH = 13, overpotential 225 mV at 10.0 mA cm^{-2}). The alkaline stable catalyst exhibits high thermal stability also [90]. As shown in Figure 7.3 (c, d), the OER performance of Fe_3-Co_2@GC, when compared with Fe_3@GC, Fe_3-Fe@GC, and Co:Fe_3@GC, is better. In addition, the OER activity of Fe_3-Co_2 coated on Cu and Ni foams has been explored. Accordingly, the results reveal an enhanced OER performance for Fe_3-Co_2 coated on Ni foam. Another work reported $[Co_2(\mu\text{-}OH)_2(bbta)]$ (H_2bbta = 1H, 5H-benzo(1,2-d:4,5-d′) bistriazole) as an efficient OER electrocatalyst in alkaline medium, accounting for the existence of more open metal sites and hydroxide ligands [91].

Two-dimensional Ni–Co bimetallic MOF ultrathin nanosheets have been reported by Zhao *et al.* for electrocatalytic OER in alkaline conditions, wherein the ultrathin nanosheet structure had a remarkably high OER activity when compared with the bulk counterparts, due to the availability of coordinatively linked unsaturated metal sites and the coupling of metal ions [18]. Recently, many similar cases have been reported [91–95]. In addition, porphyrin-based MOFs have been reported as efficient electrocatalysts for OER. Dai *et al.* demonstrated the electrocatalytic OER of $[Pb_2(H_2TCPP)]\cdot 4DMF\cdot H_2O$ (H_6TCPP = 5,10,15,20-tetra(carboxyl-phenyl) porphyrin), which owns gas adsorption properties as well in addition to its electrocatalytic activity [96]. Likewise, several MOF-based OER electrocatalysts are also reported [15, 97–99].

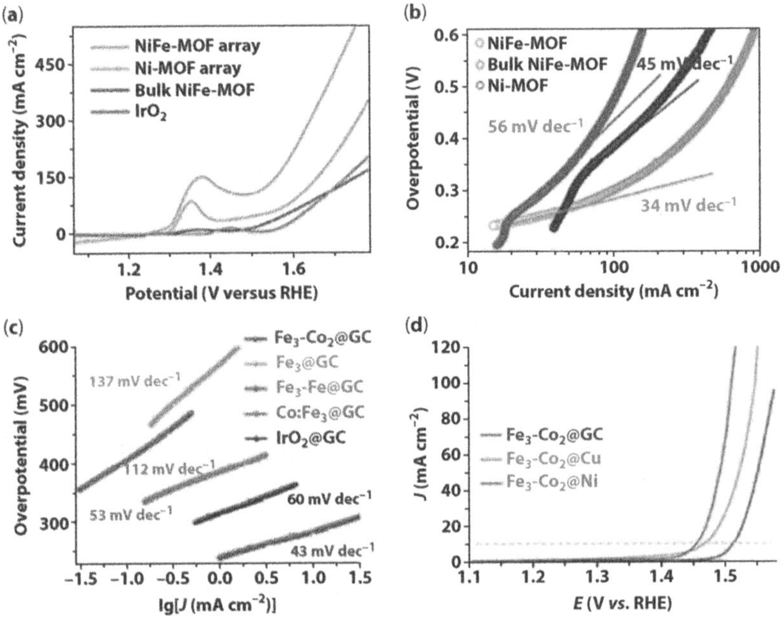

Figure 7.3 (a) Linear sweep voltammetry (LSV) plots corresponding to the OER performance of NiFe-MOF, Ni-MOF, bulk NiFe-MOF, and IrO_2 (at 10 mV s^{-1} in 0.1 M KOH). (b) Tafel plots corresponding to NiFe-MOF, Ni-MOF, and bulk NiFe-MOF. Reprinted under the terms of Creative Commons CCBY license from ref. [84]. (c) Tafel plots corresponding to Fe_3Co_2@GC, Fe_3@GC, Fe_3-Fe@GC, Co:Fe_3@GC, and IrO_2@GC in water (pH = 13). (d) LSV curves of Fe_3Co_2@GC, Fe_3-Co_2@Cu, and Fe_3-Co_2@Ni in water (pH = 13). Reprinted with permission from ref. [90].

7.4.4 ORR

Cu-BTC ($Cu_3(BTC)_2$) has been reported as the first electrocatalyst for ORR [100]. However, it lacked structural stability in aqueous solution. Thus, copper(II) 2,20-bipyridinebenzene-1,3,5-tricarboxylate, a water-soluble MOF, has been designed to catalyze the ORR efficiently, which followed a four-electron transfer pathway. Another Cu-based MOF has been subsequently reported as an ORR electrocatalyst [101], wherein the ORR is catalyzed through a two- to four-electron reduction pathway. Intrinsically conducting, $Ni_3(HITP)_2$ is a tunable and precise electrocatalyst for ORR [102, 103]. Furthermore, its ORR activity is found to be competent with the most active electrocatalysts of the non-Pt metal group. Transition metal-based porphyrin systems have been reported as effective ORR molecular catalysts. Accordingly, PCN-223-Fe, fashioned from Fe(III) porphyrin linkers and Zr_6-oxo clusters, has been demonstrated to catalyze ORR via a four-electron pathway which is highly selective [104]. In addition, MOF-based composite materials and conductive supports are reported to enable ORR. A metalloporphyrin MOF, based on graphene, was designed by Jahan *et al.* for catalyzing ORR [105]. The introduction of C_6H_5N functionalized graphene has accelerated the rate of electrochemical charge transfer in the iron–porphyrin system, hence demonstrating an enhanced electrocatalytic performance of the hybrid MOF through a facile four-electron transfer pathway. Fang *et al.* reported Co-MOF@CNTs as superior bifunctional electrocatalysts with remarkable stability toward OER and ORR activities competent with 20 wt.% Pt/C and RuO_2 electrocatalysts, respectively, owing to the synergism of Co(II), CNTs, and organic ligands [106].

7.4.5 CO_2RR

Electrocatalytic carbon dioxide reduction reaction (CO_2RR) is an effective and ecologically benign methodology for converting carbon dioxide to high-value carbon-containing products. Cu-based MOFs are the most commonly reported electrocatalysts for CO_2RR [107–109]. Hinogami *et al.* verified the competence of Cu-based MOFs, by preparing a copper rubeanate MOF for reducing CO_2 to HCOOH, with selectivity as high as 98%, wherein the Cu metal electrode produced a sequence of products consisting of formic acid, methane, ethane, and ethylene. Alb *et al.* explored several Cu-based metal–organic materials, with high proficiency for methanol and ethanol production [110]. It is a result of their high surface area, availability of electrodes, and catalytic Cu active sites. Majidi *et al.* reported a two-dimensional Cu-based conductive MOF, Cu-THQ (THQ: tetrahydroxyquinone), where a current density as high as 173 mA cm^{-2} (at −0.45 V vs. RHE), with an average Faraday efficiency reaching 91% for the CO product, and high turnover frequency as high as 20.82 s^{-1} are observed. The Cu ions and neighboring oxygen are confirmed as the catalytically active centers [111]. Metalloporphyrin-based MOF systems are also alleged as well-organized electrocatalysts for CO_2RR [12, 108, 109]. Hod *et al.* reported Fe-porphyrin-based MOF-525 (Fe_MOF-525) as an auspicious electrocatalyst for CO_2 reduction to yield hydrogen and CO with a Faraday efficiency closer to 100% [112]. In the same manner, the catalytic activity of cobalt–porphyrin–MOF: $Al_2(OH)_2$TCPP-Co was demonstrated for the selective reduction of CO_2 to CO (aqueous medium) [113], wherein CO selectivity surpasses 76% and stability lasts beyond 7 h (TOF 1,400/site). For a better understanding, an additional comparison to summarize the above-discussed applications of the selected conductive MOFs is also included in Table 7.1.

In addition, the applications of theoretically viable ORR catalysts, pioneering electrocatalytic nitrogen reduction (NRR), and various electrocatalytic reactions are yet to be investigated. In the meantime, the categories of conductive MOFs and their stability concerns under appropriate conditions should be augmented well for all the precise electrocatalytic applications.

Table 7.1 Comparison of HER, OER, ORR, and CO_2RR activities of various conductive MOFs.

Catalytic systems	Applications	Efficiency	References
Ni-THT	HER	Overpotential: 80.5 mV; Tafel slope: 333 mV/dec at 10 mA cm^{-2}	[78]
Cu-BHT	HER	Overpotential: 450 mV; Tafel slope: ~95 mV/dec at 10 mA cm^{-2}	[80]
Co-BHT	HER	Overpotential: 185 mV; Tafel slope: 88 mV/dec at 10 mA cm^{-2}	[84]
$Ni_3(Ni_3.HAHATN)_2$	HER	Overpotential: 115 mV; Tafel slope: 45.6 mV/dec at 10 mA cm^{-2}	[86]
$\{Fe_3(\mu_3\text{-}O)(bdc)_3\}_4\{Co_2(na)_4(L^T)_2\}_3$	OER	Overpotential: 225 mV; Tafel slope: 43 mV/dec at 10.0 mA cm^{-2}	[90]
MAF-X27-OH ($[Co_2(\mu\text{-}OH)_2(bbta)]$) ($H_2$bbta = 1H, 5H-benzo(1,2-d:4,5-d¢) bistriazole)	OER	Overpotential 292 mV at 10.0 mA cm^{-2}	[91]
NiCo-UMOFNs	OER	Overpotential: 250 mV; Tafel slope: 42 mV/dec at 10.0 mA cm^{-2}	[18]
$Pb_2(H_2TCPP)]\cdot 4DMF\cdot H_2O$	OER	Overpotential: 1.70 V; Tafel slope: 106.20 mV/dec at 10.0 mA cm^{-2}	[96]

(*Continued*)

Table 7.1 Comparison of HER, OER, ORR, and CO_2RR activities of various conductive MOFs. (*Continued*)

Catalytic systems	Applications	Efficiency	References
$Ni_3(HITP)_2$	ORR	Overpotential: 0.18 V relative to Pt; Tafel slope: −128 mV/dec	[102]
Metalloporphyrin MOF, based on graphene	ORR	ORR onset potential at −0.087 V vs. SCE at current density >−4.5 mA cm^{-2}	[105]
Co-MOF@CNTs (5 wt.%)	OER, ORR	Overpotential: <0.34 V; Tafel slope: 69 mV/dec at 10.0 mA cm^{-2}	[106]
H_Ru10 (Ru(III)-based metalloporphyrin-based MOF)	CO_2RR	Faraday efficiency of 47.2% at 20 mA cm^{-2}	[109]
Cu-THQ	CO_2RR	Faraday efficiency of ~91%, turnover frequency (TOF) >20.82 s^{-1} at ~173 mA cm^{-2}	[111]
Fe_MOF-525	CO_2RR	Faraday efficiency ~100%, TOF: 0.11 s^{-1} at 5.9 mA cm^{-2}	[112]
$Al_2(OH)_2$TCPP-Co	CO_2RR	CO selectivity surpasses 76% (turnover rate of 1,400 per site)	[113]

7.5 Challenges and Forthcoming Outlook

Metal–organic frameworks are hybrid inorganic–organic materials, and with the help of different kinds of linkages, metal ions, and organic compounds or metal–oxygen clusters, they are interlinked together [114]. In previous decades, there was a drastic change in the field of metal–organic framework (MOF) synthesis and structural/topological identification. With its extreme topological possibilities and compositional rigidity, the MOF synthetic route paves the way to understand and control both structural [115] and compositional complexities [116]. In comparison to conventional porous materials, MOFs have exceptionally high, ordered, and tuneable structures. Moreover, the availability of a surplus variety of organic ligands in MOFs introduces an extra degree of electronic complexity owing to its metal–organic orbital mixing at the interface. Due to these beneficial properties, the MOF can act as an emerging advanced material to exploit a wide range of applications in energy storage and conversion, drug delivery, separation, medicine, magnetics, CO_2 adsorption, and environmental and engineering sectors [117–120]. The incorporation of oxides, hydroxides, phosphides, and chalcogenides of transition metal into MOF enhances the overall electrochemical performance of the system [121]. Most of the modified MOF derivatives have similarities in their morphology with their own precursors/templates, and since each of the components in the modified MOF has unique properties, the combination of various components induces synergistic interactions within the system; thus, as compared with single-node MOFs, mixed-node MOFs can provide open metal sites and an electron-rich environment, providing efficient charge transfer and dramatically increased electrochemical activity, and hence overcome the drawbacks of single-ingredient materials by the superior performance of hybrid MOF [122]. However, the modification of the MOF layer is very difficult due to less interaction between the functional materials and metal–organic framework. The practical usage of MOFs is also difficult due to their low conductivity [123, 124]. The simple mixing of two or more metal species into the MOF matrix forms less mechanically stable crystalline coordination polymers. Slow diffusion coefficient and low electron transfer kinetics are their major challenges. Henceforth, it is important to design an appropriate synthetic route for the modification of MOF with diverse functionalities in order to arbitrate this issue.

Recently, several strategies such as spray drying, epitaxial growth, irradiation of microwave, self-templated solvothermal approach, self-sacrificing template growth, electrospinning, and interfacial engineering are opted for the synthesis of MOF. However, the high temperature associated with the major synthetic routes leads to a decrease in the number of active sites in MOF due to the destruction of its chemical structure. An alternative method to enhance the activity of MOFs is the incorporation of inorganic compounds, polymers, and conductive supporting matrix such as carbon cloth, carbon paper, and graphene layers [125, 126], inducing porosity in MOF, and some of its advantages include a simple operational approach, high rate of productivity, controllable components, adjustable morphologies, environmental friendliness, and superior electrochemical properties.

The complete exploration of the MOF derivatives to employ in various practical applications directly relies on certain parameters including the interactions of its guest–host structure, actual components present in the system, synthetic route opted, particle size, and economic viability of the material. Thus, the proper consideration of the as-mentioned parameters including the design strategy, structure, and properties provides innovative MOF derivatives with desirable functionalities for pioneering approaches toward various commercial applications.

7.6 Conclusion

MOF which holds the benefit of both organic and inorganic parts is undoubtedly a promising candidate for electrochemical transformations. Thus, designing of conductive MOF is highly desirous and can be achieved by accelerating the electron–proton transfer pathway. The three recognized electron transfer pathways in MOFs are "through-space," "through-bond," and "through-guest." The mechanism of ionic conduction in MOFs is also explored in the chapter. Based on design, conductive MOFs are classified into intrinsically conductive, guest-based, and conductive MOF composites. The chapter illustrates the application of MOF-based electrocatalysts in various electrochemical transformations (HER, HOR, OER, ORR, and CO_2 reduction). The difficulty involved in introducing functional materials into the MOF matrix leads to poor host–guest interaction put forward by the major challenge in the design of efficient MOF-based catalysts. The challenges and future direction in the area are discussed here in detail. The present chapter can definitely guide the researchers in the area.

Acknowledgments

The authors acknowledge Dr. S.M.A. Shibli, Professor, Department of Chemistry, University of Kerala for proper guidance and support.

References

1. Krishnan, A., Viswanath, S., Mohan, A.C., Panchami, R., Vishwanathan, P.V., Surface engineering of Ni-P electrode by cobalt oxide co-deposition for electrochemical hydrogen evolution reaction. *J. Environ. Chem. Eng.*, **9**, 4, 105300, 2021.

2. Joo, S.H., Choi, S.J., Oh, I., Kwak, J., Liu, Z., Terasaki, O., Ryoo, R., Ordered nanoporous arrays of carbon supporting high dispersions of platinum nanoparticles. *Nature*, **412**, 6843, 169–172, 2001.
3. Mahmood, A., Guo, W., Tabassum, H., Zou, R., Metal-organic framework-based nanomaterials for electrocatalysis. *Adv. Energy Mater.*, **6**, 17, 1600423, 2016.
4. Liu, Z., Ling, X.Y., Su, X., Lee, J.Y., Carbon-supported Pt and PtRu nanoparticles as catalysts for a direct methanol fuel cell. *J. Phys. Chem. B*, **108**, 24, 8234–8240, 2004.
5. Krishnan, A., Sha, M.A., Basheer, R., Riyas, A.H., Shibli, S.M.A., Vacancy mediated improvement in electrocatalytic HER on Nano-CeO2- Al2O3 incorporated Ni-P electrode. *Mater. Sci. Semicond. Process.*, **116**, 105138, 2020.
6. Krishnan, A., Ajay, R., Anakha, J., Namboothiri, U.S.K., Understanding defect chemistry in TMOS involved electrocatalytic OER; an analysis for advancement. *Surf. Interfaces*, **30**, 101942, 2022.
7. Mamtani, K., Jain, D., Dogu, D., Gustin, V., Gunduz, S., Co, A.C., Ozkan, U.S., Insights into oxygen reduction reaction (ORR) and oxygen evolution reaction (OER) active sites for nitrogen-doped carbon nanostructures (CNx) in acidic media. *Appl. Catal B: Environ.*, **220**, 88–97, 2018.
8. Krishnan, A., Beena, S., Shibli, S.M.A., A novel high performance Ti/Ti–W-reinforced polyaniline functionalized Ni–P electrode for high sensitive detection of dopamine from urine sample. *Mater. Chem. Phys.*, **244**, 122680, 2020.
9. Krishnan, A. and Aboobakar Shibli, S.M., Electroactive P-Ani/coreshell/TiO2/TiO2-WO3 employed surface engineering of Ni-P electrodes for alkaline hydrogen evolution reaction. *J. Ind. Eng. Chem.*, **87**, 198–212, 2020.
10. Al-Naggar, A.H., Shinde, N.M., Kim, J.-S., Mane, R.S., Water splitting performance of metal and non-metal-doped transition metal oxide electrocatalysts. *Coord Chem. Rev.*, 474, 214864, 2023.
11. Debata, U., Patra, S., Banerjee, S., Madhuri, R., Sharma, P.K., Controlled hydrothermal synthesis of graphene supported NiCo2O4 coral-like nanostructures: An efficient electrocatalyst for overall water splitting. *Appl. Surf. Sci.*, **449**, 203–212, 2018.
12. Ma, R., Lin, G., Zhou, Y. *et al.*, A review of oxygen reduction mechanisms for metal-free carbon-based electrocatalysts. *NPJ Comput. Mater.*, **5**, 78, 2019.
13. Mukhopadhyay, S., Basu, O., Nasani, R., Das, S.K., Evolution of metal organic frameworks as electro catalysts for water oxidation. *Chem. Commun.*, **56**, 79, 11735–11748, 2020.
14. Gouré, E., Gerey, B., Pécaut, J., Thomas, F., Fortage, J., Collomb, M.-N., Multireversible Redox Processes in a Self-Assembled Nickel Pentanuclear Bis (Triple- stranded Helicate): Structural and Spectroscopic Characterizations in the Ni^{II}_5 and $Ni^{I}Ni^{II}_4$ Redox States. *ChemElectroChem.*, **8**, 15, 2912–2920, 2021.

15. Qin, J.-S., Du, D.-Y., Guan, W., Bo, X.-J., Li, Y.-F., Guo, L.-P., Su, Z.-M., Wang, Y.-Y., Lan, Y.-Q., Zhou, H.-C., Ultrastable polymolybdate-based metal–organic frameworks as highly active electrocatalysts for hydrogen generation from water. *J. Am. Chem. Soc.*, **137**, 22, 7169–7177, 2015.
16. Kornienko, N., Zhao, Y., Kley, C.S., Zhu, C., Kim, D., Lin, S., Chang, C.J., Yaghi, O.M., Yang, P., Metal–organic frameworks for electrocatalytic reduction of carbon dioxide. *J. Am. Chem. Soc.*, **137**, 44, 14129–14135, 2015.
17. Kung, C.-W., Mondloch, J.E., Wang, T.C., Bury, W., Hoffeditz, W., Klahr, B.M., Klet, R.C., Pellin, M.J., Farha, O.K., Hupp, J.T., Metal–organic framework thin films as platforms for atomic layer deposition of cobalt ions to enable electrocatalytic water oxidation. *ACS Appl. Mater. Interfaces*, 7, 51, 28223–28230, 2015.
18. Zhao, S., Wang, Y., Dong, J., He, C.-T., Yin, H., An, P., Zhao, K., Zhang, X., Gao, C., Zhang, L., Lv, J., Wang, J., Zhang, J., Khattak, A.M., Khan, N.A., Wei, Z., Zhang, J., Liu, S., Zhao, H., Tang, Z., Ultrathin metal–organic framework nanosheets for electrocatalytic oxygen evolution. *Nat. Energy*, **1**, 12, 16184, 2016.
19. Liu, L., Xu, Q., Zhu, Q., Electrically conductive metal–organic frameworks for electrocatalytic applications. *Adv. Energy Sustain. Res.*, 2, 11, 2100100, 2021.
20. Zheng, W. and Lee, L.Y.S., Metal–organic frameworks for electrocatalysis: Catalyst or precatalyst? *ACS Energy Lett.*, 6, 8, 2838–2843, 2021.
21. Horike, S., Umeyama, D., Kitagawa, S., Ion conductivity and transport by porous coordination polymers and metal–organic frameworks. *Acc. Chem. Res.*, **46**, 11, 2376–2384, 2013.
22. Zhang, G., Jin, L., Zhang, R., Bai, Y., Zhu, R., Pang, H., Recent advances in the development of electronically and ionically conductive metal-organic frameworks. *Coord. Chem. Rev.*, **439**, 213915, 2021.
23. Sun, L., Campbell, M.G., Dincă, M., Electrically conductive porous metal–organic frameworks. *Angew. Chem. Int. Ed.*, **55**, 11, 3566–3579, 2016.
24. Deng, X., Hu, J.-Y., Luo, J., Liao, W.-M., He, J., Conductive metal–organic frameworks: Msechanisms, design strategies and recent advances. *Top. Curr. Chem.*, **378**, 2, 27, 2020.
25. Park, S.S., Hontz, E.R., Sun, L., Hendon, C.H., Walsh, A., Van Voorhis, T., Dincă, M., Cation-dependent intrinsic electrical conductivity in isostructural tetrathiafulvalene-based microporous metal–organic frameworks. *J. Am. Chem. Soc*, **137**, 5, 1774–1777, 2015.
26. Leong, C.F., Usov, P.M., D'Alessandro, D.M., Intrinsically conducting metal–organic frameworks. *MRS Bull.*, **41**, 11, 858–864, 2016.
27. Zhu, B., Wen, D., Liang, Z., Zou, R., Conductive metal-organic frameworks for electrochemical energy conversion and storage. *Coord. Chem. Rev.*, **446**, 214119, 2021.
28. Shan, Y., Chen, L., Pang, H., Xu, Q., Metal–organic framework- based hybrid frameworks. *Small Struct.*, **2**, 2, 2000078, 2021.

29. Zheng, F., Xiang, D., Li, P., Zhang, Z., Du, C., Zhuang, Z., Li, X., Chen, W., Highly conductive bimetallic Ni–Fe metal organic framework as a novel electrocatalyst for water oxidation. *ACS Sustain. Chem. Eng.*, **7**, 11, 9743–9749, 2019.
30. Sheberla, D., Sun, L., Blood-Forsythe, M.A., Er, S., Wade, C.R., Brozek, C.K., Aspuru-Guzik, A., Dincă, M., High electrical conductivity in Ni3 (2,3,6,7,10,11-hexaiminotriphenylene)2, a semiconducting metal–organic graphene analogue. *J. Am. Chem. Soc.*, **136**, 25, 8859–8862, 2014.
31. Zhang, G., Jin, L., Zhang, R., Bai, Y., Zhu, R., Pang, H., Recent advances in the development of electronically and ionically conductive metal-organic frameworks. *Coord. Chem. Rev.*, **439**, 213915, 2021.
32. Khatua, S., Bar, A.K., Sheikh, J.A., Clearfield, A., Konar, S., Achieving amphibious superprotonic conductivity in a Cu I metal-organic framework by strategic pyrazinium salt impregnation. *Chem. - Eur. J.*, **24**, 4, 872–880, 2018.
33. Ponomareva, V.G., Kovalenko, K.A., Chupakhin, A.P., Dybtsev, D.N., Shutova, E.S., Fedin, V.P., Imparting high proton conductivity to a metal–organic framework material by controlled acid impregnation. *J. Am. Chem. Soc.*, **134**, 38, 15640–15643, 2012.
34. Nie, X., Kulkarni, A., Sholl, D.S., Computational prediction of metal organic frameworks suitable for molecular infiltration as a route to development of conductive materials. *J. Phys. Chem. Lett.*, **6**, 9, 1586–1591, 2015.
35. Zhu, B., Wen, D., Liang, Z., Zou, R., Conductive metal-organic frameworks for electrochemical energy conversion and storage. *Coord. Chem. Rev.*, **446**, 214119, 2021.
36. Xie, X.-X., Yang, Y.-C., Dou, B.-H., Li, Z.-F., Li, G., Proton conductive carboxylate-based metal–organic frameworks. *Coord. Chem. Rev.*, **403**, 213100, 2020.
37. Li, A.-L., Gao, Q., Xu, J., Bu, X.-H., Proton-conductive metal- organic frameworks: Recent advances and perspectives. *Coord. Chem. Rev.*, **344**, 54–82, 2017.
38. Yamada, T., Sadakiyo, M., Shigematsu, A., Kitagawa, H., Proton- conductive metal–organic frameworks. *Bull. Chem. Soc. Jpn.*, **89**, 1, 1–10, 2016.
39. Agmon, N., The grotthuss mechanism. *Chem. Phys. Lett.*, **244**, 5–6, 456–462, 1995.
40. Miyake, T. and Rolandi, M., Grotthuss mechanisms: From proton transport in proton wires to bioprotonic devices. *J. Phys. Condens. Matter.*, **28**, 2, 023001, 2016.
41. Grancha, T., Ferrando-Soria, J., Cano, J., Amorós, P., Seoane, B., Gascon, J., Bazaga-García, M., Losilla, E.R., Cabeza, A., Armentano, D., Pardo, E., Insights into the dynamics of grotthuss mechanism in a proton-conducting chiral *bio* MOF. *Chem. Mater.*, **28**, 13, 4608–4615, 2016.

42. Kreuer, K.-D., Rabenau, A., Weppner, W., Vehicle mechanism, a new model for the interpretation of the conductivity of fast proton conductors. *Angew. Chem. Int. Ed. Engl.*, **21**, 3, 208–209, 1982.
43. Ludueña, G.A., Kühne, T.D., Sebastiani, D., Mixed grotthuss and vehicle transport mechanism in proton conducting polymers from ab initio molecular dynamics simulations. *Chem. Mater.*, **23**, 6, 1424–1429, 2011.
44. Sadakiyo, M., Yamada, T., Kitagawa, H., Hydrated proton- conductive metal-organic frameworks. *ChemPlusChem*, **81**, 8, 691–701, 2016.
45. Wahiduzzaman, M., Wang, S., Schnee, J., Vimont, A., Ortiz, V., Yot, P.G., Retoux, R., Daturi, M., Lee, J.S., Chang, J.-S., Serre, C., Maurin, G., Devautour-Vinot, S., A high proton conductive hydrogen-sulfate decorated titanium carboxylate metal–organic framework. *ACS Sustain. Chem. Eng.*, **7**, 6, 5776–5783, 2019.
46. Stavila, V., Talin, A.A., Allendorf, M.D., MOF-based electronic and opto-electronic devices. *Chem. Soc. Rev.*, **43**, 16, 5994–6010, 2014.
47. Okubo, T., Anma, H., Tanaka, N., Himoto, K., Seki, S., Saeki, A., Maekawa, M., Kuroda-Sowa, T., Crystal structure and carrier transport properties of a new semiconducting 2D coordination polymer with a 3,5-dimethylpiperidine dithiocarbamate ligand. *Chem. Commun.*, **49**, 39, 4316–4318, 2013.
48. Wu, Z.-L., Wang, C.-H., Zhao, B., Dong, J., Lu, F., Wang, W.-H., Wang, W.-C., Wu, G.-J., Cui, J.-Z., Cheng, P., A semi-conductive copper-organic framework with two types of photocatalytic activity. *Angew. Chem. Int. Ed.*, **55**, 16, 4938–4942, 2016.
49. Sun, L., Miyakai, T., Seki, S., Dincă, M., Mn 2 (2,5-disulfhydryl- benzene-1,4-dicarboxylate): A microporous metal–organic framework with infinite (–mn–s–) ∞ chains and high intrinsic charge mobility. *J. Am. Chem. Soc.*, **135**, 22, 8185–8188, 2013.
50. Narayan, T.C., Miyakai, T., Seki, S., Dincă, M., High charge mobility in a tetrathiafulvalene-based microporous metal–organic framework. *J. Am. Chem. Soc*, **134**, 31, 12932–12935, 2012.
51. Zeng, M.-H., Wang, Q.-X., Tan, Y.-X., Hu, S., Zhao, H.-X., Long, L.-S., Kurmoo, M., Rigid pillars and double walls in a porous metal-organic framework: Single-crystal to single-crystal, controlled uptake and release of iodine and electrical conductivity. *J. Am. Chem. Soc.*, **132**, 8, 2561–2563, 2010.
52. Hao, Z., Yang, G., Song, X., Zhu, M., Meng, X., Zhao, S., Song, S., Zhang, H., A europium(iii) based metal–organic framework: bifunctional properties related to sensing and electronic conductivity. *J. Mater. Chem. A*, **2**, 1, 237–244, 2014.
53. Neumann, T., Liu, J., Wächter, T., Friederich, P., Symalla, F., Welle, A., Mugnaini, V., Meded, V., Zharnikov, M., Wöll, C., Wenzel, W., Superexchange charge transport in loaded metal organic frameworks. *ACS Nano*, **10**, 7, 7085–7093, 2016.
54. Han, S., Warren, S.C., Yoon, S.M., Malliakas, C.D., Hou, X., Wei, Y., Kanatzidis, M.G., Grzybowski, B.A., Tunneling electrical connection to the

interior of metal-organic frameworks. *J. Am. Chem. Soc.*, **137**, 25, 8169–8175, 2015.

55. Kung, C.-W., Otake, K., Buru, C.T., Goswami, S., Cui, Y., Hupp, J.T., Spokoyny, A.M., Farha, O.K., Increased electrical conductivity in a mesoporous metal–organic framework featuring metallacarboranes guests. *J. Am. Chem. Soc.*, **140**, 11, 3871–3875, 2018.
56. Guo, Z., Panda, D.K., Maity, K., Lindsey, D., Parker, T.G., Albrecht-Schmitt, T.E., Barreda-Esparza, J.L., Xiong, P., Zhou, W., Saha, S., Modulating the electrical conductivity of metal–organic framework films with intercalated guest π-systems. *J. Mater. Chem. C*, **4**, 5, 894–899, 2016.
57. Xie, L.S., Skorupskii, G., Dincă, M., Electrically conductive metal–organic frameworks. *Chem. Rev.*, **120**, 16, 8536–8580, 2020.
58. Le Ouay, B., Boudot, M., Kitao, T., Yanagida, T., Kitagawa, S., Uemura, T., Nanostructuration of PEDOT in porous coordination polymers for tunable porosity and conductivity. *J. Am. Chem. Soc.*, **138**, 32, 10088–10091, 2016.
59. Mohmeyer, A., Schaate, A., Hoppe, B., Schulze, H.A., Heinemeyer, T., Behrens, P., Direct grafting-from of PEDOT from a photoreactive Zr-based MOF – A novel route to electrically conductive composite materials. *Chem. Commun.*, **55**, 23, 3367–3370, 2019.
60. Wang, T.C., Hod, I., Audu, C.O., Vermeulen, N.A., Nguyen, S.T., Farha, O.K., Hupp, J.T., Rendering high surface area, mesoporous metal–organic frameworks electronically conductive. *ACS Appl. Mater. Interfaces*, **9**, 14, 12584–12591, 2017.
61. Goswami, S., Ray, D., Otake, K., Kung, C.-W., Garibay, S.J., Islamoglu, T., Atilgan, A., Cui, Y., Cramer, C.J., Farha, O.K., Hupp, J.T., A porous, electrically conductive hexazirconium(iv) metal–organic framework. *Chem. Sci.*, **9**, 19, 4477–4482, 2018.
62. Alfè, M., Gargiulo, V., Lisi, L., Di Capua, R., Synthesis and characterization of conductive copper-based metal-organic framework/graphene-like composites. *Mater. Chem. Phys.*, **147**, 3, 744–750, 2014.
63. Kim, D., Kim, D.W., Hong, W.G., Coskun, A., Graphene/ZIF-8 composites with tunable hierarchical porosity and electrical conductivity. *J. Mater. Chem. A*, **4**, 20, 7710–7717, 2016.
64. Cao, C., Ma, D., Xu, Q., Wu, X., Zhu, Q., Semisacrificial template growth of self-supporting MOF nanocomposite electrode for efficient electrocatalytic water oxidation. *Adv. Funct. Mater.*, **29**, 6, 1807418, 2019.
65. Dhakshinamoorthy, A., Li, Z., Garcia, H., Catalysis and photocatalysis by metal organic frameworks. *Chem. Soc. Rev.*, **47**, 22, 8134–8172, 2018.
66. Shi, Y., Zhang, X., Liu, H., Han, J., Yang, Z., Gu, L., Tang, Z., Metalation of catechol-functionalized defective covalent organic frameworks for lewis acid catalysis. *Small*, **16**, 24, 2001998, 2020.
67. Wang, J., Hu, H., Lu, S., Hu, J., Zhu, H., Duan, F., Du, M., Conductive metal and covalent organic frameworks for electrocatalysis: Design principles, recent progress and perspective. *Nanoscale*, **14**, 2, 277–288, 2022.

68. Ukaj, D., Bunzen, H., Berger, J., Kieslich, G., Fischer, R.A., Charge-transfer-induced electrical conductivity in a tetrathiafulvalene-based metal–organic framework. *Chem. Mater.*, **33**, 7, 2532–2542, 2021.
69. Zhu, Z., Zhao, B., Hou, S., Jiang, X., Liang, Z., Zhang, B., Zhao, B., A facile strategy for constructing a carbon-particle-modified metal–organic framework for enhancing the efficiency of CO2 electroreduction into formate. *Angew. Chem.*, **133**, 43, 23582–23590, 2021.
70. Stolz, R.M., Mahdavi-Shakib, A., Frederick, B.G., Mirica, K.A., host–guest interactions and redox activity in layered conductive metal–organic frameworks. *Chem. Mater.*, **32**, 18, 7639–7652, 2020.
71. Xu, S., Chansai, S., Stere, C., Inceesungvorn, B., Goguet, A., Wangkawong, K., Taylor, S.F.R., Al-Janabi, N., Hardacre, C., Martin, P.A., Fan, X., Sustaining metal–organic frameworks for water–gas shift catalysis by non-thermal plasma. *Nat. Catal.*, **2**, 2, 142–148, 2019.
72. Yi, J., Xie, R., Xie, Z., Chai, G., Liu, T., Chen, R., Huang, Y., Cao, R., Highly selective CO_2 electroreduction to CH 4 by *In Situ* generated Cu_2O single-type sites on a conductive MOF: Stabilizing key intermediates with hydrogen bonding. *Angew. Chem. Int. Ed.*, **59**, 52, 23641–23648, 2020.
73. Huang, N., Lee, K.H., Yue, Y., Xu, X., Irle, S., Jiang, Q., Jiang, D., A stable and conductive metallophthalocyanine framework for electrocatalytic carbon dioxide reduction in water. *Angew. Chem.*, **59**, 16587, 2020.
74. Yi, J., Si, D., Xie, R., Yin, Q., Zhang, M., Wu, Q., Chai, G., Huang, Y., Cao, R., Conductive two-dimensional phthalocyanine-based metal–organic framework nanosheets for efficient electroreduction of CO_2. *Angew. Chem.*, **133**, 31, 17245–17251, 2021.
75. Cai, S., Sun, B., Li, X., Yan, Y., Navarro, A., Garzón-Ruiz, A., Mao, H., Chatterjee, R., Yano, J., Zhu, C., Reimer, J.A., Zheng, S., Fan, J., Zhang, W., Liu, Y., Reversible interlayer sliding and conductivity changes in adaptive tetrathiafulvalene-based covalent organic frameworks. *ACS Appl. Mater. Interfaces*, **12**, 16, 19054–19061, 2020.
76. Liu, D., Liu, T.-F., Chen, Y.-P., Zou, L., Feng, D., Wang, K., Zhang, Q., Yuan, S., Zhong, C., Zhou, H.-C., A reversible crystallinity-preserving phase transition in metal–organic frameworks: Discovery, mechanistic studies, and potential applications. *J. Am. Chem. Soc.*, **137**, 24, 7740–7746, 2015.
77. Shi, Y., Momeni, M.R., Chen, Y.-J., Zhang, Z., Shakib, F.A., Water-induced structural transformations in flexible two-dimensional layered conductive metal–organic frameworks. *Chem. Mater.*, **32**, 22, 9664–9674, 2020.
78. Dong, R., Pfeffermann, M., Liang, H., Zheng, Z., Zhu, X., Zhang, J., Feng, X., Large-area, free-standing, two-dimensional supramolecular polymer single-layer sheets for highly efficient electrocatalytic hydrogen evolution. *Angew. Chem. Int. Ed.*, **54**, 41, 12058–12063, 2015.
79. Clough, A.J., Yoo, J.W., Mecklenburg, M.H., Marinescu, S.C., Two-dimensional metal–organic surfaces for efficient hydrogen evolution from water. *J. Am. Chem. Soc.*, **137**, 1, 118–121, 2015.

80. Huang, X., Yao, H., Cui, Y., Hao, W., Zhu, J., Xu, W., Zhu, D., Conductive copper benzenehexathiol coordination polymer as a hydrogen evolution catalyst. *ACS Appl. Mater. Interfaces*, **9**, 46, 40752–40759, 2017.
81. Downes, C.A. and Marinescu, S.C., One dimensional metal dithiolene (M = Ni, Fe, Zn) coordination polymers for the hydrogen evolution reaction. *Dalton Trans.*, **45**, 48, 19311–19321, 2016.
82. Downes, C.A. and Marinescu, S.C., Bioinspired metal selenolate polymers with tunable mechanistic pathways for efficient H2 evolution. *ACS Catal.*, **7**, 1, 848–854, 2017.
83. Liu, L., Xu, Q., Zhu, Q., Electrically conductive metal–organic frameworks for electrocatalytic applications. *Adv. Energy Sustain. Res.*, **2**, 11, 2100100, 2021.
84. Downes, C.A., Clough, A.J., Chen, K., Yoo, J.W., Marinescu, S.C., Evaluation of the H 2 evolving activity of benzenehexathiolate coordination frameworks and the effect of film thickness on H 2 production. *ACS Appl. Mater. Interfaces*, **10**, 2, 1719–1727, 2018.
85. Dong, R., Zheng, Z., Tranca, D.C., Zhang, J., Chandrasekhar, N., Liu, S., Zhuang, X., Seifert, G., Feng, X., Immobilizing molecular metal dithiolene-diamine complexes on 2D metal-organic frameworks for electrocatalytic H 2 production. *Chem. - Eur. J.*, **23**, 10, 2255–2260, 2017.
86. Huang, H., Zhao, Y., Bai, Y., Li, F., Zhang, Y., Chen, Y., Conductive metal–organic frameworks with extra metallic sites as an efficient electrocatalyst for the hydrogen evolution reaction. *Adv. Sci.*, **7**, 9, 2000012, 2020.
87. Nohra, B., El Moll, H., Rodriguez Albelo, L.M., Mialane, P., Marrot, J., Mellot-Draznieks, C., O'Keeffe, M., Ngo Biboum, R., Lemaire, J., Keita, B., Nadjo, L., Dolbecq, A., Polyoxometalate-based metal organic frameworks (POMOFs): Structural trends, energetics, and high electrocatalytic efficiency for hydrogen evolution reaction. *J. Am. Chem. Soc.*, **133**, 34, 13363–13374, 2011.
88. Duan, J., Chen, S., Zhao, C., Ultrathin metal-organic framework array for efficient electrocatalytic water splitting. *Nat. Commun.*, **8**, 1, 15341, 2017.
89. Li, X., Ma, D., Cao, C., Zou, R., Xu, Q., Wu, X., Zhu, Q., Inlaying ultrathin bimetallic MOF nanosheets into 3D ordered macroporous hydroxide for superior electrocatalytic oxygen evolution. *Small*, **15**, 35, 1902218, 2019.
90. Shen, J.-Q., Liao, P.-Q., Zhou, D.-D., He, C.-T., Wu, J.-X., Zhang, W.-X., Zhang, J.-P., Chen, X.-M., Modular and stepwise synthesis of a hybrid metal–organic framework for efficient electrocatalytic oxygen evolution. *J. Am. Chem. Soc.*, **139**, 5, 1778–1781, 2017.
91. Lu, X.-F., Liao, P.-Q., Wang, J.-W., Wu, J.-X., Chen, X.-W., He, C.-T., Zhang, J.-P., Li, G.-R., Chen, X.-M., An alkaline-stable, metal hydroxide mimicking metal–organic framework for efficient electrocatalytic oxygen evolution. *J. Am. Chem. Soc.*, **138**, 27, 8336–8339, 2016.
92. Wang, X.-L., Dong, L.-Z., Qiao, M., Tang, Y.-J., Liu, J., Li, Y., Li, S.-L., Su, J.-X., Lan, Y.-Q., Exploring the performance improvement of the oxygen evolution

reaction in a stable bimetal-organic framework system. *Angew. Chem. Int. Ed.*, **57**, 31, 9660–9664, 2018.

93. Zheng, F., Xiang, D., Li, P., Zhang, Z., Du, C., Zhuang, Z., Li, X., Chen, W., Highly conductive bimetallic Ni–Fe metal organic framework as a novel electrocatalyst for water oxidation. *ACS Sustain. Chem. Eng.*, **7**, 11, 9743–9749, 2019.
94. Zheng, F., Zhang, Z., Xiang, D., Li, P., Du, C., Zhuang, Z., Li, X., Chen, W., Fe/Ni bimetal organic framework as efficient oxygen evolution catalyst with low overpotential. *J. Colloid Interface Sci.*, **555**, 541–547, 2019.
95. Wang, B., Shang, J., Guo, C., Zhang, J., Zhu, F., Han, A., Liu, J., A general method to ultrathin bimetal-mof nanosheets arrays via *in situ* transformation of layered double hydroxides arrays. *Small*, **15**, 6, 1804761, 2019.
96. Dai, F., Fan, W., Bi, J., Jiang, P., Liu, D., Zhang, X., Lin, H., Gong, C., Wang, R., Zhang, L., Sun, D., A lead–porphyrin metal–organic framework: Gas adsorption properties and electrocatalytic activity for water oxidation. *Dalton Trans.*, **45**, 1, 61–65, 2016.
97. Usov, P.M., Ahrenholtz, S.R., Maza, W.A., Stratakes, B., Epley, C.C., Kessinger, M.C., Zhu, J., Morris, A.J., Cooperative electrochemical water oxidation by Zr nodes and Ni–porphyrin linkers of a PCN-224 MOF thin film. *J. Mater. Chem. A*, **4**, 43, 16818–16823, 2016.
98. Zhao, Y., Zhang, S., Wang, M., Han, J., Wang, H., Li, Z., Liu, X., Engineering iridium-based metal organic frameworks towards electrocatalytic water oxidation. *Dalton Trans.*, **47**, 13, 4646–4652, 2018.
99. Lin, S., Pineda-Galvan, Y., Maza, W.A., Epley, C.C., Zhu, J., Kessinger, M.C., Pushkar, Y., Morris, A.J., Electrochemical water oxidation by a Catalyst-modified metal-organic framework thin film. *ChemSusChem*, **10**, 3, 514–522, 2017.
100. Mao, J., Yang, L., Yu, P., Wei, X., Mao, L., Electrocatalytic four-electron reduction of oxygen with Copper (II)-based metal-organic frameworks. *Electrochem. Commun.*, **19**, 29–31, 2012.
101. Jiang, M., Li, L., Zhu, D., Zhang, H., Zhao, X., Oxygen reduction in the nanocage of metal–organic frameworks with an electron transfer mediator. *J. Mater. Chem. A*, **2**, 15, 5323–5329, 2014.
102. Miner, E.M., Fukushima, T., Sheberla, D., Sun, L., Surendranath, Y., Dincă, M., Electrochemicaloxygenreductioncatalysedby Ni3(hexaiminotriphenylene)2. *Nat. Commun.*, **7**, 1, 10942, 2016.
103. Miner, E.M., Gul, S., Ricke, N.D., Pastor, E., Yano, J., Yachandra, V.K., Van Voorhis, T., Dincă, M., Mechanistic evidence for ligand-centered electrocatalytic oxygen reduction with the conductive MOF Ni_3 (hexaiminotriphenylene)2. *ACS Catal.*, **7**, 11, 7726–7731, 2017.
104. Usov, P.M., Huffman, B., Epley, C.C., Kessinger, M.C., Zhu, J., Maza, W.A., Morris, A.J., Study of electrocatalytic properties of metal–organic framework PCN-223 for the oxygen reduction reaction. *ACS Appl. Mater. Interfaces*, **9**, 39, 33539–33543, 2017.

105. Jahan, M., Bao, Q., Loh, K.P., Electrocatalytically active graphene–porphyrin MOF composite for oxygen reduction reaction. *J. Am. Chem. Soc.*, **134**, 15, 6707–6713, 2012.
106. Fang, Y., Li, X., Li, F., Lin, X., Tian, M., Long, X., An, X., Fu, Y., Jin, J., Ma, J., Self-assembly of cobalt-centered metal organic framework and multiwalled carbon nanotubes hybrids as a highly active and corrosion-resistant bifunctional oxygen catalyst. *J. Power Sources*, **326**, 50–59, 2016.
107. Tan, X., Yu, C., Zhao, C., Huang, H., Yao, X., Han, X., Guo, W., Cui, S., Huang, H., Qiu, J., Restructuring of Cu_2 O to Cu_2 O@Cu-Metal–organic frameworks for selective electrochemical reduction of CO_2. *ACS Appl. Mater. Interfaces*, **11**, 10, 9904–9910, 2019.
108. Qiu, Y.-L., Zhong, H.-X., Zhang, T.-T., Xu, W.-B., Su, P.-P., Li, X.-F., Zhang, H.-M., Selective electrochemical reduction of carbon dioxide using cu based metal organic framework for CO_2 capture. *ACS Appl. Mater. Interfaces*, **10**, 3, 2480–2489, 2018.
109. Perfecto-Irigaray, M., Albo, J., Beobide, G., Castillo, O., Irabien, A., Pérez-Yáñez, S., Synthesis of heterometallic metal–organic frameworks and their performance as electrocatalyst for CO_2 reduction. *RSC Adv.*, **8**, 38, 21092–21099, 2018.
110. Albo, J., Vallejo, D., Beobide, G., Castillo, O., Castaño, P., Irabien, A., Copper-based metal-organic porous materials for CO_2 electrocatalytic reduction to alcohols. *ChemSusChem*, **10**, 6, 1100–1109, 2017.
111. Majidi, L., Ahmadiparidari, A., Shan, N., Misal, S.N., Kumar, K., Huang, Z., Rastegar, S., Hemmat, Z., Zou, X., Zapol, P., Cabana, J., Curtiss, L.A., Salehi-Khojin, A., 2D Copper tetrahydroxyquinone conductive metal–organic framework for selective CO_2 electrocatalysis at low overpotentials. *Adv. Mater.*, **33**, 10, 2004393, 2021.
112. Hod, I., Sampson, M.D., Deria, P., Kubiak, C.P., Farha, O.K., Hupp, J.T., Fe-porphyrin-based metal–organic framework films as high-surface concentration, heterogeneous catalysts for electrochemical reduction of CO_2. *ACS Catal.*, **5**, 11, 6302–6309, 2015.
113. Dou, S., Song, J., Xi, S., Du, Y., Wang, J., Huang, Z., Xu, Z.J., Wang, X., Boosting electrochemical CO_2 reduction on metal–organic frameworks via ligand doping. *Angew. Chem.*, **131**, 12, 4081–4085, 2019.
114. O'Keeffe, M. and Yaghi, O.M., Deconstructing the crystal structures of metal–organic frameworks and related materials into their underlying nets. *Chem. Rev.*, **112**, 2, 675–702, 2012.
115. Li, P., Vermeulen, N.A., Malliakas, C.D., Gómez-Gualdrón, D.A., Howarth, A.J., Mehdi, B.L., Dohnalkova, A., Browning, N.D., O'Keeffe, M., Farha, O.K., Bottom-up construction of a superstructure in a porous uranium-organic crystal. *Science*, **356**, 6338, 624–627, 2017.
116. Zhou, C., Longley, L., Krajnc, A., Smales, G.J., Qiao, A., Erucar, I., Doherty, C.M., Thornton, A.W., Hill, A.J., Ashling, C.W., Qazvini, O.T., Lee, S.J., Chater, P.A., Terrill, N.J., Smith, A.J., Yue, Y., Mali, G., Keen, D.A., Telfer,

S.G., Bennett, T.D., Metal-organic framework glasses with permanent accessible porosity. *Nat. Commun.*, **9**, 1, 5042, 2018.

117. Liao, X., Fu, H., Yan, T., Lei, J., Electroactive metal–organic framework composites: Design and biosensing application. *Biosens. Bioelectron.*, **146**, 111743, 2019.
118. Razavi, S.A.A. and Morsali, A., Linker functionalized metal- organic frameworks. *Coord. Chem. Rev.*, **399**, 213023, 2019.
119. Rada, Z.H., Abid, H.R., Sun, H., Wang, S., Bifunctionalized metal organic frameworks, UiO-66-NO_2 -N (N = -NH_2, -$(OH)_2$, -$(COOH)_2$), for Enhanced Adsorption and Selectivity of CO_2 and N_2. *J. Chem. Eng. Data*, **60**, 7, 2152–2161, 2015.
120. Chowdhury, M.A., Metal-organic-frameworks for biomedical applications in drug delivery, and as MRI contrast agents: applications of MOFs in drug delivery and as contrast agents. *J. Biomed. Mater. Res. A*, **105**, 4, 1184–1194, 2017.
121. Wu, Y., Huang, Z., Jiang, H., Wang, C., Zhou, Y., Shen, W., Xu, H., Deng, H., Facile synthesis of uniform metal carbide nanoparticles from metal–organic frameworks by laser metallurgy. *ACS Appl. Mater. Interfaces*, **11**, 47, 44573–44581, 2019.
122. Wang, L., Zhu, Y., Du, C., Ma, X., Cao, C., Advances and challenges in metal–organic framework derived porous materials for batteries and electrocatalysis. *J. Mater. Chem. A*, **8**, 47, 24895–24919, 2020.
123. Forgan, R.S., Modulated self-assembly of metal–organic frameworks. *Chem. Sci.*, **11**, 18, 4546–4562, 2020.
124. Yang, P., Zhao, W., Shkurenko, A., Belmabkhout, Y., Eddaoudi, M., Dong, X., Alshareef, H.N., Khashab, N.M., Polyoxometalate–cyclodextrin metal–organic frameworks: From tunable structure to customized storage functionality. *J. Am. Chem. Soc*, **141**, 5, 1847–1851, 2019.
125. Li, X., Ma, W., Li, H., Zhang, Q., Liu, H., Sulfur-functionalized metal-organic frameworks: Synthesis and applications as advanced adsorbents. *Coord. Chem. Rev.*, **408**, 213191, 2020.
126. Dong, J., Han, X., Liu, Y., Li, H., Cui, Y., Metal–covalent organic frameworks (MCOFs): A bridge between metal–organic frameworks and covalent organic frameworks. *Angew. Chem. Int. Ed.*, **59**, 33, 13722–13733, 2020.

8

Carbonaceous Materials for Supercapattery

J.R. Low[1], H.N. Lim[1,2*], I. Ibrahim[2,3†], C. Y. Foo[4] and Z. Zainal[1]

[1]*Department of Chemistry, Faculty of Science, University Putra Malaysia, Serdang, Selangor, Malaysia*
[2]*Foundry of Reticular Materials for Sustainability (FORMS) Laboratory, Institute of Nanoscience and Nanotechnology, Universiti Putra Malaysia, Serdang, Selangor, Malaysia*
[3]*Functional Nanotechnology Devices Laboratory (FNDL), Institute of Nanoscience and Nanotechnology, Universiti Putra Malaysia, Serdang, Selangor Darul Ehsan, Malaysia*
[4]*School of Energy and Chemical Engineering, Xiamen University Malaysia, Selangor Darul Ehsan, Malaysia & College of Chemistry and Chemical Engineering, Xiamen University, Xiamen, China*

Abstract
The demand for energy storage is exponentially increasing with the growth of population and technological innovations. In this regard, supercapattery shows great promise due to its potential to achieve high energy and power density with long cyclability, which is higher in magnitude than supercapacitors and batteries. The next-generation electrochemical energy storage device of supercapattery couples both supercapacitor (high power density and long cyclability) and battery (high energy density) into a single device. Carbonaceous electrode materials attract tremendous attention in supercapattery owing to their excellent properties of high specific surface area, electrical conductivity, and mechanical stability. These properties are responsible for enhancing the charged storage capacity and act as conducive support to boost the electrical properties of battery-like material. Therefore, carbonaceous materials, particularly, carbon nanotubes, graphene, mesoporous carbon, and their hybrids, have been widely investigated as efficient electrode materials in supercapattery. This chapter summarizes recent advances in high-performance supercapattery based on carbonaceous material with an

**Corresponding author*: hongngee@upm.edu.my
†*Corresponding author*: izwaharyanie@upm.edu.my

Sudheesh K. Shukla, Chaudhery Mustansar Hussain, Santanu Patra, Meenakshi Choudhary (eds.) Electrocatalytic Materials for Renewable Energy, (195–228) © 2024 Scrivener Publishing LLC

emphasis on the fabrication of electrode structures and the elucidation of charge storage mechanisms. In addition, a new class of carbonaceous materials, metal–organic framework-derived porous carbon material was also discussed, which promises great potential in not only improving the supercapattery performances but also unraveling the related mechanisms.

***Keywords*:** Supercapattery, battery-supercapacitor hybrid, activated carbon, electrochemical energy storage, carbon nanotubes, graphene, metal-organic framework-derived carbon

8.1 Introduction

Environmental pollution issues and energy crises have become critical nowadays due to the excessive utilization of fossil fuel resources [1]. To overcome this problem, developing new energy sources is expeditious [2]. Electrochemical energy storage (EES) such as supercapacitors, regenerative fuel cells, and rechargeable batteries has drawn much attention due to their high performance and affordable price. Commonly, EES is classified into two types based on their energy and power density which are rechargeable batteries and supercapacitors. Rechargeable batteries possess high energy density due to faradaic charge storage mechanisms accompanied by high redox potential windows [3]. Lithium-ion battery (LIB) has surpassed the performance of all rechargeable batteries because of the lightest metal of lithium and most electropositive metallic elements (–3.04 V vs. standard hydrogen electrode) [4–6]. Therefore, lithium ions (Li^+) are easy to accommodate and possess rapid ion mobility among most metal ions [5]. Meanwhile, supercapacitors suffer from limited energy density because of their finite ion adsorption and desorption processes at the electrode and electrolyte interface [7].

The development of supercapattery is to pursue both high power and energy density with a long lifespan that could provide a huge amount of energy in a short period. In this regard, the next generation of EES started to develop in the past few years to hybridize a rechargeable battery and supercapacitor into a single device. As a result, the generic name "supercapattery" was given later. Supercapattery could overcome the low specific power of the battery and low specific energy of supercapacitors without sacrificing their excellent cyclability. In addition, supercapattery utilizes nanomaterials by hybridizing between capacitive electrode from supercapacitor and Nernstian electrode from LIB in a single electrode or device [8]. Therefore, the novel supercapattery is expected to possess the best features of both supercapacitors and LIBs. The capacitive electrode used in supercapattery can be either EDLC type or pseudocapacitive type.

Generally, the EDLC type can be a carbonaceous material, while the pseudocapacitive type of material is a negative potential metal oxide such as Fe_2O_3, Bi_2O_3, and MnO_3. Moreover, the EDLC type of material can be either anode or cathode [9, 10]. Carbon-based materials are widely used in energy storage applications, especially supercapacitors owing to their high specific surface area (SSA) and porous structure where the electrostatic double layer is formed at the electrode–electrolyte interface, while the porous structure allows electrolyte ions to pass through the whole surface of the electrode to provide high power density. Carbon-based material is inexpensive due to its abundance, impeccable corrosion resistance, and properties of high electrical conductivity and chemical stability [11]. Moreover, it is easy to handle and fabricate with controllable porosity. Various allotropes of carbon (e.g., graphite, activated carbon, carbon nanotubes, graphene) can occur in different forms from 1D to 3D including sheets, fibers, aerogels, composites, foams, and tubes [12]. In the energy storage application, the specific capacitance is affected by the SSA, pore size, surface functionality, and electrical conductivity of electrode materials [10].

This chapter summarizes recent advances in high-performance supercapattery based on carbonaceous material with an emphasis on the material synthesis and elucidation of charge storage mechanisms of electrode material toward supercapattery. In addition, a new class of carbonaceous materials, metal–organic framework (MOF)-derived porous carbon material, has been managed into the supercapattery field, which promises great potential in not only improving the supercapattery performances but also unraveling the related mechanisms.

8.2 Mechanism and the Fundamental of Supercapattery

The charge storage mechanism of supercapattery relies on the adsorption–desorption of anions on the surface of the cathode and the intercalation–deintercalation of Li^+ in the anode during the reversible charging and discharging process [13]. The capability of capacitive cathode materials to absorb and release the charge quickly results in high power density, while the lithiation ability of the battery-type anode material leads to high energy density. However, the intercalation reaction of Li^+ at the anode is far less than solvated ion adsorption on the cathode causing the phenomenon of decreasing energy density, especially at a high charging rate, thus restricting energy storage. Thus, modification and improvement of the electrode material to utilize capacitive carbon material as anode and battery-type material (high SSA, high conductivity, and high capacitance performances) as cathode material will accelerate the rate of the process for supercapattery (Figure 8.1) [14]. Anions and cations move to or are separated from two electrodes, respectively, during the charging and discharging process. Rapid charge transfer or ion accumulation and separation occurred at the capacitive electrode, while bulk redox reaction happened at the battery-type electrode. In the meantime, electrons flow across the external circuit [9].

Capacitive carbon anode and battery-type cathode bring the advantages of not consuming electrolytes and generally have high output voltage due to the high potential plateau of the cathode. Moreover, the charge storage capability of a battery-type cathode is higher than a capacitive cathode which can improve the energy density of a supercapattery [9]. Therefore,

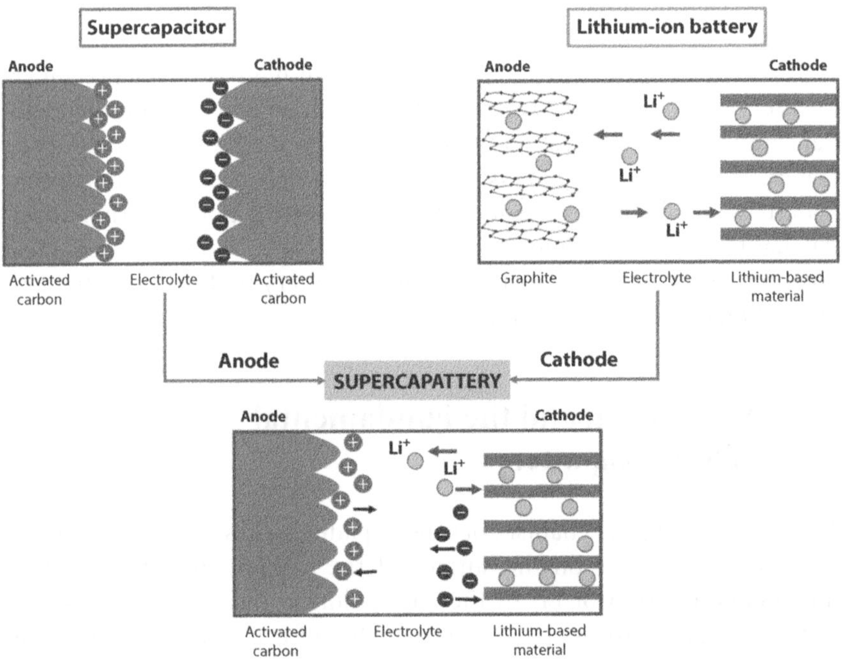

Figure 8.1 Schematic illustrating the comparison between supercapacitor, LIB, and supercapattery.

a supercapattery that has the combination of EDLC and battery-type working mechanism can achieve high energy density, moderate power density, and stable cyclability. For instance, dendrite-structured FeF_2 as a battery-type cathode and activated carbon (AC) anode exhibits a high energy density value of 152 Wh kg^{-1} and a high power density of 4,900 W kg^{-1} [15]. Furthermore, a $LiFePO_4$ as battery-type cathode and a free-standing reduced graphene oxide/carbon nanotube (rGO/CNT) nanocarbon membrane as anode indicated superior performance with both high energy density (180 Wh kg^{-1}) and high power density (218 W kg^{-1}) [16]. By contrast, a 3D carbon-modified Li_2TiSiO_5 anode and a lignin-derived AC cathode delivered an energy density of 115.3 Wh kg^{-1} at 163.5 W kg^{-1}, and Fe_3O_4 nanocrystallites anchored on the 3D graphene layers cathode and AC anode possess energy density of 120 Wh kg^{-1} at 45.4 W kg^{-1} [17, 18]. The literature indicated the former (capacitive carbon as an anode) to show higher energy density compared with the latter (capacitive carbon as a cathode).

The energy density of a supercapattery can be improved through two different approaches which are capacity improvement and voltage expansion. The high capacity of battery-type electrode material can increase the performance of a supercapattery approximately two-fold (or even more) than a symmetrical supercapacitor [19]. Beyond that, selecting an appropriate battery-type electrode material that works in the separated potential range enables the utilization of the full capacitance of capacitive electrode material. Thus, the output voltage of the full cell of the supercapattery can be enlarged. Figure 8.2 further comprehends the remarkable capacity extension using the electrode potential profile versus specific capacity plots

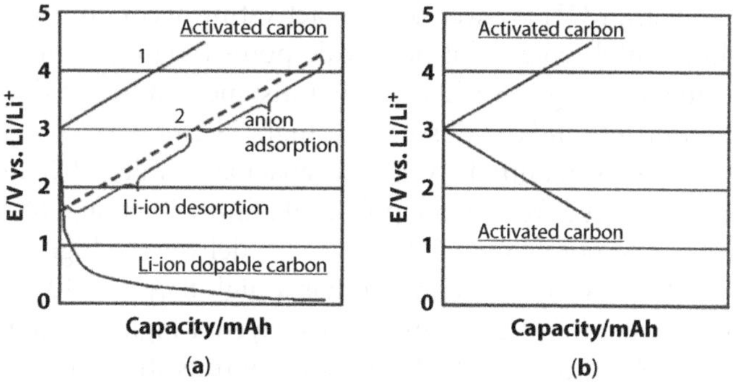

Figure 8.2 Typical voltage profiles for (a) supercapattery of activated carbon // lithium-doped carbon and (b) symmetric supercapacitor of activated carbon. Reproduced from ref. [20].

for both symmetric supercapacitor and supercapattery configurations. In Figure 8.2a, a wide potential window of 1.5 to 4.5 V versus Li/Li^+ can be used for the cathode in supercapattery based on the activated carbon cathode and intercalated type carbon anode. By contrast, only a narrow potential from 3.0 to 4.5 V versus Li/Li^+ has been used for activated carbon in symmetrical supercapacitors (Figure 8.2b) [9, 20, 21]. Thus, the energy density of the supercapattery was improved.

8.3 Utilization of Carbonaceous Materials in Supercapattery Application

8.3.1 Activated Carbon (AC)

Activated carbon is an amorphous carbon that is widely used in supercapattery devices due to its good electrical conductivity, porous structure, availability of raw material, and cost-effectiveness. A good porous structure with a high SSA provides numerous charge storage sites and facilitates electrolyte ion transportation to enhance the capacitances. A carbon piece-rich material can be converted into pure carbon by the carbonation process by heating (also known as pyrolysis), followed by an activation process to enlarge the pore size and SSA.

The chemical activation process is the most popular activation method used to produce the AC with a high SSA (around 1,000–2,000 $m^2\ g^{-1}$) and a wide pore distribution (about 0.4–4 nm) [22, 23]. Moreover, it takes place at a slightly lower temperature compared with physical activation. The chemical activation process involved the dehydration of chemical agents such as phosphorus acid (H_3PO_4), potassium hydroxide (KOH), and zinc chloride ($ZnCl_2$). KOH is a well-known activating agent due to KOH etching building a high volume of nanoscale pores ranging from 2.67 nm to 3.20 nm and a very high SSA [24, 25]. For instance, AC derived from neem leaves (*Azadirachta indica*) possesses a high SSA of 3,315 $m^2\ g^{-1}$ and an average pore size of 3.4 nm delivering a capacitance value of 262 $F\ g^{-1}$ [26]. A similar AC derived from waste polyimide film with a high SSA of 2,898 $m^2\ g^{-1}$ and a total pore volume of 1.4 $cm^3\ g^{-1}$ delivered a high capacitance of 482 $F\ g^{-1}$ [27]. Another AC which has a high SSA of 3,405 $m^2\ g^{-1}$ and a total pore structure of 2.01 $cm^3\ g^{-1}$ displayed a high capacitance value of 335 $F\ g^{-1}$ [28]. Thus, high SSA and good porous structure promote the electrolyte ion storage to enhance the electrochemical performances in supercapattery.

Apart from the type of activating reagent, carbonization temperature influenced the SSA and porosity characteristics of the AC produced by chemical activation. The optimum carbonization temperature was 800°C. The SSA was increased from 600.05 to 618.54 m^2 g^{-1} when the temperature was raised from 600°C to 800°C [29]. This is because of the vigorous reaction offered by thermal energy resulting in more carbon atoms on active sites reacting with activating agents which contributed to a large SSA. A large amount of potassium vapor gets into the existing pores at 800°C to develop an abundant pore structure. However, the features then started to decline after 800°C as the temperature continued to rise due to the excessive reaction of the activating agent and carbon. Moreover, the microporous volume of AC started to reduce, while the mesoporous volume can be observed at above 650°C [30]. The specific capacitance value of AC was reported as 174 F g^{-1} at 800°C and 156 F g^{-1} and 135 F g^{-1} at 600°C and 900°C, respectively [31].

Generally, AC was used as an anode electrode in supercapattery applications. The high SSA and porous structure of AC promote the large number of active sites, which hence allow the insertion/desertion of Li^+ to provide high capacity and outstanding rate performance after assembly with the battery-type cathode. Furthermore, the high accessible surface area and the pore size of the porous AC reduced the equivalent series resistance, thus enabling high power and energy density. For instance, porous zinc cobalt sulfide nanosheet array as a cathode and AC as an anode ($Zn_{0.5}Co_{0.5}S$//AC) delivered an energy density of 49 Wh kg^{-1} at 957 W kg^{-1} with 88% retention over 10,000 cycles [32]. A high energy density of 45.6 Wh kg^{-1} at a power density of 2,250 W kg^{-1} was achieved by employing nickel-iron phosphate as a positive electrode and AC as the negative electrode (NiFe-P//AC) [33]. Polyanionic lithium manganese silicate as a cathode and AC as an anode (Li_2MnSiO_4//AC) showed an energy density of 38.4 Wh kg^{-1} at a power density of 516 W kg^{-1} [34]. Likewise, various supercapattery devices were fabricated using AC as a negative electrode such as $FeCoCuS_2$//AC [35], $Co_3(PO_4)_2$//AC [36], $Ni_{0.75}Mn_{0.25}(PO_4)_2$//AC [37], $Co_{0.5}Mn_{0.5}S$//AC [38], Co_3O_4//AC [39], and $NiMnO_4$//AC [40], and those devices exhibited energy density in the range of 20–65 Wh kg^{-1}.

In addition, AC can also be used as a conductive agent hybridized with electroactive material to promote electron and ionic diffusion, thus improving the energy density. The high SSA of AC particles connecting is beneficial to support $Ni(OH)_2$ growth and increase the active sites. A $Ni(OH)_2$/AC/CNT composite was reported to deliver 232 F g^{-1} in the half cell at 1 A g^{-1}. Moreover, a supercapattery device using a $Ni(OH)_2$/AC/CNT composite as cathode with AC anode showed an energy density of 32.3 Wh

kg^{-1} at 504.8 W kg^{-1} with 83.5% capacitance retention after 1,000 cycles [41]. Another literature reported that AC was converted from starch at a high temperature indicating a high surface-to-volume ratio which enables more active sites involved in a redox reaction. It promoted long cycle life and reduced the dissolution of manganese ions in the electrolyte after combining with manganese oxide. The hybrid electrode material of Ag-Mn_3O_4/AC was reported to obtain a high specific capacitance of 981 F g^{-1} at 1 A g^{-1}. When the Ag-Mn_3O_4/AC cathode assembled with the AC anode, it delivered a specific cell capacitance of 180 F g^{-1} at a high current rate of 10 A g^{-1} and yielded a high energy density of 81 Wh kg^{-1} at 4,486 W kg^{-1} [42]. NiS nanoparticles loaded on activated carbon (NiS/AC) were synthesized via a hydrothermal method to exhibit a high specific capacity of 677 C g^{-1} at 0.5 A g^{-1}. The study revealed that the pore structure of AC is conducive to the rapid transportation of the electrolyte ions in NiS/AC. Moreover, a high energy density of 45.6 Wh kg^{-1} at a power density of 509.5 W kg^{-1} and an excellent stability of 92.6% retention after 20,000 cycles were reported after being assembled with the AC anode [43].

8.3.2 Carbon Nanotubes (CNTs)

Carbon nanotubes were discovered by Iijima in 1991 as 1D carbon columns structured with a diameter of 1–20 nm obtained by rolling a single or multilayer graphene into a tubular structure with a large aspect ratio [44]. The high mechanical and conductivity properties of CNT are based on its high SSA and 1D tubular structure that can support the rapid movement of electrons, especially under high-speed charging and discharging [10]. Therefore, CNT is one of the attractive candidate electrode materials applied in supercapattery applications. However, CNT is mostly limited to laboratory research in energy storage applications due to its low dispersion and poor electrochemical performance including low specific capacitance. Hence, surface modification with a functional group is needed to enhance its dispersion [45].

Different synthesis routes of CNTs were suggested such as the arc discharge method, laser ablation method, and chemical vapor deposition (CVD) (Table 8.1) [46–48]. High energy input is necessary for the synthesis of CNTs by using arc discharge and laser ablation methods such as arc discharge or laser beams due to the high degree of graphitization in the CNTs [43]. Furthermore, vacuum conditions are required in these systems, as well as continuous graphite target replacement for continuous large-scale production. In comparison, CVD is a promising method used to synthesize CNTs due to its high product yield and simplicity compared with arc discharge and laser ablation techniques. Moreover, the process of CVD can be done at low temperatures.

A catalyst plays an important role in the CVD technique as it decides the end products to be either CNT or graphene. Moreover, it promotes the growth of carbon materials. Nickel (Ni), iron (Fe), and cobalt (Co) catalysts were the common catalysts used to synthesize CNT owing to the strong adhesion with grown CNTs, which allows for the formation of SWCNT [49]. A comparison study of catalysts was conducted and found that the Fe catalyst indicated the best performances due to its capability of

Table 8.1 Summary of the CNT synthesis method.

Synthesis technique	Arc discharge	Laser ablation	Chemical vapor deposition
Process	CNTs were grown on the graphite cathode by DC power supply under inert condition	CNTs formed after condensation of vaporized graphite by a pulse laser inside quartz under high temperature and inert gas	CNTs formed involving catalyst decomposition or carbon precursor on catalyst substrate
Yield	~70% (moderate)	80%~85% (above moderate)	95%~99% (high)
Products	SWCNT and MWCNT	SWCNT	SWCNT and MWCNT
Advantages	Simple, low cost	Relative high purity CNTs	Simple, large-scale production, high purity and yield, low cost
Disadvantages	High temperature, less pure product, low production scale, tangled nanotube	Limited to lab-scale production	MWCNTs with defects

producing high-ordered crystalline CNT structure with the least structural defect. Moreover, it also performed at a fast-growing rate owing to its high carbon solubility and small particle size of Fe [50, 51]. High-ordered crystalline CNT not only acts as a mechanical support for composite materials but also provides a good conductive path for electron and ionic transport; hence, high capacitance and excellent rate capability can be obtained at supercapattery [52]. The formation of CNT in the CVD technique can be explained in "vapor–liquid–solid" (VLS) and "vapor–solid–solid" (VSS) mechanisms. In the VLS mechanism, the hydrocarbon is absorbed on the catalyst substrate and dissociated carbon atom into liquid metastable carbide followed by solid carbon precipitates at the end of the catalyst particle to form CNT. In the VSS mechanism, the hydrocarbon precursor is first dissociated to active carbon atoms with subsequent diffusion in catalyst particles and then precipitates in the form of CNT [53].

In the case of graphene synthesis, copper (Cu) was the most common and best catalyst due to its low carbon solubility [54]. High carbon solubility of Ni catalyst was also reported in the literature to synthesize graphene, but the formation of graphene was observed to begin during the cooling step. This is because the nucleation of graphene was initiated after the bulk Ni catalyst reached the threshold carbon concentration [49]. Figure 8.3 illustrates the mechanism of graphene formation using Ni and Cu catalysts [53]. Five steps of the mechanism are needed using Ni catalyst which involved the absorption of hydrocarbon on the catalyst substrate (step 1), dissociation of hydrocarbon through dehydrogenation (step 2), and chemisorption of carbon atom in the bulk catalyst (step 3). The chemisorption is facilitated by the empty D-shell of the catalyst that is ready to receive electrons. The excess carbon will be diffused to the catalyst surface (step 4) and graphene is formed through segregation (step 5) (Figure 8.3a). The segregation process does not stop until the concentration in the catalyst is reduced to equilibrium even if the supply of hydrocarbon is stopped. On the contrary, the growth of graphene would start during hydrocarbon dissociation and terminate immediately when the supply of hydrocarbon is stopped in the existence of Cu catalyst (Figure 8.3b) [53].

CNT generally acts as the conducting agent due to its high conductivity features to fabricate the nanocomposite with a variety of electroactive materials to enhance Li^+ diffusion and energy density. For instance, a ternary nanocomposite (MWCNT-Co_3O_4-Ag) was formed by the *in-situ* growing of Co_3O_4 nanograin on multiwalled CNT (MWCNT) and anchored with silver (Ag) nanoparticles. The ternary nanocomposite delivered a maximum specific capacity of 83.88 C g^{-1} at 0.6 A g^{-1} which is substantially higher than MWCNT-Co_3O_4 (55.33 C g^{-1}) and Co_3O_4

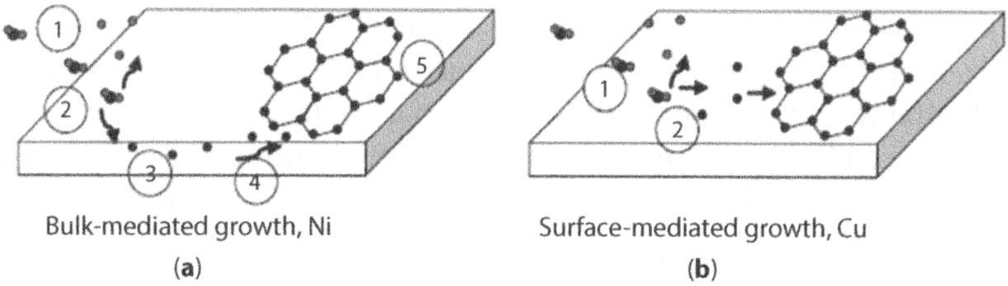

Figure 8.3 Formation mechanism of graphene through (a) high carbon solubility of Ni catalyst (step 1: absorption, 2: dehydrogenation, 3: chemisorption, 4: out-diffusion, 5: surface segregation) and (b) low carbon solubility of Cu catalyst (step 1: dissociation, 2: segregation). Reproduced from ref. [53].

nanograins (39.24 C g^{-1}) [55]. CNT has a unique structure and remarkable physiochemical properties such as ballistic electron transport, high porosity, and hollow structure. These characteristic features enabled an ideal platform for Co_3O_4 nanograins to be grafted onto it to reduce particle aggregation and boost rapid electron shuttling throughout the composite network. Thus, intercalation of Co_3O_4 nanograins onto the CNT network prevents capacity fading, structural distortion, and particle aggregations. The supercapattery devices were fabricated in a configuration of MWCNT-Co_3O_4-Ag//AC to demonstrate an energy density of 16.5 Wh kg^{-1} at 297.5 W kg^{-1} and remained at 93% retention after 3,000 cycles. Furthermore, a binary composite of an anode (rGO/CNT) delivered a capacity of 230 mA h g^{-1} at 1 A g^{-1}. The remarkable electrochemical performance of rGO/CNT was associated with the effect of expanded surface area imparted by the 3D nanocarbon network which provides a sufficient transport pathway for Li^+ during the charge–discharging process. After being constructed into a supercapattery device, rGO/CNT anode and $LiFePO_4$ (LFP) cathode achieved a high energy density of 180 Wh kg^{-1} at 218 W kg^{-1} and retained over 97% from initial capacity after 100 cycles [16]. Highly interconnected Cu_2SnS_3 nanospheres encapsulated in the carbon nanotube framework (Cu_2SnS_3/CNT) were fabricated as an anode to indicate a superior Li^+ storage capability. The benefits from the composite materials of Cu_2SnS_3 and CNT, with an exceptional high-capacity value of 600 mA h g^{-1} at 0.5 A g^{-1}, were reported. This is due to the presence of CNT providing efficient charge transfer and ionic diffusion path and, hence, enhancing the reaction kinetics of Cu_2SnS_3. Moreover, it restores the structural architecture of Cu_2SnS_3 by avoiding volume expansion. Thus, Cu_2SnS_3/CNT anode and AC cathode device achieved an exceptionally high electrochemical performance in terms of high energy density (158.77 Wh kg^{-1}) and power density (12.5 kW kg^{-1}) with superior stability of 84% retention after 10,000 cycles [56]. CNT-coated lithium manganese phosphate composite (CNT/LMP) attained a high capacity of 159 mA h g^{-1} at 0.1A g^{-1}. Furthermore, it delivered excellent stability with 93% capacity retention over 5,000 cycles. Figure 8.4 revealed that the morphology of LMP was identified as a uniform plate-like structure, while the porous and interconnected networks of tubes were observed in the CNT/LMP structure [57]. Porous and interconnected networks of CNT/LMP with high SSA promote electrolyte ion diffusion. Furthermore, CNT/LMP reduced the side reaction that occurred at the electrode–electrolyte interface and Li^+ decomposition. 3D hierarchical MnO_2 nanoflakes grown on vertically aligned carbon nanotubes (MnO_2/VACNT) were reported to reveal its specific capacitances of 1,131 F g^{-1} which was significantly higher than VACNT (218 F g^{-1}). A similar trend was observed in the initial discharged capacity where MnO_2/VACNT obtained 1,289 mA h g^{-1}, while the vertically aligned carbon nanotubes (VACNT) exhibited 533 mA h g^{-1} at 0.25 A g^{-1}. The fabricated electrode demonstrated an extraordinarily high cycling stability of 98.4% coulombic efficiency over 150 cycles revealing an improvement in stability and lithium storage capability compared with VACNT [58]. A 3D structure formed by a scaffold of metal oxide on VACNTs creates extra active sites to facilitate the movement of electrolyte ions to react with metal oxide. Furthermore, the growth of VACNTs on the conductive material creates

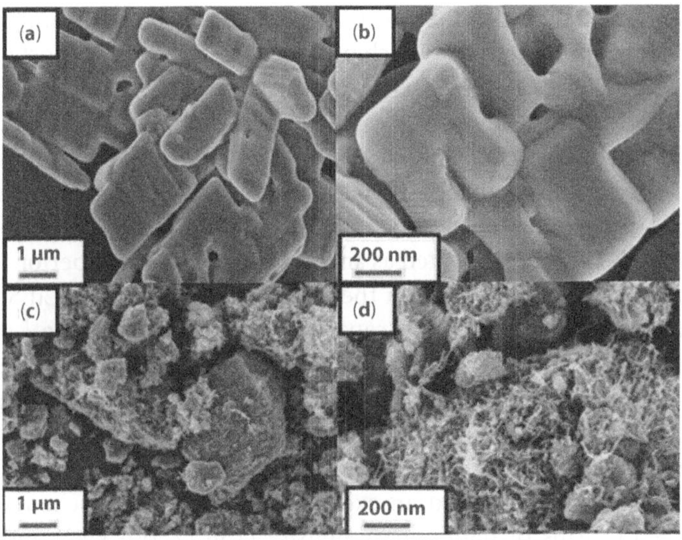

Figure 8.4 SEM image of (a, b) pure LMP and (c, d) CNT/LMP. Reproduced from ref. [57].

low contact resistances of the electrode. In conclusion, the hybridization of electroactive material with CNT enhanced the electrochemical performances of supercapattery by stimulating more active sites to increase the electron transfer efficiency and Li^+ storage capacity [59].

Heteroatom-doped 1D CNT-induced structural defects have aroused widespread interest in energy storage due to their effective ion-accessible surface with low manufacturing cost [60]. It affected the morphologies of CNTs and created surface reactivity, thereby exerting an impact on their capacity value. Moreover, Li^+ diffuses more easily in the CNTs with structure defects (which have holes in their sidewall) compared with defect-free CNTs. Element doping such as boron or nitrogen creates structural defects [61]. For instance, nitrogen-doped multiwalled CNT contains wall defects that expand the storage region through an interwall space. This was proven through the HRTEM image in Figures 8.5a, b, where the interwall spacing increased from 3.4 to 3.7 (wall thickness from 15.7 to 17.8 nm) after fully charged. Furthermore, the Fourier transformed (FT) patterns (in the right box of Figures 8.5a, b) show the spot-to-spot distance reduced to keep the interwall spacing increased. In addition, the capacity of the

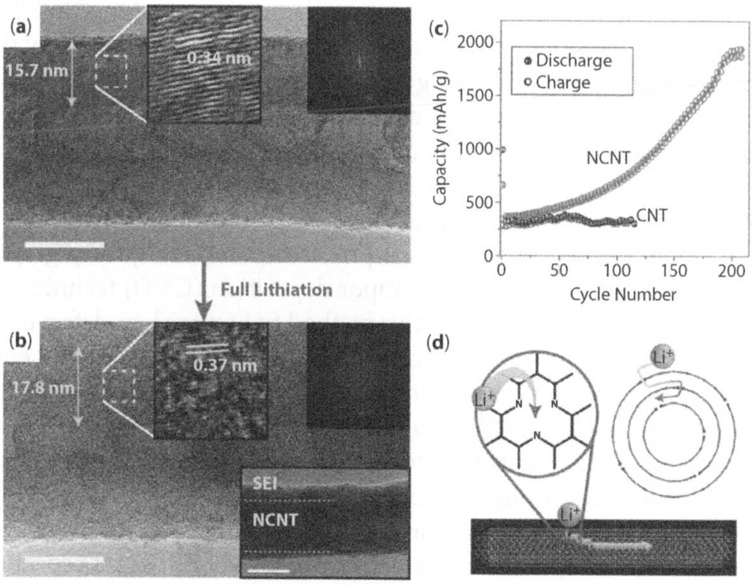

Figure 8.5 Schematic illustrating the TEM image and the FT pattern of N-doped CNT (a) before and (b) after full charging; (c) gravimetric capacities and cycling performances of N-doped CNT with commercial CNT; (d) lithium ions insert into the interwall space of NCNTs through wall defects. Reproduced from ref. [62].

N-doped CNT electrode increases significantly with cycling compared with undoped CNT (Figure 8.5c). The capacity increase can be explained by the schematic illustrated in Figure 8.5d where a large amount of Li^+ is favorable to diffuse into the interwall space through surface defects due to their high electronegativity to attract a large number of Li^+ [62]. In addition, B-doped CNT was synthesized via thermal reduction of boron acid with CNT in the mixed H_2/Ar gas. The specific capacity of B-doped CNT (993 mA h g^{-1}) is higher than pristine CNT (800 mA h g^{-1}) and retained a reversible capacity of 169 mA h g^{-1} after 5,000 cycles [63]. This is because the B-CNT displayed a favorable surface capacitance-controlled mechanism to provide fast lithium storage kinetics and fascinating rate capability. Moreover, good mechanical integrity structure and the least resistance (R_{ct}) of B-doped CNT (69.8 Ω) than CNT (86.0 Ω) enhance the electrochemical performances.

8.3.3 Graphene

Graphene was first discovered by Novoselov *et al.* in 2004 and attracted widespread attention in energy conversion due to its unique properties [11, 64–66]. It has excellent electrical properties of 1×10^{-8} S m^{-1} due to each carbon atom in graphene forming a large π-bond of multiple atoms penetrating the entire layer. Furthermore, it has superior mechanical properties and high SSA (2,630 m^2 g^{-1}) which is beneficial to supercapattery [67]. Various strategies can be used to synthesize graphene and can be further classified as "top-down" or "bottom-up" (Figure 8.6). The "top-down" approach is preferred because this technique makes it more convenient to synthesize nanoparticle graphene. Some of the techniques involved complex procedures and long chemical processes, and low-quality graphene was generated [68]. The chemical vapor deposition (CVD) technique was identified as a powerful and effective method to prepare low-defect density and large SSA of graphene in a recent study due to its simplicity and energy efficiency. Moreover, it is also applicable in commercial-scale production. Plasma-enhanced CVD (PECVD) is the most popular one as it can grow graphene rapidly at low temperatures with better control in nanostructure which is beneficial for high productivity [69].

Graphene was reported to have a high specific capacitance of 550 F g^{-1} when it was fully utilized. In fact, it could only achieve a capacitance value of about 247 F g^{-1} due to the layer stacking of graphene sheets from the Van der Waals attraction. Thus, a layer of graphene is not suitable for energy storage applications because of its low SSA. Moreover, narrow interlayer distances between graphene sheets (0.34 nm) limit the storage of

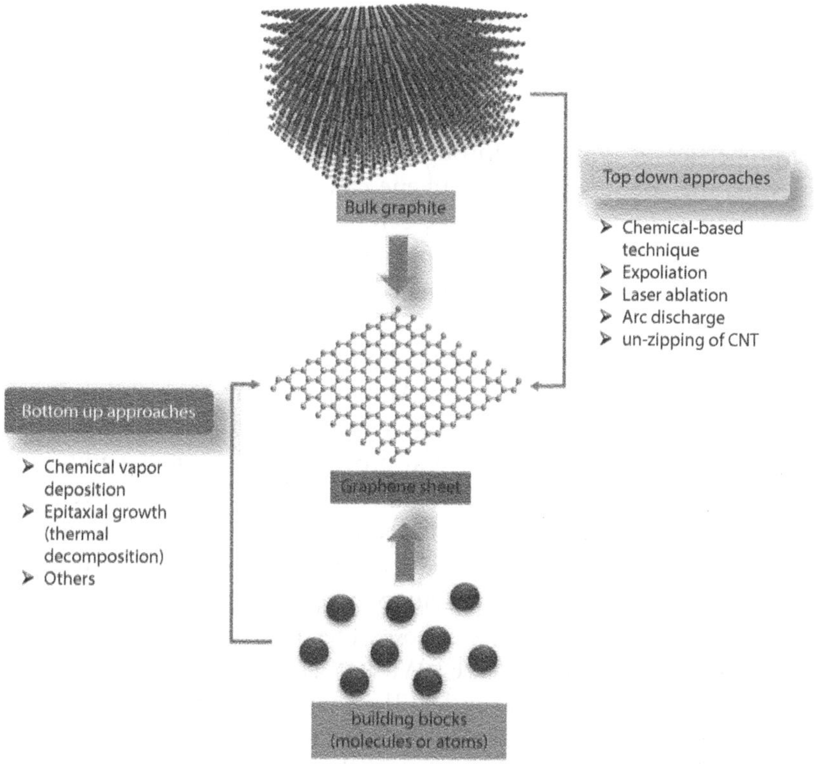

Figure 8.6 Graphene synthesis process.

Li^+. Therefore, 3D graphene was introduced, which increases the electron transfer kinetics for fast Li^+ diffusion, and it can inhibit aggregation [70, 71]. For instance, 3D porous graphene delivered an excellent volumetric performance of 299 F cm^{-3} at 0.1 A g^{-1}, high stability with 85% retention after 30,000 cycles, and volumetric energy density up to 118 Wh L^{-1} at 116 W L^{-1} [72]. 3D graphene hydrogel achieved 196 F g^{-1} at 1 A g^{-1} and only 8.4% capacitance decay over 10,000 cycles [73]. Another 3D sponge-like structure of graphene aerogel (GA) was reported to achieve 148.45 F g^{-1} with an SSA of 103.35 m^2 g^{-1}. Similar to all carbon allotropes, graphene can be combined with various materials to enhance its electrochemical performance. Azure A (Az) was hybridized with GA to facilitate its storage capacity. Compared with GA, Az-GA with relatively high SSA (128.22 m^2 g^{-1}) and pore size distribution (2–5 nm) possesses a more active site to increase electrode/electrolyte interaction and achieve high specific capacitance value (439.04 F g^{-1}) [74].

Furthermore, graphene cannot be decomposed due to thermodynamics instability because of the presence of strong interatomic bonds between the carbon atom which prevented crystal dislocation and defects [68]. Immobilization of graphene with heteroatom was used to improve the properties of graphene through surface or structural modification. Likewise, the conductivity of graphene can further be improved. Doped (e.g., boron and nitrogen) graphene can be observed to have high performance compared with pristine graphene due to increasing defects on the graphene lattice. The defects happened because of the chemical doping process that incorporates nitrogen and boron atoms into the graphene lattice. Aside from the large SSA for Li^+ storage, structural defects also allow more ion channels for Li^+ diffusion. Although the presence of the element created structural defects in the graphene lattice, it does not disrupt the conductivity of doped graphene and provides a continuous pathway for electron transmission [70]. Furthermore, elemental doping prevents the agglomeration of graphene and induces a p-type conductive network enhancing reversibility [75, 76]. B-doped graphene was created and delivered a high initial discharge capacity of 1,400 mA h g^{-1} at 1 A g^{-1}. After being assembled with an AC electrode, the supercapattery device delivered a high energy density of 162 Wh kg^{-1} at 590 W kg^{-1} and remained at 65% retention after 25,000 cycles. The energy, power, and stability performance of the assembled system were found to be at par or even better compared with the results from other studies [77]. A 3D honeycomb-like $CoMn_2O_4$ constructed on N-doped graphene was also reported to assemble with N-doped graphene as a supercapattery application. The application delivered an energy density of 54.37 Wh kg^{-1} at 420.46 W kg^{-1} and retained 92% at 3,000 cycles. The outstanding performance of supercapattery is caused by the stable honeycomb architecture of $CoMn_2O_4$ providing a short ionic diffusion pathway and increased electrode stability during continuous charge–discharge processes even at high current density. Moreover, the introduction of N-doped graphene provides excess active sites for the reaction and strengthens the binding bond with electroactive materials, enabling better capacity and cycling stability [78].

Oxidation of graphene to form graphene oxide (GO) or graphene oxide-based composites was found to effectively provide structural deficiencies of graphene. Graphene oxide was then further reduced to reduced graphene oxide (rGO) to facilitate Li^+ transmission. Graphene oxide contains functional groups (epoxy, hydroxyl, carbonyl, and carbonyl groups) on the graphene lattices that enhance the surface area and provide a nucleation site for the growth of metal ions. Therefore, ion adsorption and ion transport are enhanced by high surface polarity along the graphene surface [67].

The solvothermal method or chemical method (also known as modified Hummer's process) was found as the most popular method to convert the GO from graphite powder and then chemically reduce it to rGO. This method produced a large volume of graphene. Typically, a mixture (9:1) of sulfuric acid (H_2SO_4) and phosphorus acid (H_3PO_4) in the first step was used to initiate the intercalation between the graphite layers followed by the surface oxidation using potassium permanganate ($KMnO_4$) to produce graphene oxide. GO was prepared from graphene with a mixture of H_2SO_4 and H_3PO_4 with a slow addition of $KMnO_4$ as oxidant. There is a close relationship between the degree of oxidation and the detrimental defect of GO. The study revealed that partial oxidation occurred under low $KMnO_4$ concentrations resulting in defects on the edges of GO, whereas defects appeared on the surface of GO at high concentrations of $KMnO_4$ [79, 80]. The advantages of this technique are safety through the elimination of explosive exothermic reaction and eco-friendliness due to non-toxic gas emission during the reaction [81]. Moreover, it is suitable for large-scale production or commercial use [82].

$LiMnPO_4$/rGO was fabricated from a facile solvothermal method as a cathode with rGO anode delivering an excellent specific energy of 16.46 Wh kg^{-1} at 0.38 W kg^{-1}. Moreover, the device maintained over 91.2% of the initial capacity after 10,000 cycles at 2 A g^{-1}, demonstrating high-rate performance and long cycle life [83]. The *in-situ* growth of Co_3O_4 nanoflakes on rGO-wrapped Ni foam was reported to exhibit a considerably high specific capacity of 1,328 C g^{-1} at 2 A g^{-1} and good stability (82.2% retention after 2,000 cycles) compared with Co_3O_4 nanoflakes on bare Ni foam (82.2% retention after 1,000 cycles). After being assembled with an AC anode, it achieved a specific capacitance of 80 F g^{-1} at 0.1 A g^{-1} and excellent stability with 94.5% retention after 10,000 cycles and a high energy density of 20 Wh kg^{-1} at 1,200 W kg^{-1} [84]. The TiO_2/rGO nanocomposite was successfully synthesized by a facile hydrothermal technique. Various amounts of GO were studied and the optimum amount of GO provides the highest specific capacity of 116.7 mA h g^{-1} at 0.2 A g^{-1}. The result inferred that high ionic diffusion efficiency was obtained with low charged transfer resistance between rGO and TiO_2 nanocrystals. Furthermore, a high-performance supercapattery with TiO_2/rGO nanocomposites and AC electrode revealed a superior energy density of 54.37 Wh kg^{-1} at 420.48 W kg^{-1} and 92% retention after 3,000 cycles at 1 A g^{-1} [85]. Copper niobate ($CuNb_2O_6$) nanowires immobilized on rGO nanosheets and assembled with AC as a supercapattery device were conducted. In the half-cell study of $CuNb_2O_6$/rGO, the electrode exhibits a high specific capacity of 312.2 mA h g^{-1} at 0.1 A g^{-1}. The high-capacity result was proposed by the formation of Li_3NbO_4 and

Cu metal during the lithiation process. The intercalation type of Li_3NbO_4 serves as the host for Li^+ storage, and the inactive Cu metal precipitation acts as the conductive network to enhance electron transportation. The energy density of the full cell achieved an energy density of 92.1 Wh kg^{-1} and a power density of 9,475 W kg^{-1}. In addition, a higher stability of the device resulted in 82.3% retention after 10,000 cycles [86].

8.3.4 Metal–Organic Framework (MOF)-Derived Carbon

In recent years, a novel electrode material of MOF-derived carbon materials has drawn great interest in supercapattery due to its excellent electrochemical performance based on its favorable nature, such as high porosity, high SSA, and excellent physiochemical properties [10]. The high SSA with large pore volume (micro-/mesoporous structure) of the MOF-derived carbon provides an abundant active site for electrode/electrolyte interaction and assures fast ionic and electron diffusion. Therefore, these porous carbon materials provide high capacitance values [87]. Interestingly, the structure and morphology of the MOF-derived carbon can be controlled by varying carbonization temperatures. Figure 8.7a illustrates the relationship between surface areas and carbonization temperatures for non-porous carbon derived from MOF-5. The highest surface area was found at a carbonization temperature of 530°C. However, it provides an unusually low capacitance value, which is attributed to the poor conductivity of nano-porous carbon at low carbonization temperature (Figure 8.7b) [88]. A sonochemistry strategy to produce 2D graphene nanoribbons (GNRibs) accompanied by chemical activation of 1D solid carbon nanorod (CNRod) derived from MOF-74 was reported. The results revealed that the higher specific capacitance of GNRibs (193 F g^{-1}) than CNRod (164 F g^{-1}) was due to the morphology of MOF-derived porous carbon, which is beneficial to the electrochemical performance [89]. Porous carbon has been applied widely in energy storage applications due to its low cost, high SSA, and good stability. Nevertheless, its relatively low specific capacitance and energy density greatly limit its application. Thus, heteroatom doping and hybridization of MOF-derived carbon materials with other electrode materials improved their electrochemical performances effectively [90].

The hybridization of two or more electrode materials optimized their electrochemical properties due to their synergistic effects. Carbon-coated copper sulfide polyhedron connected to carbon nanotubes (CuS/CNTs) was fabricated from the HKUST-1 and delivered a high capacitance value of 606 F g^{-1} at 1 A g^{-1}. The remarkable performances of porous HKUST-1-derived CuS polyhedrons were attributed to the large SSA of 126.4

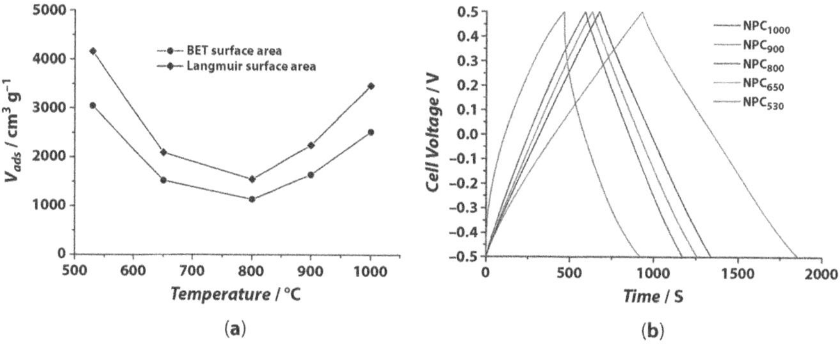

Figure 8.7 (a) The relationship between SSA (BET and Langmuir) and the carbonization temperatures for non-porous carbon (NPC). (b) Comparisons of the galvanostatic charge/discharge curves at a current density of 100 mA g^{-1}. Reproduced from ref. [88].

m^2 g^{-1} enhanced the reactivity and improved the specific capacitances. Furthermore, the *in-situ* growth of CuS polyhedrons maximized the integral SSA that furnishes a nice transport environment for the electrons and ions of the electrolyte. In addition, the interwoven network of CNTs further boosts the electrical conductivity and stability of the thin film. The as-prepared material was assembled in a supercapattery device with nitrogen-doped carbon polyhedrons interspersed into carbon nanotube films that possess 87% retention up to 6,000 cycles with an energy density of 38.4 Wh kg^{-1} at 750 W kg^{-1} [91]. Porous carbon derived from ZTF obtained a super large porous volume (3.504 cm^3 g^{-1}) and SSA (3132 m^2 g^{-1}) due to the unique 3D hollow structure. The large SSA and porous structures of carbon derived from ZTF allow the electrode material to contact the electrolyte completely resulting in a larger electric double-layer effect. Thus, the porous carbon was reported to have a high specific capacity (123.4 mA h g^{-1}) and excellent cycling stability (10,000 cycles) in a half cell with an organic electrolyte. After being assembled as a supercapattery device with a sucrose hard carbon electrode, the device possesses an energy density of 157 Wh kg^{-1} at 40 kW kg^{-1} and the capacities remained at 85% after 4,000 cycles [92]. Another high SSA (2,714 m^2 g^{-1}) of 3D carbon cuboids synthesized from MOF-5 indicated similar trends as previous literature and was reported to obtain about 150 F g^{-1} at 0.15 A g^{-1}. After being assembled with $Li_4Ti_5O_{12}$, the density of the cell was 65 Wh kg^{-1} which was significantly higher than the commercial AC (~36 Wh kg^{-1}). Furthermore, the assembled material also demonstrated good cycle stability with 82% retention after 10,000 cycles [93]. Thus, the high SSA and controllable size of the inerratic morphology of the MOF-derived carbon hybridized with transition metal oxide maximized the electrochemical properties of both components.

The introduction of heteroatom doping modified the properties of the materials to further enhance their electrochemical performances. For instance, nitrogen-doped MOF-derived carbon materials can provide even higher performance due to the added wettability and conductivity provided by nitrogen. Moreover, boron creates additional active sites for boron-doped MOF-derived carbon. Heteroatom-doped MOF-derived carbon can be created by two different strategies: (1) direct pyrolysis of the heteroatom-containing MOF precursor and (2) mixed heteroatom sources (e.g., urea, melamine, polyaniline, thiourea, and boric acid) with MOF precursor during pyrolysis. Zeolitic imidazolate frameworks (ZIFs) are a typical MOF precursor used to prepare nitrogen-doped MOF-derived carbon as they contain both carbon-rich and nitrogen elements [90]. Furthermore, zinc can be eliminated readily at high-temperature pyrolysis as metallic vapor [94]. Nitrogen-doped porous carbon polyhedral (NC) derived from zeolite imidazolate framework-8 (ZIF-8) assembled into flexible nanopapers assisted with rGO was reported. The intimate contacts between the NC and rGO facilitate rapid electron transfer, which greatly improves conductivity, thus enhancing the capacitance value. As a result, NC/rGO nanopapers delivered a high specific capacitance of 280 F g^{-1} at 1 A g^{-1} which is higher than NC (104 F g^{-1}) and rGO (88 F g^{-1}) [95]. The study of MOF-derived carbon on supercapattery devices is still limited. So, the development of electrode material using MOF precursors is still required to improve the performance of supercapattery. All the electrochemical performances of the supercapattery discussed in this chapter are summarized and tabulated in Table 8.2.

8.4 Conclusion and Outlook

Supercapattery is a new generation of energy storage device characterized by both high power and energy density with a stable cycling process. This chapter provides insight into the need for supercapattery in the modern world, starting from the basic understanding of the mechanism of supercapattery. The operating voltages, energy, and power density of supercapattery can be improved by using capacitive carbonaceous material as the anode and battery-type materials as the cathode. Moreover, the carbonaceous material can act as a conducting network that combines with a battery-type cathode, further enhancing the

Table 8.2 A summary of the supercapattery performances based on carbonaceous materials.

Cathode electrode	Anode electrode	Electrolyte	Specific capacity (mA h g^{-1})	Specific capacitance	Capacity retention cycles (%)	Energy density (Wh kg^{-1})	Ref.
$Na_{0.76}V_6O_{15}$/AC	Graphite	1 M $LiPF_6$ (EC:DMC)	–	–	5,000 (70)	43.7	[96]
$Ni(OH)_2$/AC/CNT	AC	6 M KOH	–	82.1 F g^{-1}	1,000 (83.5)	32.3	[41]
Ag-Mn_3O_4/AC	AC	1 M Na_2SO_4	–	180 F g^{-1}	1,000 (96)	81	[42]
NiS/AC	AC	3 M KOH	–	102 F g^{-1}	20,000 (92.6)	45.6	[43]
MWCNT-Co_3O_4-Ag	AC	1 M KOH	–	–	3,000 (93)	16.5	[55]
Cu_2SnS_3/CNT	AC	1 M $LiPF_6$ (EC:DC)	–	272 F g^{-1}	10,000 (84)	158.77	[56]
$LiMnPO_4$/CNT	AC	1 M $LiPF_6$ (EC:DMC)	159	–	100 (92)	–	[57]
$LiMnPO_4$/rGO	rGO	1 M LiOH	–	–	10,000 (91.2)	16.46	[83]
Co_3O_4/rGO	AC	1 M KOH	–	80 F g^{-1}	10,000 (94.5)	20	[84]
AC	TiO_2/rGO	1 M KOH	–	–	3,000 (92)	54.37	[85]
AC	$CuNb_2O_6$/rGO	1 M $LiPF_6$ (EC:DMC:DEC)	–	–	10,000 (82.3)	92.1	[86]

(Continued)

Table 8.2 A summary of the supercapattery performances based on carbonaceous materials. (*Continued*)

Cathode electrode	Anode electrode	Electrolyte	Specific capacity (mA h g^{-1})	Specific capacitance	Capacity retention cycles (%)	Energy density (Wh kg^{-1})	Ref.
AC	B-doped graphene	1 M $LiPF_6$ (EC:DEC)	–	–	25,000 (65)	162	[77]
Porous carbon	SHC	1 M $LiPF_6$ (EC:DMC:DEC)	–	–	4,000 (85)	157	[92]
3D carbon cuboid	$Li_4Ti_5O_{12}$	1 M $LiPF_6$ (EC:DMC)	–	–	10,000 (82)	65	[93]
$Zn_{0.5}Co_{0.5}S$	AC	2 M KOH	–	614 C g^{-1}	10,000 (88)	49.0	[32]
NiFe-P	AC	1 M KOH	–	326 C g^{-1}	1,000 (87.6)	45.6	[33]
Li_2MnSiO_4	AC	1 M Na_2SO_4	–	108.2 F g^{-1}	3,000 (99.6)	38.4	[34]
$FeCoCuS_2$	AC	3 M KOH	–	133.5 F g^{-1}	–	48.2	[35]
$Co_3(PO_4)_2$	AC	1 M KOH	–	275 C g^{-1}	5,000 (95.1)	63.9	[36]
$Ni_{0.75}Mn_{0.25}(PO_4)_2$	AC	1 M KOH	–	–	2,000 (119)	64.2	[37]
$Co_{0.5}Mn_{0.5}S$	AC	1 M KOH	–	260 C g^{-1}	2,000 (90.2)	61.3	[38]
Co_3O_4	AC	1 M KOH	–	108.8 C g^{-1}	2,500 (88.5)	23.7	[39]

(*Continued*)

Table 8.2 A summary of the supercapattery performances based on carbonaceous materials. (*Continued*)

Cathode electrode	Anode electrode	Electrolyte	Specific capacity (mA h g^{-1})	Specific capacitance	Capacity retention cycles (%)	Energy density (Wh kg^{-1})	Ref.
$NiMnO_4$	AC	6 M KOH	–	341 C g^{-1}	5,000 (102)	64.07	[40]
$LiFePO_4$	rGO/CNT	Gel polymer	–	–	100 (97)	180	[16]
CuS	NC/C	6 M KOH	–	–	6,000 (87)	38.4	[91]
MnO_2/VACNT	–	1 M $LiPF_6$ (EC:DEC)	1,289	–	150 (98.4)	–	[58]
N-CNT	–	1 M $LiPF_6$ (EC:DEC)	–	340 F g^{-1}	200	–	[62]
B-CNT	–	1 M $LiPF_6$ (EC:DMC:DEC)	993	–	5,000	–	[63]
3D graphene hydrogel	–	H_2SO_4-PVA gel	–	196 F g^{-1}	10,000 (91.6)	–	[73]
3D graphene aerogel (GA)	–	3 M KOH	–	148.45 F g^{-1}	–	–	[74]
NC/rGO	–	6 M KOH	280	–	10,000	–	[95]
Cu_7S_4/C	–	1 M H_2SO_4	–	321.9 F g^{-1}	3,000 (78.1)	–	[97]

electrochemical performances. The capacitance of carbonaceous materials is highly dependent on their characteristics, such as SSA and porosity. Li^+ storage occurred primarily at the micropores, while the mesopores provided fast Li^+ transport channels, which enhanced the electrode–electrolyte interaction. In addition, carbonaceous material accommodates a huge volume expansion and contraction under continuous lithiation and delithiation processes due to its excellent mechanical integrity. To improve the performance of supercapattery, an electrode material with nanostructuring must be fabricated as the nanostructured electrode material reduces ionic and electronic transport channels in the electrochemical process. Furthermore, an effective synthesis method needs to be chosen to provide high-precision control of characteristics, while cost-effectiveness and scalability need to be considered for commercialization. MOF-derived carbon is a promising electrode material for supercapattery applications due to its well-regulated morphology with flexible structure, high SSA, and large pore volume. Although MOF-derived carbon exhibits an outstanding electrochemical performance of supercapattery, it still faces some difficulty because of the limited selection and complex synthesis method of the MOF precursor. Moreover, the low yield production of MOF-derived carbon hinders its commercialization. Thus, many efforts on production strategies, exceptional morphology and structure, and reaction chemistry are still required to achieve high performance in supercapattery applications. In conclusion, a supercapattery can be a more feasible and sustainable energy storage device after optimizing its synthesis process. Therefore, the field of research pertaining to supercapattery is wide open for exploration and has great potential in the future.

Acknowledgments

The authors acknowledge the financial support from the Fundamental Research Grant Scheme (FRGS) FRGS/1/2021/STG05/UPM/01/1 awarded by the Ministry of Higher Education of Malaysia (MOHE).

References

1. Wu, H., Liu, Q., Guo, S., Composites of graphene and LiFePO4 as cathode materials for lithium-ion battery: A mini-review. *Nano-Micro Lett.*, 6, 4, Shanghai Jiaotong University, 316–326, Oct. 01, 2014, doi: 10.1007/s40820-014-0004-6.
2. Wang, J., Lyu, X., Wang, L., Yu, S., Zhu, W., Han, C., Cao, X., Preparation and electrochemical performance of hierarchical CuS-rGO composite. *J. Alloys Compd.*, 694, 1067–1072, 2017, doi: 10.1016/j.jallcom.2016.10.155.
3. Matheswaran, P., Karuppiah, P., Thangavelu, P., Contribution of different charge storage mechanisms in cobalt pyrophosphate-based supercapattery. *Ionics (Kiel)*, 27, 1769–1780, 2021, doi: 10.1007/s11581-021-03908-2/Published.
4. Xie, J., Yang, P., Wang, Y., Qi, T., Lei, Y., Li, C.M., Puzzles and confusions in supercapacitor and battery: Theory and solutions. *J. Power Sources*, 401, 213–223, Oct. 2018, doi: 10.1016/J.JPOWSOUR.2018.08.090.
5. Wu, F., Maier, J., Yu, Y., Guidelines and trends for next-generation rechargeable lithium and lithium-ion batteries. *Chem. Soc. Rev.*, 49, 5, Royal Society of Chemistry, 1569–1614, Mar. 07, 2020, doi: 10.1039/c7cs00863e.
6. Theerthagiri, J., Lee, S.J., Shanmugam, P., Choi, M.Y., Basic principles in energy conversion and storage, in: *Nanostructured, Functional, and Flexible Materials for Energy Conversion and Storage Systems*, pp. 1–14, Jan. 2020, doi: 10.1016/B978-0-12-819552-9.00001-4
7. Yu, F., Zhang, C., Wang, F., Gu, Y., Zhang, P., Waclawik, E.R., Du, A., Ostrikov, K.K., Wang, H., A zinc bromine 'supercapattery' system combining triple functions of capacitive, pseudocapacitive and battery-type charge storage. *Mater. Horiz.*, 7, 2, 495–503, Feb. 2020, doi: 10.1039/c9mh01353a.
8. Yu, L. and Chen, G.Z., Supercapatteries as high-performance electrochemical energy storage devices. *Electrochem. Energy Rev.*, 3, 2, Springer Science and Business Media B.V, 271–285, Jun. 01, 2020, doi: 10.1007/s41918-020-00063-6.
9. Zuo, W., Li, R., Zhou, C., Li, Y., Xia, J., Liu, J., Battery-supercapacitor hybrid devices: Recent progress and future prospects. *Adv. Sci.*, 4, 7, 1–21, 2017, doi: 10.1002/advs.201600539.
10. Balasubramaniam, S., Mohanty, A., Balasingam, S.K., Kim, S.J., Ramadoss, A., Comprehensive insight into the mechanism, material selection and performance evaluation of supercapatteries. *Nano-Micro Lett.*, 12, 1, 1–46, Apr. 01, 2020, doi: 10.1007/s40820-020-0413-7.
11. Choi, H.J., Jung, S.M., Seo, J.M., Chang, D.W., Dai, L., Baek, J.B., Graphene for energy conversion and storage in fuel cells and supercapacitors. *Nano Energy*, 1, 4, 534–551, Jul. 2012, doi: 10.1016/J. NANOEN.2012.05.001.
12. Li, X. and Wei, B., Supercapacitors based on nanostructured carbon. *Nano Energy*, 2, 2, 159–173, Mar. 2013, doi: 10.1016/J. NANOEN.2012.09.008.

13. Gu, H., Zhu, Y.E., Yang, J., Wei, J., Zhou, Z., Nanomaterials and technologies for lithium-ion hybrid supercapacitors. *Chem. Nano Mat.*, 2, 7, Wiley-VCH Verlag, 578–587, Jul. 01, 2016, doi: 10.1002/ cnma.201600068.
14. Liu, X., Sun, Y., Tong, Y., Wang, X., Zheng, J., Wu, Y., Li, H., Niu, L., Hou, Y., Exploration in materials, electrolytes and performance towards metal ion (Li, Na, K, Zn and Mg)-based hybrid capacitors: A review. *Nano Energy*, 86, 106070, Aug. 2021, doi: 10.1016/J.NANOEN.2021.106070.
15. Liang, H., Hu, Z., Zhao, Z., Chen, D., Zhang, H., Wang, H., Wang, X., Li, Q., Guo, X., Li, H., Dendrite-structured FeF_2 consisting of closely linked nanoparticles as cathode for high-performance lithium-ion capacitors. *J. Energy Chem.*, 55, 517–523, Apr. 2021, doi: 10.1016/J. JECHEM.2020.07.031.
16. Dai, J., Fu, K., Palanisamy, R., Gong, A., Pastel, G., Kornfeld, R., Zhu, H., Sanghadasa, M., Bekyarova, E., Hu, L., A solid state energy storage device with supercapacitor-battery hybrid design. *J. Mater. Chem. A Mater*, 5, 29, 15266–15272, 2017, doi: 10.1039/c7ta02638b.
17. Jin, L., Gong, R., Zhang, W., Xiang, Y., Zheng, J., Xiang, Z., Zhang, C., Xia, Y., Zheng, J.P., Toward high energy-density and long cycling-lifespan lithium ion capacitors: A 3D carbon modified low-potential Li_2TiSiO_5 anode coupled with a lignin-derived activated carbon cathode. *J. Mater Chem. A*, 7, 14, 8234–8244, 2019, doi: 10.1039/C8TA12000E.
18. Zhang, S., Li, C., Zhang, X., Sun, X., Wang, K., Ma, Y., High performance lithium-ion hybrid capacitors employing Fe_3O_4–graphene composite anode and activated carbon cathode. *ACS Appl. Mater. Interfaces*, 9, 20, 17136–17144, May 2017, doi: 10.1021/acsami.7b03452.
19. Numan, A., Zhan, Y., Khalid, M., Hatamvand, M., Introduction to supercapattery, in: *Advances in Supercapacitor and Supercapattery: Innovations in Energy Storage Devices*, pp. 45–61, Elsevier, Elsevier Inc., Amsterdam, Netherlands, 2020, doi: 10.1016/ B978-0-12-819897-1.00008-2.
20. Aida, T., Yamada, K., Morita, M., An advanced hybrid electrochemical capacitor that uses a wide potential range at the positive electrode. *Electrochem. Solid-State Lett.*, 9, 12, A534, 2006, doi: 10.1149/1.2349495.
21. Weng, Z., Li, F., Wang, D.-W., Wen, L., Cheng, H.-M., Controlled electrochemical charge injection to maximize the energy density of supercapacitors. *Angew. Chem. Int. Ed.*, 52, 13, 3722–3725, 2013, doi: https://doi.org/10.1002/anie.201209259.
22. Davies, A. and Yu, A., Material advancements in supercapacitors: From activated carbon to carbon nanotube and graphene. *Can. J. Chem. Eng.*, 89, 6, 1342–1357, Dec. 2011, doi: 10.1002/ cjce.20586.
23. Gu, W. and Yushin, G., Review of nanostructured carbon materials for electrochemical capacitor applications: Advantages and limitations of activated carbon, carbide-derived carbon, zeolite-templated carbon, carbon aerogels, carbon nanotubes, onion-like carbon, and graphene. *Wiley Interdiscip. Rev.: Energy Environ.*, 3, 5, 424–473, 2014, doi: 10.1002/wene.102.

24. Xiong, R., Zhang, Y., Zhou, W., Xia, K., Sun, Q., Chen, G., Han, B., Gao, Q., Zhou, C., Chemical activation of carbon materials for supercapacitors: Elucidating the effect of spatial characteristics of the precursors. *Colloids Surf. A Physicochem. Eng. Asp.*, 597, 124762, Jul. 2020, doi: 10.1016/J.COLSURFA.2020.124762.
25. Peng, C., X. bin Yan, Wang, R.T., Lang, J.W., Ou, Y.J., Xue, Q.J., Promising activated carbons derived from waste tea-leaves and their application in high performance supercapacitors electrodes. *Electrochim. Acta*, 87, 401– 408, Jan. 2013, doi: 10.1016/J.ELECTACTA.2012.09.082.
26. Sundriyal, S., Shrivastav, V., Dubey, P., Singh, M., Deep, A., Dhakate, S.R., Highly porous carbon from azadirachta indica leaves and Uio-66 derived metal oxide for asymmetrical supercapacitors. *IEEE Trans. Nanotechnol.*, 21, 60–65, 2022, doi: 10.1109/TNANO.2022.3144367.
27. Liu, Y., Wang, H., Li, C., Wang, S., Li, L., Song, C., Wang, T., Hierarchical flaky porous carbon derived from waste polyimide film for high-performance aqueous supercapacitor electrodes. *Int. J. Energy Res.*, 46, 1, 370–382, 2022, doi: 10.1002/er.7106.
28. Xiao, Y., Long, C., Zheng, M.T., Dong, H.W., Lei, B.F., Zhang, H.R., Liu, Y.L., High-capacity porous carbons prepared by KOH activation of activated carbon for supercapacitors. *Chin. Chem. Lett.*, 25, 6, 865–868, Jun. 2014, doi: 10.1016/J.CCLET.2014.05.004.
29. Tay, T., Ucar, S., Karagöz, S., Preparation and characterization of activated carbon from waste biomass. *J. Hazard Mater.*, 165, 1–3, 481–485, Jun. 2009, doi: 10.1016/J.JHAZMAT.2008.10.011.
30. Sudaryanto, Y., Hartono, S.B., Irawaty, W., Hindarso, H., Ismadji, S., High surface area activated carbon prepared from cassava peel by chemical activation. *Bioresour. Technol.*, 97, 5, 734–739, Mar. 2006, doi: 10.1016/J.BIORTECH.2005.04.029.
31. Farma, R., Putri, A., Taer, E., Awitdrus, A., Apriwandi, A., Synthesis of highly porous activated carbon nanofibers derived from bamboo waste materials for application in supercapacitor. *J. Mater. Sci.: Mater. Electron.*, 32, 6, 7681–7691, 2021, doi: 10.1007/ s10854-021-05486-5.
32. Li, Y., Luo, Z., Liang, S., Qin, H., Zhao, X., Chen, L., Wang, H., Chen, S., Two-dimensional porous zinc cobalt sulfide nanosheet arrays with superior electrochemical performance for supercapatteries. *J. Mater. Sci. Technol.*, 89, 199–208, Oct. 2021, doi: 10.1016/J.JMST.2021.01.085.
33. How, Y.Y., Bibi, F., Numan, A., Walvekar, R., Jagadish, P., Khalid, M., Iqbal, J., Mubarak, N.M., Fabrication of binary metal phosphate-based binder-free electrode for new generation energy storage device. *Surf. Coat. Technol.*, 429, 127924, Jan. 2022, doi: 10.1016/J.SURFCOAT.2021.127924.
34. Ndipingwi, M.M., Ikpo, C.O., Nwanya, A.C., Januarie, K.C., Ramoroka, M.E., Uhuo, O.V., Nwambaekwe, K., Yussuf, S.T., Iwuoha, E., II, Engineering the chemical environment of lithium manganese silicate by Mn ion substitution to boost the charge storage capacity for application in high efficiency

supercapattery. *Electrochim. Acta*, 414, 140180, May 2022, doi: 10.1016/J.ELECTACTA.2022.140180.

35. Amiri, M., Moosavifard, S.E., Davarani, S.S.H., Shamsipur, M., Novel rugby-ball-like $FeCoCuS_2$ triple-shelled hollow nanostructures with enhanced performance for supercapattery. *Energy Fuels*, 35, 18, 15108–15117, Sep. 2021, doi: 10.1021/acs. energyfuels.1c01807.
36. Iqbal, M.Z., Khan, J., Siddique, S., Afzal, A.M., Aftab, S., Optimizing electrochemical performance of sonochemically and hydrothermally synthesized cobalt phosphate for supercapattery devices. *Int. J. Hydrog. Energy*, 46, 29, 15807–15819, Apr. 2021, doi: 10.1016/J.IJHYDENE.2021.02.077.
37. Alzaid, M., Iqbal, M.Z., Alam, S., Almoisheer, N., Afzal, A.M., Aftab, S., Binary composites of nickel-manganese phosphates for supercapattery devices. *J. Energy Storage*, 33, 102020, Jan. 2021, doi: 10.1016/J. EST.2020.102020.
38. Iqbal, M.Z. and Khan, J., Optimization of cobalt-manganese binary sulfide for high performance supercapattery devices. *Electrochim. Acta*, 368, 137529, Feb. 2021, doi: 10.1016/J.ELECTACTA.2020.137529.
39. Numan, A., Khalid, M., Ramesh, S., Ramesh, K., Shamsudin, E.M., Zhan, Y., Jagadesh, P., Facile sonochemical synthesis of 2D porous Co_3O_4 nano-flake for supercapattery. *J. Alloys Compd.*, 819, 153019, Apr. 2020, doi: 10.1016/J.JALLCOM.2019.153019.
40. Raj, C.J., Manikandan, R., Yu, K.H., Nagaraju, G., Park, M.S., Kim, D.W., Park, S.Y., Kim, B.C., Engineering thermally activated $NiMoO_4$ nanoflowers and biowaste derived activated carbon-based electrodes for high-performance supercapatteries. *Inorg. Chem. Front.*, 7, 2, 369–384, 2020, doi: 10.1039/C9QI01085H.
41. Sui, L., Tang, S., Chen, Y., Dai, Z., Huangfu, H., Zhu, Z., Qin, X., Deng, Y., Haarberg, G.M., An asymmetric supercapacitor with good electrochemical performances based on $Ni(OH)_2$/AC/CNT and AC. *Electrochim. Acta*, 182, 1159–1165, Nov. 2015, doi: 10.1016/J.ELECTACTA.2015.09.111.
42. Nagamuthu, S., Vijayakumar, S., Muralidharan, G., Ag incorporated Mn_3O_4/AC nanocomposite based supercapacitor devices with high energy density and power density. *Dalton Trans.*, 43, 46, 17528–17538, 2014, doi: 10.1039/C4DT02287D.
43. Cao, S., Li, H., Zhou, X., Guo, H., Chen, Y., NiS/activated carbon composite derived from sodium lignosulfonate for long cycle-life asymmetric supercapacitors. *J. Alloys Compd.*, 900, 163546, Apr. 2022, doi: 10.1016/J.JALLCOM.2021.163546.
44. Iijima, S., Helical microtubules of graphitic carbon. *Nature*, 354, 6348, 56–58, 1991, doi: 10.1038/354056a0.
45. Senthil, T., Divakaran, N., Kale, M.B., Mubarak, S., Dhamodharan, D., Wu, L., Bensingh, R.J., Kader, M.A., Dutta, K., Low-dimensional carbon-based nanomaterials for energy conversion and storage applications, in: *Nanostructured, Functional, and Flexible Materials for Energy Conversion and*

Storage Systems, pp. 15–68, Elsevier, Elsevier Inc., Amsterdam, Netherlands, 2020, doi: 10.1016/b978-0-12-819552-9.00002-6.

46. Rathinavel, S., Priyadharshini, K., Panda, D., A review on carbon nanotube: An overview of synthesis, properties, functionalization, characterization, and the application. *Mater. Sci. Eng. B*, 268, 115095, Jun. 2021, doi: 10.1016/J.MSEB.2021.115095.
47. Gupta, N., Gupta, S.M., Sharma, S.K., Carbon nanotubes: Synthesis, properties and engineering applications. *Carbon Lett.*, 29, 5, 419–447, 2019, doi: 10.1007/s42823-019-00068-2.
48. Baddour, C.E. and Briens, C., Carbon nanotube synthesis: A review. *Int. J. Chem. Reactor Eng.*, 3, 1–20, 2005, doi: 10.2202/1542-6580.1279.
49. Shudin, N.H., Mukri, M. 'A. R., Aziz, M., Othman, M.H.D., Tanemura, M., Yusop, M.Z.M., The role of solid, liquid and gaseous hydrocarbon precursors on chemical vapor deposition grown carbon nanomaterials' growth temperature. *Synth. Met.*, 274, 116735, Apr. 2021, doi: 10.1016/J.SYNTHMET.2021.116735.
50. Labbaye, T., Kovacevic, E., Lecas, T., Ammar, M.R., Canizarès, A., Raimboux, N., Strunskus, T., Jaeger, C., Simon, P., Boulmer-Leborgne, C., Enhancement of catalytic effect for CNT growth at low temperature by PECVD. *Appl. Surf. Sci.*, 453, 436–441, Sep. 2018, doi: 10.1016/J.APSUSC.2018.05.059.
51. Lin, J., Yang, Y., Zhang, H., Li, F., Huang, G., Wang, J., Effect of catalyst on carbon nanotubes synthesis on titanium diboride via chemical vapor deposition. *J. Am. Ceram. Soc.*, 103, 8, 4691–4699, 2020, doi: 10.1111/jace.17128.
52. Chen, T. and Dai, L., Carbon nanomaterials for high-performance supercapacitors. *Mater. Today*, 16, 7–8, 272–280, Jul. 2013, doi: 10.1016/J.MATTOD.2013.07.002.
53. Seah, C.M., Chai, S.P., Mohamed, A.R., Mechanisms of graphene growth by chemical vapour deposition on transition metals. *Carbon*, 70, 1–21, Apr. 2014, doi: 10.1016/J.CARBON.2013.12.073.
54. Addou, R., Dahal, A., Sutter, P., Batzill, M., Monolayer graphene growth on Ni(111) by low temperature chemical vapor deposition. *Appl. Phys. Lett.*, 100, 2, 21601, 2012, doi: 10.1063/1.3675481.
55. Iqbal, J., Numan, A., Rafique, S., Jafer, R., Mohamad, S., Ramesh, K., Ramesh, S., High performance supercapattery incorporating ternary nanocomposite of multiwalled carbon nanotubes decorated with Co_3O_4 nanograins and silver nanoparticles as electrode material. *Electrochim. Acta*, 278, 72–82, Jul. 2018, doi: 10.1016/J.ELECTACTA.2018.05.040.
56. Lokhande, A.C., Hussain, T., Shelke, A.R., Babar, P.T., Kim, J.H., Choi, D., High performance Li-ion capacitor based on novel carbon-coated chalcogenide anode. *J. Energy Storage*, 50, 104251, Jun. 2022, doi: 10.1016/J.EST.2022.104251.
57. Sharmila, V. and Parthibavarman, M., Lithium manganese phosphate associated with MWCNT: Enhanced positive electrode for lithium

hybrid batteries. *J. Alloys Compd.*, 858, 157715, Mar. 2021, doi: 10.1016/J.JALLCOM.2020.157715.
58. Abdollahi, A., Abnavi, A., Ghasemi, F., Ghasemi, S., Sanaee, Z., Mohajerzadeh, S., Facile synthesis and simulation of MnO_2 nanoflakes on vertically aligned carbon nanotubes, as a high-performance electrode for Li-ion battery and supercapacitor. *Electrochim. Acta*, 390, 138826, Sep. 2021, doi: 10.1016/J.ELECTACTA.2021.138826.
59. Gao, C., Liu, S., Yan, P., Zhu, M., Qiu, T., Enhanced electrochemical kinetics and three dimensional architecture lithium iron phosphate/carbon nanotubes nanocomposites for high rate lithium-ion batteries. *Colloids Surf. A Physicochem. Eng. Asp.*, 643, 128718, Jun. 2022, doi: 10.1016/J.COLSURFA.2022.128718.
60. Kumar, R., Joanni, E., Sahoo, S., Shim, J.J., Tan, W.K., Matsuda, A., Singh, R.K., An overview of recent progress in nanostructured carbon-based supercapacitor electrodes: From zero to bi-dimensional materials. *Carbon*, 193, 298–338, Jun. 2022, doi: 10.1016/J.CARBON.2022.03.023.
61. Kumar, S., Nehra, M., Kedia, D., Dilbaghi, N., Tankeshwar, K., Kim, K.H., Carbon nanotubes: A potential material for energy conversion and storage. *Prog. Energy Combust. Sci.*, 64, 219–253, Jan. 2018, doi: 10.1016/J.PECS.2017.10.005.
62. Ho Shin, W., Mo Jeong, H., Gon Kim, B., Ku Kang, J., Wook Choi, J., Nitrogen-doped multiwall carbon nanotubes for lithium storage with extremely high capacity. *Nano Lett.*, 12, 5, 2283–2288, Apr. 2012, doi: 10.1021/nl3000908.
63. Zhu, G., Tang, C., Jiang, M., Du, A., Zhang, H., Yang, J., Regulating the interfacial behavior of carbon nanotubes for fast lithium storage. *Electrochim. Acta*, 388, 138591, Aug. 2021, doi: 10.1016/J.ELECTACTA.2021.138591.
64. Novoselov, K.S. *et al.*, Electric field effect in atomically thin carbon films. *Science (1979)*, 306, 5696, 666–669, 2004, doi: 10.1126/ science.1102896.
65. Bonaccorso, F., Colombo, L., Yu, G., Stoller, M., Tozzini, V., Ferrari, A.C., Ruoff, R.S., Pellegrini, V., Graphene, related two-dimensional crystals, and hybrid systems for energy conversion and storage. *Science*, 347, 6217, 1–9, American Association for the Advancement of Science, Jan. 02, 2015, doi: 10.1126/science.1246501.
66. Eskusson, J., Rauwel, P., Nerut, J., Jänes, A., A hybrid capacitor based on Fe_3O_4-graphene nanocomposite/few-layer graphene in different aqueous electrolytes. *J. Electrochem. Soc.*, 163, 13, A2768–A2775, Oct. 2016, doi: 10.1149/2.1161613jes.
67. Li, L., Zhang, D., Deng, J., Gou, Y., Fang, J., Cui, H., Zhao, Y., Cao, M., Carbon-based materials for fast charging lithium-ion batteries. *Carbon*, 183, 721–734, Oct. 2021, doi: 10.1016/J.CARBON.2021.07.053.
68. Ajala, O.J., Tijani, J.O., Bankole, M.T., Abdulkareem, A.S., A critical review on graphene oxide nanostructured material: Properties, synthesis, characterization and application in water and wastewater treatment. *Environ.*

Nanotechnol. Monit. Manag., 18, 100673, Dec. 2022, doi: 10.1016/J.ENMM.2022.100673.

69. Bo, Z., Yang, Y., Chen, J., Yu, K., Yan, J., Cen, K., Plasma-enhanced chemical vapor deposition synthesis of vertically oriented graphene nanosheets. *Nanoscale*, 5, 12, 5180–5204, 2013, doi: 10.1039/C3NR33449J.
70. Basirun, W.J., Saeed, I.M., Rahman, M.S., Mazari, S.A., Nickel oxides/hydroxides-graphene as hybrid supercapattery nanocomposites for advanced charge storage materials – A review. *Crit. Rev. Solid State Mater. Sci.*, 46, 6, 553–586, 2021, doi: 10.1080/10408436.2021.1886040.
71. Patra, S., Roy, E., Tiwari, A., Madhuri, R., Sharma, P.K., 2-dimensional graphene as a route for emergence of additional dimension nanomaterials. *Biosens. Bioelectron.*, 89, Elsevier Ltd, 8–27, Mar. 15, 2017, doi: 10.1016/j.bios.2016.02.067.
72. Xu, X., Zhao, X., Yang, Z., Lin, Q., Jian, B., Li, N., Zheng, C., Lv, W., High-density three-dimensional graphene cathode with a tailored pore structure for high volumetric capacity zinc-ion storage. *Carbon*, 186, 624–631, Jan. 2022, doi: 10.1016/J.CARBON.2021.10.060.
73. Xu, Y., Lin, Z., Huang, X., Liu, Y., Huang, Y., Duan, X., Flexible solid-state supercapacitors based on three-dimensional graphene hydrogel films. *ACS Nano.*, 7, 5, 4042–4049, Apr. 2013, doi: 10.1021/nn4000836.
74. Renani, T.S., Khoshfetrat, S.M., Arjomandi, J., Shi, H., Khazalpour, S., Fabrication and design of new redox active azure A/3D graphene aerogel and conductive trypan blue–nickel MOF nanosheet array electrodes for an asymmetric supercapattery. *J. Mater. Chem. A*, 9, 21, 12853–12869, 2021, doi: 10.1039/D1TA02850B.
75. Wang, X., Sun, G., Routh, P., Kim, D.-H., Huang, W., Chen, P., Heteroatom-doped graphene materials: Syntheses, properties and applications. *Chem. Soc. Rev.*, 43, 20, 7067–7098, 2014, doi: 10.1039/C4CS00141A.
76. Naresh, V., Bhattacharjee, U., Martha, S.K., Boron doped graphene nanosheets as negative electrode additive for high-performance lead-acid batteries and ultracapacitors. *J. Alloys Compd.*, 797, 595–605, Aug. 2019, doi: 10.1016/J.JALLCOM.2019.04.311.
77. Bhattacharjee, U., Bhowmik, S., Ghosh, S., Vangapally, N., Martha, S.K., Boron-doped graphene anode coupled with microporous activated carbon cathode for lithium-ion ultracapacitors. *Chem. Eng. J.*, 430, 132835, Feb. 2022, doi: 10.1016/J.CEJ.2021.132835.
78. Liu, G., Xie, J., Sun, Y., Zhang, P., Li, X., Zheng, L., Hao, L., Shanmin, G., Constructing 3D honeycomb-like CoMn2O4 nanoarchitecture on nitrogen-doped graphene coating Ni foam as flexible battery-type electrodes for advanced supercapattery. *Int. J. Hydrog. Energy*, 46, 73, 36314–36322, Oct. 2021, doi: 10.1016/J.IJHYDENE.2021.08.159.
79. Liou, Y.J., da Tsai, B., Huang, W.J., An economic route to mass production of graphene oxide solution for preparing graphene oxide papers. *Mater. Sci. Eng., B*, 193, C, 37–40, Mar. 2015, doi: 10.1016/J.MSEB.2014.11.007.

80. Ikram, R., Jan, B.M., Ahmad, W., An overview of industrial scalable production of graphene oxide and analytical approaches for synthesis and characterization. *J. Mater. Res. Technol.*, 9, 5, 11587–11610, Sep. 2020, doi: 10.1016/J.JMRT.2020.08.050.
81. Santamaría-Juárez, G., Gómez-Barojas, E., Quiroga-González, E., Sánchez-Mora, E., Quintana-Ruiz, M., Santamaría-Juárez, J.D., Safer modified Hummers' method for the synthesis of graphene oxide with high quality and high yield. *Mater. Res. Express*, 6, 12, 125631, Jan. 2020, doi: 10.1088/2053-1591/ab4cbf.
82. Agudosi, E.S., Abdullah, E.C., Numan, A., Mubarak, N.M., Khalid, M., Omar, N., A review of the graphene synthesis routes and its applications in electrochemical energy storage. *Crit. Rev. Solid State Mater. Sci.*, 45, 5, 339–377, 2020, doi: 10.1080/10408436.2019.1632793.
83. Xu, L., Wang, S., Zhang, X., He, T., Lu, F., Li, H., Ye, J., A facile method of preparing $LiMnPO_4$/reduced graphene oxide aerogel as cathodic material for aqueous lithium-ion hybrid supercapacitors. *Appl. Surf. Sci.*, 428, 977–985, Jan. 2018, doi: 10.1016/J.APSUSC.2017.09.247.
84. Raj, S., Srivastava, S.K., Kar, P., Roy, P., *In situ* growth of Co_3O_4 nanoflakes on reduced graphene oxide-wrapped Ni-foam as high performance asymmetric supercapacitor. *Electrochim. Acta*, 302, 327–337, Apr. 2019, doi: 10.1016/J.ELECTACTA.2019.02.010.
85. Heng, I., Low, F.W., Lai, C.W., Juan, J.C., Amin, N., Tiong, S.K., High performance supercapattery with rGO/TiO_2 nanocomposites anode and activated carbon cathode. *J. Alloys Compd.*, 796, 13–24, Aug. 2019, doi: 10.1016/J.JALLCOM.2019.04.347.
86. Zhang, H., Zhang, X., Li, H., Gao, Y., Yan, J., Zhu, K., Ye, K., Cheng, K., Wang, G., Cao, D., Copper niobate nanowires immobilized on reduced graphene oxide nanosheets as rate capability anode for lithium ion capacitor. *J. Colloid Interface Sci.*, 583, 652–660, Feb. 2021, doi: 10.1016/J. JCIS.2020.09.076.
87. Mehtab, T., Yasin, G., Arif, M., Shakeel, M., Korai, R.M., Nadeem, M., Muhammad, N., Lu, X., Metal-organic frameworks for energy storage devices: Batteries and supercapacitors. *J. Energy Storage*, 21, 632–646, Feb. 2019, doi: 10.1016/J.EST.2018.12.025.
88. Liu, B., Shioyama, H., Jiang, H., Zhang, X., Xu, Q., Metal–organic framework (MOF) as a template for syntheses of nanoporous carbons as electrode materials for supercapacitor. *Carbon*, 48, 2, 456–463, Feb. 2010, doi: 10.1016/J.CARBON.2009.09.061.
89. Pachfule, P., Shinde, D., Majumder, M., Xu, Q., Fabrication of carbon nanorods and graphene nanoribbons from a metal–organic framework. *Nat. Chem.*, 8, 7, 718–724, 2016, doi: 10.1038/nchem.2515.
90. Chu, X., Meng, F., Deng, T., Zhang, W., Metal organic framework derived porous carbon materials excel as an excellent platform for high-performance packaged supercapacitors. *Nanoscale*, 13, 11, 5570–5593, 2021, doi: 10.1039/D1NR00160D.

91. Niu, H., Liu, Y., Mao, B., Xin, N., Jia, H., Shi, W., *In-situ* embedding MOFs-derived copper sulfide polyhedrons in carbon nanotube networks for hybrid supercapacitor with superior energy density. *Electrochim. Acta*, 329, 135130, Jan. 2020, doi: 10.1016/J.ELECTACTA.2019.135130.
92. Xu, Z., Li, J., Li, X., Chen, Z., Chen, C., Shah, S.A.A., Wu, M., An instantaneous metal organic framework to prepare ultra-high pore volume porous carbon for lithium ion capacitors. *Appl. Surf. Sci.*, 565, 150528, Nov. 2021, doi: 10.1016/J.APSUSC.2021.150528.
93. Banerjee, A., Upadhyay, K.K., Puthusseri, D., Aravindan, V., Madhavi, S., Ogale, S., MOF-derived crumpled-sheet-assembled perforated carbon cuboids as highly effective cathode active materials for ultra-high energy density Li-ion hybrid electrochemical capacitors (Li-HECs). *Nanoscale*, 6, 8, 4387–4394, 2014, doi: 10.1039/C4NR00025K.
94. Xi, K., Cao, S., Peng, X., Ducati, C., Kumar, R.V., Cheetham, A.K., Carbon with hierarchical pores from carbonized metal–organic frameworks for lithium sulphur batteries. *Chem. Commun.*, 49, 22, 2192–2194, 2013, doi: 10.1039/C3CC38009B.
95. Lu, H., Liu, S., Zhang, Y., Huang, Y., Zhang, C., Liu, T., Nitrogen-doped carbon polyhedra nanopapers: An advanced binder-free electrode for high-performance supercapacitors. *ACS Sustain. Chem. Eng.*, 7, 5, 5240–5248, Feb. 2019, doi: 10.1021/ acssuschemeng.8b06159.
96. Lu, R., Ren, X., Wang, C., Zhan, C., Nan, D., Lv, R., Shen, W., Kang, F., Huang, Z.H., $Na_{0.76}V_6O_{15}$/activated carbon hybrid cathode for high-performance lithium-ion capacitors. *Materials*, 14, 1, 1–10, 2021, doi: 10.3390/ma14010122.
97. Li, L., Liu, Y., Han, Y., Qi, X., Li, X., Fan, H., L., Metal-organic framework-derived carbon coated copper sulfide nanocomposites as a battery-type electrode for electrochemical capacitors. *Mater. Lett.*, 236, 131–134, Feb. 2019, doi: 10.1016/J.MATLET.2018.10.093.

9

Graphene-Based Electrocatalytic Materials Toward Electrochemical Water Splitting

Prasanta Pattanayak[1], Paulomi Singh[1,2], Nitin Kumar Bansal[1], Snehangshu Mishra[1] and Trilok Singh[1]*

[1]*Functional Materials and Device Laboratory, School of Energy Science and Engineering, Indian Institute of Technology, Kharagpur, India*
[2]*School of Nano Science and Technology, Indian Institute of Technology, Kharagpur, India*

Abstract

Graphene-based electrocatalytic materials have attracted great interest toward electrochemical water splitting for producing green hydrogen. Moreover, it is expected that graphene-based metal and their nanohybrids could further enhance the hydrogen evolution reaction (HER) and oxygen evolution reaction (OER) of a designed electrochemical system. This chapter will focus on the state-of-the-art technological developments of graphene-based electrocatalytic materials for water splitting in the recent past. This includes the synthesis and fabrication of graphene/metal hybrids, graphene/conducting polymer hybrids, and functionalized graphene-based electrodes for improved HER and OER activity. The future prospects for issues like cost-effectiveness, efficiency, and potential loss recovery in the system are also discussed in this chapter.

Keywords: Renewable energy, graphene, electrocatalytic materials, hydrogen evolution reaction (HER), oxygen evolution reaction (OER), electrochemical

**Corresponding author*: trilok@iitkgp.ac.in

Sudheesh K. Shukla, Chaudhery Mustansar Hussain, Santanu Patra, Meenakshi Choudhary (eds.) Electrocatalytic Materials for Renewable Energy, (229–270) © 2024 Scrivener Publishing LLC

9.1 Introduction

In the present century, rising environmental issues have prompted extensive investigations in the areas of renewable energy sources. However, the world's maximum energy is still dependent on different non-renewable resources like coal, oil, and natural gas. These sources are limited, depleting rapidly with growing industrialization and producing harmful greenhouse gases, and they have a major contribution to global warming and climate change. As an alternative, people are investigating electrochemical systems that can transform chemical energy into electrical energy, using bifunctional electrocatalytic materials. These systems cover a novel route for sustained green energy production that can serve the dual purpose of pure hydrogen and oxygen evolution with zero pollutants from the widely known technique "water splitting" [1–7]. In general, water electrolysis is primarily dependent on two key half reactions, i.e., hydrogen evolution reaction (HER) at the cathode and oxygen evolution reaction (OER) at the anode. However, the slower reaction kinetics with larger overpotentials at both ends of OER and HER becomes the major limitation for industrial production of the systems [8].

Commercially, the most active state-of-the-art electrocatalyst for the water-splitting process is Pt/C which leads to HER with an overpotential of ~0 mV at the cathode. Similarly, Ru with an overpotential of ~130 mV or Ir with an overpotential of ~210 mV serves for OER at the anode side. However, these electrocatalysts have shown promising current density and minimal overpotentials, yet their large-scale implementation is limited because of the scarcity and relatively exorbitant cost. Accordingly, cheap, earth-abundant, and highly efficient electrode materials are necessary for water electrolysis to substitute this state-of-the-art electrocatalyst. In this regard, studies have been extended toward high-performance electrode materials and robust support materials for water-splitting processes. In the past few years, different combinations of electrocatalysts like transition metal oxides, noble metals, metal (alloys), metal (sulfides, phosphides, nitrides, and carbides), chalcogenides, and various composites have been tested as an alternative for HER and OER [8, 9]. However, the numbers of such positive combinations that have responded well toward both OER and HER in the same electrolytic medium are quite low [10–13]. Moreover, different types of metal-free electrocatalysts and efficient support materials like carbon nanotubes (CNTs), graphene (0D, 2D, 3D, rGO, GQD), functionalized graphene, conducting polymers, modified conducting polymers, and their composites have equally become an area of foremost interest due to their excellent stability and unique electroactive properties. Accordingly, graphene can be used in a variety of electrochemical systems, such as solar panels [14], fuel cells [15–17], energy storage devices, chemical detection sensors, supercapacitors [7], and composites (Figure 9.1).

Various carbon family configurations are exploited for efficient water splitting; for example, direct or electrode support based on graphene could be one of the game-changing electrocatalytic materials that can provide high efficiency in electrochemical water-splitting systems with its superior electron transferability. Another new carbon family member, graphene (0D, 2D, 3D, rGO, GQD), and its functionalized forms are considered promising due to superior chemical and physical properties (such as larger surface area, enhanced electroconductivity, optical transparency, high mobility, tunable photoluminescence, high absorption capacity, strong quantum confinement effect, slow hot carrier relaxation, biocompatibility, and good dispersion ability in various solvents). Moreover, it contains -OH, -COOH, and $-NH_2$ groups at the surface, which could provide increased active sites for heteroatom doping. The latest addition to the family is the graphene quantum dots (GQDs), which are small nanocarbon graphene sheets approximately 20 nm long and about 0.4–2.0 nm thick [18], covering a broad spectrum of applications, having efficiency in designing different electrocatalysts for water-splitting applications. The reason is attributed

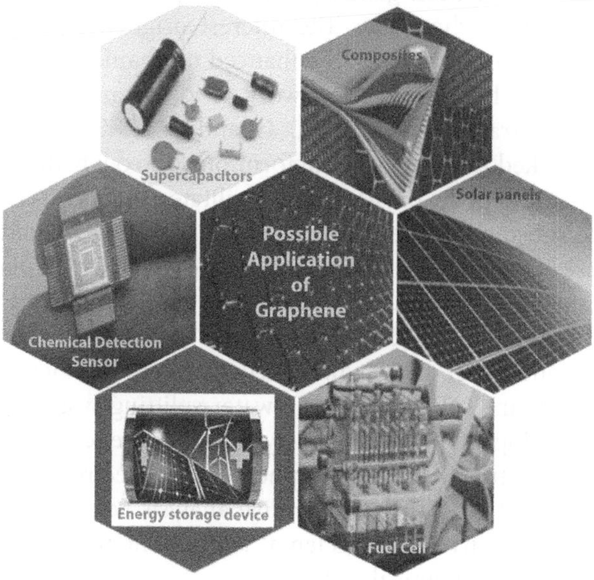

Figure 9.1 Diagram illustrating the possible applications of graphene.

to its unique properties like 0D planar structure, higher edge fractions and defects, tunable functionalities for doping heteroatoms and chemical moieties, higher dispersibility, and electron transfer ability [19]. Likely, in a single-atom 2D crystalline, each carbon atom gets sp^2-hybridized with the benzene ring [20]. With a larger specific area and enhanced electrical conductivity, 2D graphene holds a promising support and reaction center for different catalyst configurations, which not only prevents the catalyst from clogging but also allows higher active regions for enhanced electrocatalysis. Moreover, the 3D graphene network provides a more robust solution for inhibiting the aggregation of graphene sheets. Additionally, it also facilitates advanced intrinsic properties of the structure, such as fairly reduced mass densities, enhanced specific surface area, and better mechanical stability with interlinked hierarchical structure. The size range of 3D graphene can vary between microns to centimeters, bestowing their general intrinsic nanoscale feature in the framework. The 3D networks have been used as a physical spacer to inhibit the agglomeration of graphene sheets. With such excellent intrinsic features, 3D graphene has been widely applied in different streams of photo- and electrocatalytical technologies [20]. Another increasingly used support material is reduced graphene oxide (RGO), whose coupling with earth-abundant metals resulted in higher charge transfer in the combined moiety via the basal planes of the incorporated metals [10]. Recently, the combinations of GO with different polymeric materials have been gaining wider acceptance in different streams of electro-analytical chemistry. The interest originates with the aim to tailor conductive composites with enhanced mechanical strength because of the nascent sp^2-hybridized framework of RGO carbon atoms [21–25]. GO consists of a framework of sp^2 (C–C) and sp^3 (C–O) hybridized atoms (up to 55%) linked to oxygen in the form of carboxyl, hydroxyl, or epoxy groups [26]. These polar oxygen units facilitate an aqueous dispersion of GO but also substantially decrease the overall conductivity of the unit. To remove these oxygen species, GO is reduced to RGO, which plays a crucial role in enhancing the conductivity within the overall polymeric nanocomposite framework [21, 27–30]. Moreover, it is expected that graphene-based metal and other nanohybrid structures can further enhance the stability of HER and OER catalysts in water-splitting systems. This would include fabricating graphene/metal hybrids, graphene/conducting polymer hybrids, and functionalized graphene-based electrodes for improved HER and OER activity. The comparative analysis would address relative issues like cost-effectiveness, efficiency, and potential loss recovery from the employed electrochemical system.

9.2 Electrochemical Water Splitting: Principles and Mechanism

9.2.1 Overview of Electrochemical Water Splitting

As mentioned earlier, electrochemical water splitting is an electrolytic reaction, leading toward the evolution of hydrogen at the cathode and oxygen at the anode side of the system. Both pathways are different and depend on the chemical and electronic nature of the used electrolyte and electrode-catalyst morphology. A minor change can alter the corresponding reaction kinetics and stability of the system, which is the most important thing to understand and regulate for optimum efficiency. Theoretically, at 1.23 V, redox potential water splitting should occur, under standard conditions (25°C and 1 atm) regardless of the electrolyte medium (acidic or alkaline) [31]. However, in real terms, voltage above the theoretical value is needed to initiate the water-splitting reaction. This difference between the voltage needed and the thermodynamically estimated value is called the overpotential. These overpotentials (η) originate from the activation barrier of the electrode reactivity as well as other systemic resistances (e.g., contact and solution resistances). Thus, the overall real potential (EP) for water splitting is therefore denoted as [32–35]:

$$EP = 1.23\ V + \eta_a + \eta_c + \eta_{other}$$

where η_a, η_c, and η_{other} represent the anodic, cathodic, and other resistive overpotentials, respectively. The efficiency of the employed catalysts reduces these undesired overpotential values originating at both terminals of the system. In the next segments, reaction steps like the fundamental processes of HER and OER will be discussed, which will highlight the requisites for high-performance electrocatalyst preparation for hydrogen and oxygen evolution [32].

9.2.2 Mechanism of Electrochemical Water Splitting

9.2.2.1 Reaction Mechanism of Hydrogen Evolution Reaction (HER)

HER is the most elementary electrochemical reaction that holds the benchmark for studying other complicated multiple-step reactions for electron–proton transfer. It undergoes a sequence of reactions following adsorption, reduction, and desorption processes at the electrode surface.

The sequence and rate vary depending on the electrode properties; however, the generally accepted mechanisms in acidic and alkaline electrolytes are given in Figure 9.2 and Table 9.1. The first step is the Volmer reaction which allows the adsorption of hydrogen (H) atoms on the active sites of the electrode surface. This results in a one-electron transfer step that forms M–H* bond between the adsorbed H atom and the electrode surface. The adsorbed H* atoms are then reduced into H_2 molecules via either Heyrovsky or Tafel reaction, which finally desorbs at the electrode surface as H_2 gas. The rate-determining step (RDS) is calculated from the Tafel slope of the HER polarization curve (Table 9.1) [32].

9.2.2.2 Reaction Mechanism of Oxygen Evolution Reaction (OER)

OER, on the other hand, is a $4e^-$ transfer reaction with different reaction mechanisms (Figure 9.2 and Table 9.2) [32, 36]. There is a formation of a series of reaction intermediates such as OH*, O*, and OOH* for both the acidic and alkaline media. Initially, intermediates of M–OH* and M–O* are formed at the electrode surface in a two-step reaction process. Next, O_2 molecules are formed either from a) the direct recombination of M–OH* or b) another intermediate M–OOH* is formed with M–OH* reacting with

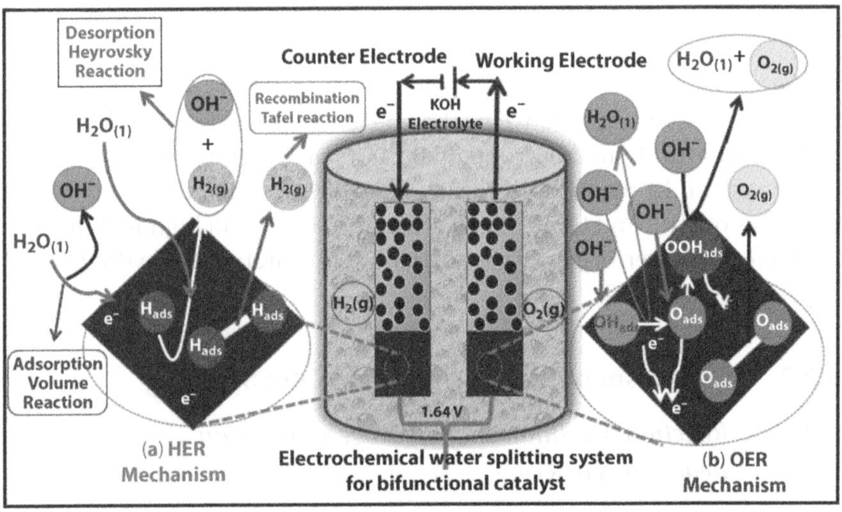

Figure 9.2 Bifunctional electrocatalyst for electrochemical water splitting in a full-cell setup. (a) Hydrogen evolution reaction (HER) mechanism on the surface of the bifunctional electrocatalyst. (b) Oxygen evolution reaction (OER) mechanism on the surface of the bifunctional electrocatalyst.

Table 9.1 A comparison of HER reaction pathways in acidic and alkaline electrolytes.

Reaction **Acidic solution**		**Electrolyte**	
		Acidic solution	**Alkaline solution**
Overall reaction		$2\,H^+ + 2\,e^- \leftrightarrow H_2$	$2\,H_2O + 2\,e^- \leftrightarrow H_2 + 2\,OH^-$
Volmer reaction **Here,** $\boldsymbol{b}$ **= Tafel slope** $= \dfrac{\mathbf{2.3\,RT}}{\boldsymbol{\alpha F}}$	b ~120 mV dec^{-1}	$M + H^+ + e^- \leftrightarrow M - H^*$	$M + H_2O + e^- \leftrightarrow M - H^* + OH^-$
Heyrovsky reaction **Here,** $\boldsymbol{b}$ **= Tafel slope** $= \dfrac{\mathbf{2.3\,RT}}{\boldsymbol{\alpha F}}$	b ~40 mV dec^{-1}	$M - H^* + H^+ + e^- \leftrightarrow M + H_2$	$M - H^* + H_2O + e^- \leftrightarrow M + OH^- + H_2$
Tafel reaction **Here,** $\boldsymbol{b}$ **= Tafel slope** $= \dfrac{\mathbf{2.3\,RT}}{\boldsymbol{\alpha F}}$	b ~30 mV dec^{-1}	$2M\text{-}H^* \leftrightarrow 2M + H_2$	

H_2O or OH. This indirect route requires much higher energy that results in slower kinetics and, thus, higher overpotential in the system. In addition to that, the binding strength of the M–O bond of M–OH*, or M–O*, or M–OOH* becomes a crucial factor in such intermediate steps. The rate-determining steps (RDS) are calculated from the Tafel slope, where a value of 120 mV dec^{-1} indicates the RDS of the first electron transfer reaction in the OER electrochemical reaction. For the second electron transfer reaction, the Tafel slope corresponds to approx. 60 mV dec^{-1}, whereas for the third electron transfer reaction, it is around 30 mV dec^{-1}. A smaller Tafel slope value indicates faster reaction kinetics, where the RDS is the next step in this multistep OER reaction [32, 36].

9.2.2.3 *Parameters of Electrochemical Water Splitting*

There are several essential parameters for rating the electrocatalysts in the system. This includes mainly the onset potentials/overpotentials, Tafel slope, exchange current density, turnover frequency, stability, and faradaic efficiency.

Table 9.2 Acidic and alkaline electrolytes and the OER reaction pathways.

Reaction	Electrolyte	
	Acidic solution	Alkaline solution
Overall reaction	$2H_2O \rightarrow O_2 + 4H^+ + 4e^-$	$4OH^- \rightarrow O_2 + 2H_2O + 4e^-$
Intermediate reactions	$M + H_2O \rightarrow M - OH^* + H^+ + e^-$ $M - OH^* + OH^- \rightarrow M - O^* + H_2O + e^-$ Then $2\,M - O^* \rightarrow 2M + O_2$ or $M - O^* + H_2O \rightarrow MOOH^* + H^+ + e^-$ $MOOH^* + H_2O \rightarrow M + O_2 + H^+ + e^-$	$M + OH^- \rightarrow M - OH^*$ $M - OH^* + OH^- \rightarrow M - O^* + H_2O$ Then $2\,M - O^* \rightarrow 2M + O_2$ or $M - O^* + OH^- \rightarrow M - OOH^* + e^-$ $M - OOH^* + OH^- \rightarrow M + O_2 + H_2O$

i. Onset Potentials and Overpotentials (η)

The overpotential (η) is defined as the difference between the applied potential and the theoretical thermodynamic value under equilibrium conditions [34, 37–39]. It is depicted as follows:

$$\eta = E_{applied} - E_0 - iR$$

Where $E_{applied}$ is the actual applied potential, E_0 is the theoretical equilibrium value, and iR is the compensation (ohmic correction) under the same equilibration conditions.

ii. Tafel Plot

It is a plot of the polarized current density versus the overpotentials (e.g., linear sweep voltammogram as a plot of log(j) vs. η). The obtained slope on the linear portion is termed as the Tafel slope, which is expressed as:

$$\mathbf{d(\log \mathit{J})/d\eta = 2{:}303\mathit{RT\alpha nF}}$$

Where α—charge transfer coefficient, R—the ideal gas constant, T—temperature, F—Faraday constant, and n—number of electrons transferred (four in the case of OER and two in the case of HER). A small Tafel slope indicates the catalyst's high charge-transfer ability, which is considered as a prime factor to distinguish and determine the electrocatalytic activity.

iii. Turnover Frequency (TOF)

Turnover frequency (TOF) depicts the intrinsic activity of the employed catalysts. It is defined as the number of molecules that reacted at each available catalytic site per time in a set given environment [40, 41]. This can be calculated as follows [42]:

$$\mathbf{TOF} = (j_A)/(\alpha F n)$$

where j is the corresponding current density at a known overpotential, A denotes the surface area of the working electrode, α represents the number of electrons of the catalyst (electrons mol^{-1}), F is the Faraday constant (96,485.3 C mol^{-1}), and n is the moles of the catalyst's metal atoms on the electrode, which is derived from the mass (g) divided by the molality (g mol^{-1}) [43, 44].

iv. Stability

Electrocatalytic stability means the durability of the employed catalyst in the system. This includes the capability to withstand parasitic effects like corrosion and fouling while maintaining its catalytical electroactivity for a longer duration. In general, two such techniques are employed to illustrate the stability in electrochemistry. One is to look for LSV curves before and after taking repetitive thousands of cycles of CV at higher scan rates (over 3,000 CV cycles). The other is the chronoamperometry or chronopotentiometry scans for a longer time (>10 h), which indicates the alterations in current density or of potentials with time under controlled potentiodynamic conditions [37].

v. Faradaic Efficiency

Faradaic efficiency (FE) tells the selectivity of the catalyst, which is the ratio of the amount of produced gas to the estimated theoretical amount in the experiment. Two well-known methods are usually applied to calculate the FE of OER or HER. Water–gas displacement is the first method of collecting the evolved gas. Second, under a constant current density or overpotential, the generated gases can be measured using gas chromatography (GC) [37].

9.3 Synthesis Methods of Graphene

Graphene is an allotrope of carbon having a two-dimensional structure organized in a honeycomb crystal lattice in which every atom has an

sp^2-bonding with its other three atoms. Thus, it shows extraordinary electrical properties where three sigma (σ) electrons of C–C bonds give a delocalized system of π-electrons perpendicular to the honeycomb plane [15, 45–49].

Nowadays, graphene has become the center of attraction for many researchers whose area of interest is the synthesis of graphene. There are various methods reported on the exfoliation of graphite into graphene, in which the top-down approach and the bottom-up approach are the two main methods shown in Figure 9.3(a) [50–52]. Graphite is the superimposition of graphene layers bonded with van der Waals forces [53]. The top-down approach includes oxidative exfoliation–reduction, arc discharge, mechanical exfoliation, unzipping of CNT, and liquid-phase exfoliation [50]. These processes convert graphite into single, bi-, or few layer graphene by delaminating and then isolating it. To form nanosized graphene, these methods are used to break larger precursors like other carbon precursors or graphite. The top-down approach seems to be a high-quality as well as a high scalability method, but it has drawbacks in forming consistent property products, has low yield, and anticipates only fewer graphite precursors. While in the bottom-up approach, other carbon materials are used instead of graphite to synthesize graphene and its derivatives. Various methods include chemical vapor deposition (CVD) [54], epitaxial growth, substrate-free gas-phase synthesis (SFGP) [55], template routes, and total organic synthesis [56]. All these methods synthesize graphene from atomic-sized precursors. The graphene product with this approach has a large surface area and is defect-free, but it comes up with some demerits such as having a sophisticated operational setup and high production cost [56–58].

Depending on chemical or physical procedures, different categories are classified for the production of different graphene as shown in Figure 9.3(b); these are the most widely used techniques for synthesizing graphene. Various methods include arc discharge, graphite liquid-phase exfoliation, oxidative exfoliation–reduction, micromechanical exfoliation, and chemical vapor deposition (CVD) of graphene oxide exfoliation [59].

9.3.1 Mechanical Exfoliation

There are different approaches for mechanical exfoliation such as shear force vectors and normal force, and the shear force methods include the fluid dynamics method and ball milling method shown in Figure 9.4(a). The ball milling method is usually used for ceramic, cement, fertilizer, and grinding minerals, to generate mechanical force with impact and attrition

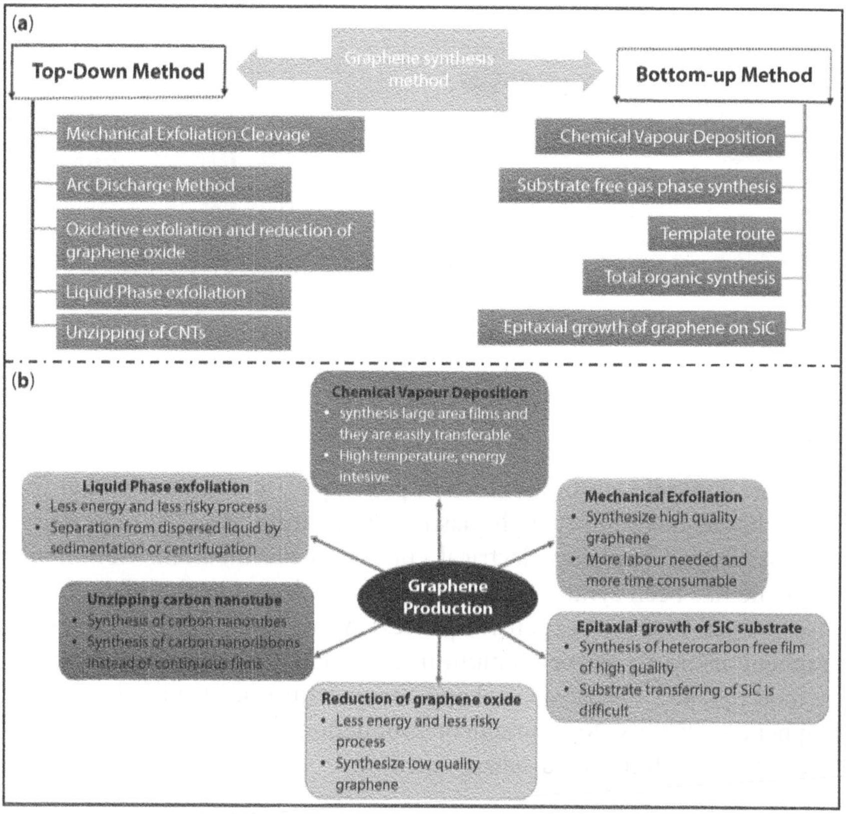

Figure 9.3 (a) Graphene synthesis methods. (b) Summary of graphene synthesis.

using grinding media, i.e., balls to convert graphite into graphene flakes. This is the basic technique to form nanosized graphene [60–63]. We can also produce composites with graphene and graphene functionalization. This process depends on a few parameters, i.e., speed, matrix, time, type of grinding media, and initial size of precursor. The process usually takes longer time, between 24 and 48 h, which is the biggest demerit for large-scale production. This factor must be overcome to make the ball milling process suitable for continuous and batch process [50, 60].

9.3.2 Arc Discharge

The arc discharge method is a resilient process for synthesizing graphene, which is a conventional method to prepare CNT and fullerene [64]. As shown in Figure 9.4(b), the reaction chamber, which consists of a cathode

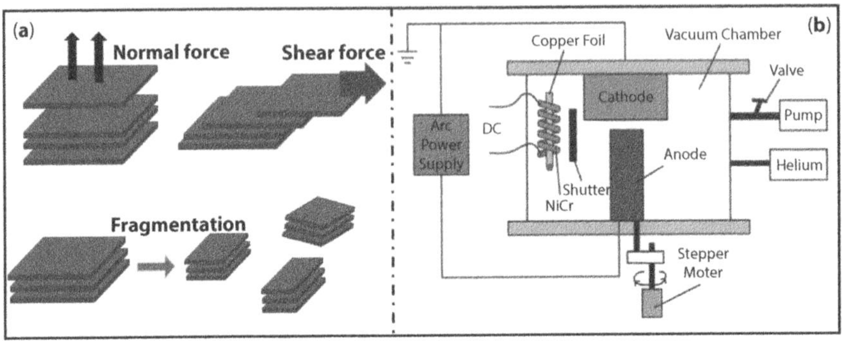

Figure 9.4 (a) Methods for the mechanical exfoliation of graphene. (b) Arc discharge method for graphene synthesis [67].

(graphite rod) and an anode (carbon precursor), is immersed either in a gas or liquid medium. A high plasma temperature (3,727°C–5,727°C) is generated upon applying electrical current in the medium. Commonly, graphene has been synthesized in vacuum conditions with different inert gases, which is a very expensive process. Wang *et al.* prepared low-cost graphene using air as the medium instead of H_2/He in the arc discharge method [65, 66]. Precise control and energy nature are required to hinder graphene scalability. Now, graphene production can be done at an affordable cost through the arc discharge method.

9.3.3 Oxidative Exfoliation Reduction

Graphene sheets or rGOs are reduced from oxidative exfoliation of graphite to graphene oxide (GO). There are four routes to the synthesis of GO, which are Brodie, Staudenmaier, Hofmann, and Hummers. Figure 9.5 shows the pathways of reaction of different methods which begin below the temperature of 100°C [68, 69]. For low-cost production, low-temperature synthesis is advantageous, but toxic gases like nitrogen dioxide (NO_2) or dinitrogen tetroxide (N_2O_4) get produced through these processes [70]. The Hummers method is used to synthesize GO as it is a faster and safe process; also, it does not generate harmful gases like chlorine dioxide (ClO_2), potassium permanganate ($KMnO_4$), or sodium nitrate ($NaNO_3$) [71]. To decrease the production cost and make it environment-friendly, the modified Hummers method is used for the synthesis of GO where only $NaNO_3$ is not used in the existing process; hence, we can obtain GO.

In a modified Hummers method, when we add a limited quantity of water to the reaction, highly oxidized GO is synthesized having good structural integrity [72]. In addition, at 95°C, the carboxyl/hydroxyl group on GO sheets gets modified by the limitation of the quantity of water in the reactions or oxidized destructive of graphite oxide when the remaining high-water content and Mn(VII) exist. At low temperatures, we can obtain a high-quality GO by this method. Strong oxidative radicals (atomic oxygen or hydroxyl radical) are formed by O_3 decomposition of Mn catalyzation, which can be generated by water oxidation having Mn(VII) in the H_2SO_4 solution of $KMnO_4$ (Figure 9.6).

9.3.4 Liquid-Phase Exfoliation (LPE)

LPE is one of the largely used methods for the synthesis of graphene. It came into consideration in 2008 [73]. LPE synthesis has three main steps: first, graphite should be dispersed in a suitable solvent, second is

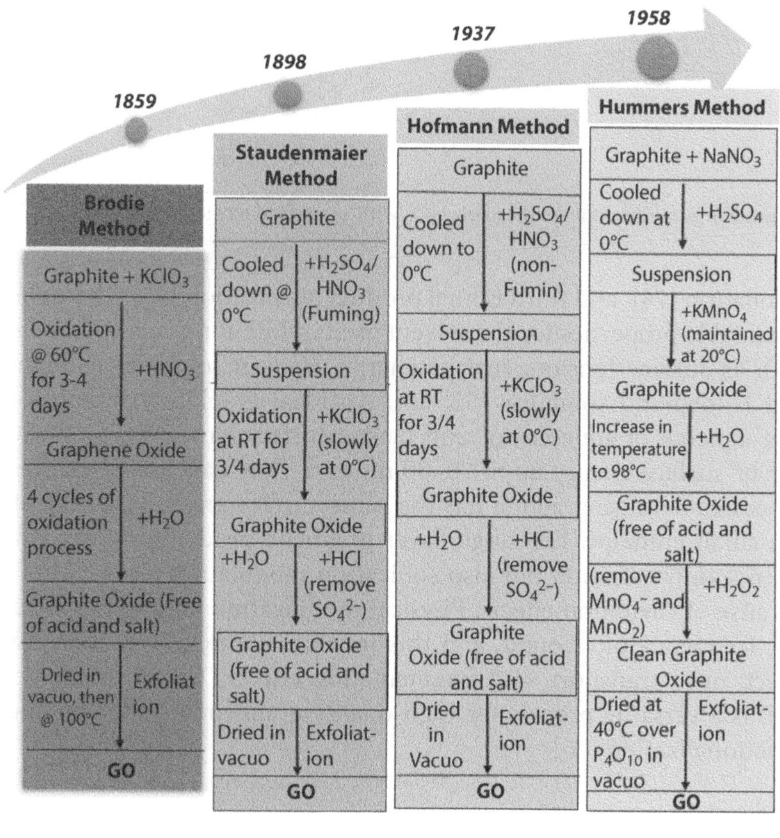

Figure 9.5 Schematic routes for graphite oxidation.

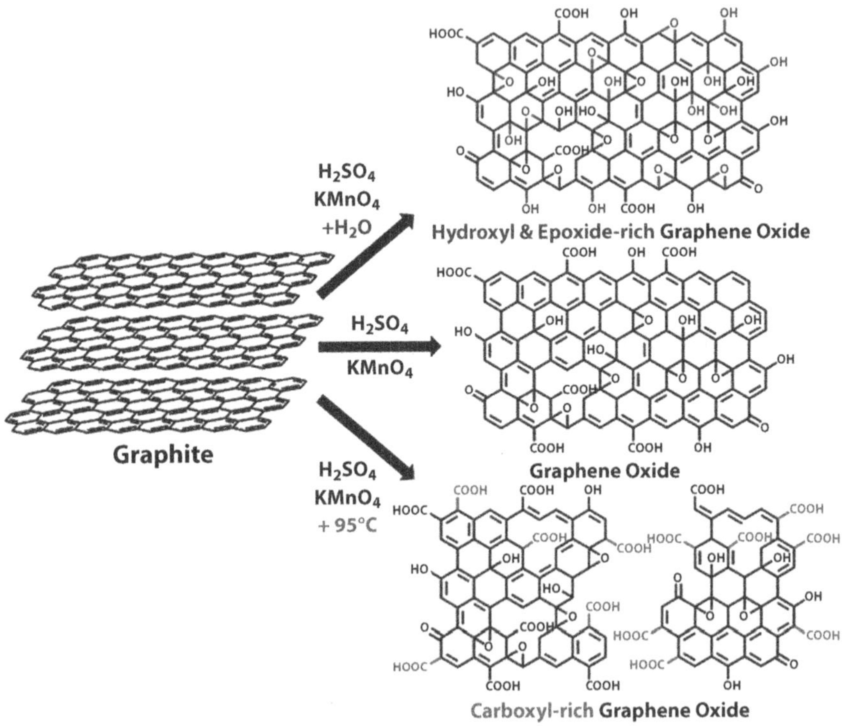

Figure 9.6 A method for synthesizing GOs with controlled oxygen group species [72].

exfoliation [74], and third is final product purification [75]. There should be suitable properties for the solvent used in LPE for graphite exfoliation such as surface tension, surface energy, Hansen solubility parameters, and Hildebrand solubility [52, 60]. An ideal graphite exfoliation uses non-aqueous or aqueous solvents having surface tension up to 40–50 mJ m^{-2} or surface energy up to 70–80 mJ m^{-2}. They have easier exfoliation processes because graphite has the same surface energy and, therefore, has small enthalpy. For single-layer graphene production from graphite, not only selection but also sonication modulation plays a vital role because of cavitation effects. Physical modification of graphite is due to low-frequency ultrasound, and this includes shock waves, well-mixing effect, mass transport, and fragmentation effect. Despite the demerits, LPE is a versatile process for high-quality graphene production and its functionalization [76].

9.3.5 Chemical Vapor Deposition (CVD)

To grow a single sheet of graphene on metallic catalysts like Cu and Ni films at high temperatures (650°C–1,000°C), CVD involves the decomposition of hydrocarbon gases like methane, ethylene, hexane, and other biomass materials [77, 78]. When the carbon precursor gets in contact with a metal catalyst's hot surface, it dissociates into hydrogen atoms and free carbon. This carbon atom diffuses from the surface and bulk metal catalyst, thus forming a graphene sheet at a certain carbon solubility limit on the metal surface. Chen *et al.* worked on the improvement of CVD processes where substrates like glass, quartz, silicon nitride, boron nitride, and silicon oxide were used [79]. CVD is advantageous as it synthesizes higher-quality graphene with high interconnected structure, large surface area, and low defects; however, it also has a few disadvantages like further purification needed to remove the remaining catalyst, high production cost, low throughput, and difficulty in transferring graphene to other substrates [80, 81].

9.4 Graphene as Electrocatalysts for Water Splitting

Carboxyl, epoxy, and hydroxy groups are found on the base plane of materials containing graphene, such as GO, whereas carboxylic groups are found toward the edges. Due to the presence of these oxide groups, diffusion of GO in different solvents increases. Under this phenomenon, GO can form bonds with other materials and possess the desired properties like high conductivity and stability. The presence of these properties motivates us to perform various modification techniques on graphene-based materials for improving HER and OER [82].

9.4.1 Heteroatom-Doped Graphene-Based Electrocatalyst

It is already known that graphene is inherently inert to obtain electrocatalytic applications, and doping or chemical modifications must be needed. The basic idea is that in a hexagonal lattice, carbon atoms can be replaced by heteroatoms. Hence, due to doping, the graphene lattice structure and composition change, making it suitable for various applications [83]. There are various techniques for tuning the chemical and electronic characteristics of graphene-based substances [84]. Graphene can be doped either by the carbonization of heteroatoms containing compounds in the presence of an indolent environment or by post-treatment techniques. Suitable gases like NH_3 and SO_2 for N doping and S doping are required for the post-treatment. Many heteroatoms, like N, P, Cl, B, Br, I, etc., are utilized as dopants to increase the electrocatalytic capabilities of graphene-based nanomaterials [82, 85]. For the replacement of expensive metallic catalysts from fuel cells and water-splitting cells, metal-free nitrogenous carbon catalysts have become a popular choice and are extensively being used [86]. It is quite possible due to the remarkable properties of metal-free nitrogenous carbon catalysts such as being cheap, having a huge surface area, having a better electrical conductivity, being CO poisoning resistant, having the required chemical stability, and being multifunctional [87]. Doping with heteroatoms (elements like N, S, P, B, I, and others) is thought to be the most effective way to introduce multiactive sites [88–90]. N doping of graphene is only done when the bonding and configuration are determined for electrocatalytic activity, and it also depends on the N-species nature [91]. *In-situ* polymerization was used by Navadeepthy *et al.* to create bifunctional electrocatalysts made of $NiFe_2O_4$ distributed on N-doped graphene [92]. The overall surface area of N-doped graphene has increased, resulting in improved electrocatalytic performance. A simple technique for synthesizing nitrogen-doped vertical graphene (N-VG) array for catalyzing HER in the acidic electrolyte was disclosed by Li *et al.* [93]. At 10 mA cm^{-2} with 290 mV overpotential and strong cycle stability, the N-VG metal-free catalyst demonstrates improved HER performance. Because of the various activity of N atoms resulting from the geometrical features (i.e., pyridinic-N, graphitic-N, and pyrrolic-N) of N-doped carbon catalysts, N doping is reported as the most effective method for creating multiactive catalysts thus far. By doping carbon-based materials with N atoms, asymmetric charge distributions on surrounding carbon atoms can be induced, which can change the electrical character of the material. As a result, it is possible to develop a significant attraction for H atoms. Furthermore, N doping may impact electron transport as well as the energy levels of the valence orbitals, making it easier to convert adsorbed H* species to molecular H_2. Although the actual cause of the catalysis is unknown, N doping has a clear favorable impact on the OER. The adsorption of negative hydroxide ions from H_2O is facilitated by the surrounding carbon atoms due to the presence of electronegative N atoms in carbon catalysts. Because of the reduction in activation energy by the charge transfer process, the breaking of the O–H bond takes place and ultimately results in the formation of O_2 during OER by the recombination of intermediate ions like O^{2-} and $-O_2^{2-}$. [94]. S and B atoms, in addition to N atoms, are utilized as dopants in GO and GQDs to modify their optoelectrical properties. Further studies support the co-doping (e.g., N-S or N-B) strategy over doping with a single atom. Due to the formation of much more active sites and the strong synergistic impact of doped heteroatoms, co-doping is preferred. Hao *et al.* carefully explored the stability and distinctive role of two-dimensional electrocatalyst for HER after theoretically designing a C vacancy and N-, S-, and P-doped penta-graphene (PG). They discovered that defection due to the addition of dopant outperformed pristine PG in terms of HER performance. Importantly, perfect HER activity can be attained with C_2 vacancies and S, P heteroatom doping, implying that defected and doped PG's catalytic characteristics may be simply and effectively controlled. Graphene can also be doped with more electropositive heteroatoms in comparison with carbon such as P and B. Graphene doped with P [95] and B improves electrocatalytic activity with unique electronic properties and its structure [96]. Another doping technique that comes into account is the dual doping of graphene which is more active than single doping of graphene. Many studies have been reported on this like dual doping of boron–nitrogen and sulfur–nitrogen. Qiao *et al.* reported on co-doping of boron and nitrogen which showed better electrocatalytic efficiency and good stability in comparison to single-doped graphene [97]. For better HER and OER, a variety of graphene-abbreviated transition metals and nanoparticles of metal alloys are shown in Table 9.3.

9.4.2 Non-Precious Transition Metal Graphene Catalysts

It has been shown that substitutional doping with transition metals can change the electrical characteristics of graphene and carbon nanotubes, which broadens their uses. It has also been observed that doping of transition metals in GO and GQDs affects their various properties such as electronic properties, surface chemistry, etc. However, doping/encapsulation also covers the active regions of the catalysts, but the true method of enhancing catalytic activity is still mysterious [18]. Moreover, transition metals have been investigated as a catalyst to upgrade the electronic characteristics of graphene-based materials [82]. Interestingly, graphene layers prevent metals from rusting in acidic and alkaline environments in this style of construction. With the use of transition metals and doping with nitrogen, graphene layers can modify electron spin density and charge distribution, increasing the number of active sites for HER. Chen *et al.* demonstrated the use of a nickel ferrite: N-graphene nanocomposite as a multifunctional electrocatalyst to improve the mechanism of O_2 and H_2 evolution in alkaline water splitting is shown in Figure 9.7. The development and implementation of metal oxide-based multifunctional electrocatalysts with long-term stability and higher activation in water splitting is a major focus of energy research. By using an *in-situ* polymerization approach, multifunctional electrocatalysts made of $NiFe_2O_4$ distributed on

Table 9.3 Overview of graphene hybrid electrocatalysts used in electrochemical water splitting.

Catalysts	Electrolyte	Tafel slope [mV dec^{-1}]	Onset potential (mV)	Overpotential at 10 mA cm^{-2} [mV]	Stability	Application	Ref.
$NiFe_2O_4$/N-graphene	0.5 M and 1 M KOH	2.9 and 93.2 for HER and OER	1.46 V vs. RHE	184 and 340 mV for HER and OER	Long-term stability	HER and OER	[92]
Nitrogen-doped vertical graphene (N-VG)	0.5 M H_2SO_4	121 for HER	–	290 for HER	Long-term stability	HER	[93]
HCNB@G-700	1.0 M KOH	66 and 58 for HER and OER	–	108 and 217 mV for HER and OER	Good stability	HER and OER	[94]
$MnFe_2O_4$/graphene	0.5 M H_2SO_4	106.4	0.79	313	Long-term stability	HER	[98]
$ZnFe_2O_4$/graphene	0.5 M H_2SO_4	114.3	0.82	315	Long-term stability	HER	[98]
MoS_2/N-RGO-220	0.5 M H_2SO4	126.6	0.205	445	Long-term stability	HER	[107]

(*Continued*)

Table 9.3 Overview of graphene hybrid electrocatalysts used in electrochemical water splitting. (*Continued*)

Catalysts	Electrolyte	Tafel slope [mV dec^{-1}]	Onset potential (mV)	Overpotential at 10 mA cm^{-2} [mV]	Stability	Application	Ref.
$NiFe_2O_4$-G	0.5 M H_2SO_4	121	0.98	259	Long-term stability	HER	[108]
$CoFe_2O_4$-G	0.5 M H_2SO_4	116.6	0.97	248.3	Long-term stability	HER	[108]
PtNPs/ $PEDOT_{0.3}SG_{0.7}$	0.5 M H_2SO_4	28.4	–	−0.040	Long-term stability	HER	[101]
$CoSe_2$/GO	0.5 M H_2SO_4	38.2	–	184	Good cycle stability	HER	[102]
Gr/AME-$NiFe_{10\%}$	0.1M KOH	46	–	519	Good stability	OER	[103]
$mCo_{0.5}Fe_{0.5}P$/rGO	1.0 M KOH	53.8 and 42.4 for HER and OER	–	190 and 250 for HER and OER	Longtime stability	HER and OER	[104]
G-INSCP	0.5 M H_2SO_4	67	–	0.09 V_{RHE}	Excellent stability	HER	[105]

(*Continued*)

Table 9.3 Overview of graphene hybrid electrocatalysts used in electrochemical water splitting. (*Continued*)

Catalysts	Electrolyte	Tafel slope [mV dec^{-1}]	Onset potential (mV)	Overpotential at 10 mA cm^{-2} [mV]	Stability	Application	Ref.
CuNi@Ni(ON)/CNT-Gr	1.0 M KOH	60 and 257 for HER and OER	1.51 V	42.1 and 410 mV for HER and OER	Good stability	HER and OER	[106]
Ni/rGO	1 M KOH	79	55 mV	–	Remarkable cycle stability	HER	[109]
0D@2D Ni_2P@NSG	1 M KOH	43 and 47 for HER and OER	1.5 V	110 and 240 for HER and OER	Good stability	HER and OER	[110]
N,S-G@CB	1 M KOH	110.5 and 90.4 for HER and OER	1.9 V	0.29 and 330 for HER and OER	Excellent stability	HER and OER	[111]

N-doped graphene were created [92]. At an alkaline electrolyte, the electrocatalyst performed admirably as a cathode in HER and an anode in OER. The nanocomposite N-doped graphene/$NiFe_2O_4$ (NGNF) achieved a current density of 10 mA cm^{-2} with a low overpotential of 184 mV in HER and 340 mV in OER, significantly superior to their immaculate counterparts. The Tafel slopes for HER and OER were reported to be 82.9 mV dec^{-1} and 93.2 mV dec^{-1}, respectively. In the form of an electrocatalyst, NGNF showed better performance compared with pure nickel ferrite and graphene/$NiFe_2O_4$ (GNF) because it possesses the desired overpotential and Tafel slopes. This demonstrates the role of graphene and N doping in graphene in pure NF activity. The N dopants and graphene in the composite induced catalytic activity and adjusted electron transport and contact with the electrolyte [92]. According to Nivetha *et al.*, solvothermal $MnFe_2O_4$/graphene and ZFe_2O_4/graphene nanocomposites (MFG and ZFG) were synthesized and tested as favorable electrocatalysts for HER [98]. During the hydrogen generation process by the hydrogen reduction technique, $MnFe_2O_4$/graphene and $ZnFe_2O_4$/graphene played a key role as electrocatalysts. When the abovementioned nanocomposites have been used in an acidic medium, they performed brilliantly with low overpotential and Tafel slope (i.e., 313 mV and 315 mV, 106.4 mV dec^{-1} and 114.3 mV dec^{-1}, respectively). MFG and ZFG had exchange current densities of 60.11 mA

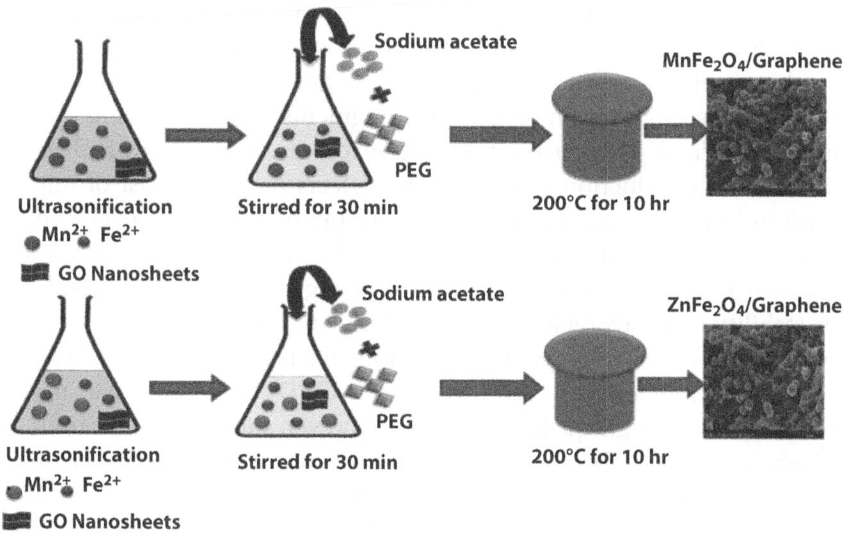

Figure 9.7 Process flow diagram of the preparation of $MnFe_2O_4$/graphene and $ZnFe_2O_4$/graphene [98].

cm^{-2} and 47 mA cm^{-2} and TOFs of 1.47 H_2 s^{-1} and 1.09 H_2 s^{-1}, respectively. Thus, the results proposed that both the composite provided a lower onset potential and fast-growing current density. A reduction in onset potential shows that the composites have strong electrocatalytic action against HER, which could be owing to the synergistic impact of ferrite/graphene composites, which promotes rapid electron transfer. For better HER and OER, a variety of graphene-abbreviated transition metals and nanoparticles of metal alloys are shown in Table 9.3.

9.4.3 Functionalized Graphene as Electrocatalysts

GO can be easily functionalized by both covalent and non-covalent strategies, but graphene is a carbon allotrope with sp^2 configuration, which makes it comparably inert. Hence, the functionalization of graphene is possible by the formation of pi–pi stacking complexation or the insertion of specific moieties onto the edge or basal planes of it [99]. Metal nanocrystals are generally deposited on graphene oxides with a large number of functional groups [82, 100]. Noble metal formation via nucleation occurs due to the presence of these functional groups and surface defects of graphene, which work as active sites or nucleation sites. Besides increasing the robustness of the hybrid nanocatalysts, these functional groups also increase the electrocatalytic activity of the catalysts through their strong interaction between graphene and the metal nanocrystals. Most of the functional groups have oxygen, and thus, graphene oxide easily gets dispersed in water as well as in other organic solvents, and this property is very important for electrocatalysts as it hinders the formation of aggregates at the time of the preparation of the catalysts. Small defective fragments of graphene and defects are unavoidable as reduced graphene oxides are derived from thermal and chemical reduction processes. The defects mainly consist of imperfections in the lattices and chemical impurities which affect the electronic properties of the reduced graphene oxide. Corrosion of graphene creates another problem as reduced graphene oxides get oxidized and, thus, loses its stability in unfavorable electrochemical conditions. Hence, for preparing the nanocrystals, graphene with minimal surface defects is necessary. In addition to chemical vapor deposition (CVD), there are other methods for yielding high-quality graphene. However, surface modifications of graphene are necessarily required to tune its surface properties as processing of graphene generally yields hydrophobic surfaces which are not suitable for anchoring of metal particles. Hence, efficient surface modification is required to improve its affinity for metal nanoparticle attachment and dispersion ability. The absence of influent functional groups, excluding those on the edges, causes a weak interaction between the graphene sheets and metal nanoparticles [101]. However, placing the nanoparticles between the graphene nanosheets and poly(3,4-ethylenedioxythiophene) (PEDOT) layer, leading to the formation of bonds, can resolve the issue to an extent. Hence, efficient hydrogen gas production can be ensured through the improvement in HER by these stabilized homogeneous metal nanoparticles. Besides obstructing the growth of the nanoparticle's agglomerates, by providing proton diffusion on the surface of the electrocatalyst, PEDOT can assist HER very well. More functionalization of the graphene nanosheets is possible by forming the sulfonate groups to enhance the performance of the electrocatalysts. There are various sulfonating agents such as sulfuric acid (SA), chlorosulfonic acid (CSA), 2-chloroethane sulfonic acid, and 4-diazoniobenzenesulfonate. CSA causes less rupture in the graphene structure than other sulfuric agents, and hence, it is a suitable candidate to form edge-selective sulfonated graphene sheets (SG) without harming its aromatic system [101]. Though platinum is an efficient electrocatalyst for HER in an acidic medium, its scarcity and high cost make it impracticable for applications. Hence, to reduce the platinum level, Abdolmaleki *et al.* decorated platinum nanoparticles (PtNPs) on poly(3,4-ethylenedioxythiophene, PEDOT) and sulfonated graphene nanosheets (SG) [101]. The synthesized nanocomposite presented low overpotential, good efficiency, and long-term durability, making it a suitable electrocatalyst for HER. After optimizing the PtNPs/$PEDOT_x$-$SG_{(1-x)}$ (x is between 0.0 and 1.0) composite, PtNPs/$PEDOT_{0.3}SG_{0.7}$ achieved the best results with low overpotential, high reduction current, and low charge transfer resistance. Yang *et al.* prepared $CoSe_2$ nanoparticles by a facile hydrothermal route. After that, a few layers of these sea-urchin-shaped $CoSe_2$ particles were attached to graphene oxide (GO) nanosheets which construct a $CoSe_2$/GO conductive network (CSGN). Different catalytic activities for the HER are observed with the change in doping concentration of graphene oxide. With 30 mg of GO doping in $CoSe_2$ (CSGN-30), the Tafel slope of CSGN-30 was 38.2 mV dec^{-1}. The obtained Tafel slope was not only much lesser than the pristine CoSe2 (98 mV dec^{-1}) but also lower than many detailed $CoSe_2$ or $NiSe_2$-based catalysts. CSGN-30 presented superior electrocatalytic movement with faster energy in hydrogen evolution reaction in comparison with the pristine $CoSe_2$. The high conductivity of graphene and the high electrochemically active surface area, along with better charge transfer between the electrolyte and the electrocatalyst, resulted in excellent HER performance [102]. Santos *et al.* achieved an efficient oxygen evolution reaction (OER) due to the efficient catalytic activity of polylactic acid (PLA)/graphene additive manufactured (3D-printed) electrodes

(Gr/AMEs) electrodeposited with Ni–Fe (oxy)hydroxide. For optimizing the Gr/AMEs' catalytic performance to achieve highly efficient OER in 0.1 M of KOH, the Fe content in the Ni–Fe (oxy)hydroxide coating was varied (5%–40%). OER catalysis happened most efficiently when the Fe content was 10%, and the OER presented an onset potential of +1.47 V and overpotential of 519 mV, respectively. The Gr/AMEs without the electrodeposited Ni–Fe (oxy)hydroxide presented negligible OER activity [103]. Huang *et al.* prepared an efficient and robust electrocatalyst for water-splitting applications. This novel nanocomposite was designed as mesoporous cobalt ferrite phosphides hybridized on reduced graphene oxide (rGO) [i.e., $mCo_{0.5}Fe_{0.5}P$/rGO]. Besides its good long-term stability, the synthesized $mCo_{0.5}Fe_{0.5}P$/rGO showed excellent catalytic activity as it resulted in a low OER overpotential of 250 mV with a current density of 10 mA cm^2. This bifunctional electrocatalyst has an abundant interface for providing large active sites, and its mesoporous structure obstructs the formation of aggregates at active sites and facilitates the fast escape of bubbles during the reaction. Additionally, the hybrid nature of rGO provides fast electron transfer and ion diffusion, and thus, the overall catalytic performance was improved [104].

9.4.4 Other Carbon Materials/Graphene Hybrids as Electrocatalysts

Nowadays, graphite is copiously available in the environment; hence, graphene materials are used as a substitute catalyst. Also, for practical execution, bifunctional catalysts are required for highly efficient water-splitting proceedings [105]. To modify the electrical properties of nanomaterials, structural engineering is a very effective method. Thus, it is possible to reveal more active centers by creating synergistic effects between components [106]. For hydrogen generation, polymer-based graphene composites were utilized. In selective media, polymers with various solventphilic units ease graphene composite dispersion, and the polymers on graphene also enhance the surface stability of graphene composites [105]. Perumal *et al.* reported poly[2-methacryloyloxy)ethyl phosphorylcholine] (MPC) stabilization with graphene iron oxide for water-splitting applications [105]. Graphene composites are extensively used for water-splitting electrocatalytic activity. MPC and poly(ethylene glycol) monomethacrylate (PEG) were polymerized on graphene surfaces with iron oxide (Fe_2O_3) NPs in the presence of *in-situ* polymerization of graphene and Fe_2O_3. In addition, composites can be synthesized using block copolymer, i.e.,

PMPC-b-PEG, and copolymer, i.e., PMPC-co-PEG (CP), called G-BCP and G-CP, respectively, shown in Figure 9.8. Then, CP and BCP polymers were used to disperse graphene in water in the appearance of IONPs. Thus, for a long period of time (2 months), the dispersion was stable for G-CP and G-INSCP. Hence, in an acidic medium, these were used as HER electrocatalysts. LSV confirmed the involvement in HER by Tafel slopes of 73 and 67 mV dec^{-1} for G-CP and G-INSCP composites, which are related to the "Volmere Heyrovsky" mechanism. Therefore, in an acidic medium, the composite G-INSCP showed exceptional electrocatalytic activity due to the existence of large active sites allocated to IONPs, charge transfer facilitated by Fe^{3+} ion synergic effect, dispersed graphene composite, composite conductivity, and polymer–graphene sheets. Also, the G-INSCP composite indicated that in acidic media it has exceptional stability. Tran *et al.* reported the synthesis of a quirky nanohybrid which includes core@shell CuNi@Ni NPs dual coordinated with heteroatoms O_2 and N_2 (CuNi@Ni(ON)NPs) and evenly gathered on porous 3D CNT graphene (CNT-Gr) [106]. HER and OER operated in 1 M of KOH have bifunctional activity when the material CuNi@Ni(ON)/CNT-Gr having electronic tunability attributes and conductive attributes showed high absorption energy for the reactant. Along with this, to improve the catalytic performance, high conductive and 3D porous CNT-Gr substrate upgraded the active material

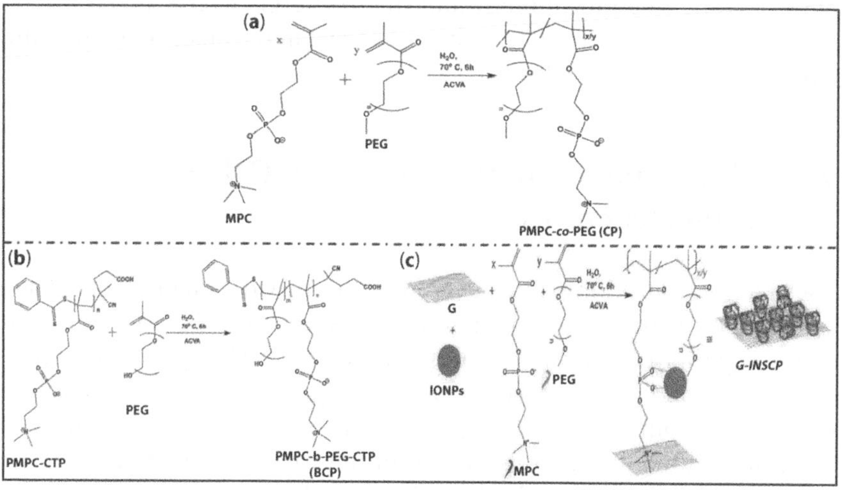

Figure 9.8 (a) Preparation of PMPC-co-PEG (CP). (b, c) Preparation of PMPC-b-PEG-CTP (BCP) and preparation of dispersion of G-INSCP using MPC and PEG monomers, respectively [105].

stabilization and the transfer of heterocharge improvement as well as the transfer of mass acceleration. For low operation cell, ~1.51 V to gain 10 mA cm^{-1} electrolyzer of alkaline used CuNi@Ni(ON)/CNT-Gr, higher than the commercially available Pt/C/RuO_2 system. In this research, CuNi@Ni(ON)/CNT-Gr has proven to be an effective catalyst for producing green H_2 gas through high-performance H_2O splitting. Oh *et al.* indicated that enhanced water splitting is produced on carbon cloth heterostructure by 3D sea-grape-like MoS_2/graphene. Because of certain properties like tunable band gap, earth abundance, and near-zero Gibbs free energy, MoS_2 is mostly used as HER electrocatalyst, but the electrochemical inert basal plane and low electrical conductivity of MoS_2 lower the HER performance. Graphene synthesized on Gr/CC and sea-grape-like morphology was produced synchronously at high temperatures using CVD and nickel electrodeposition. Thus, using CVD two-zone (MoS_2/Gr/CC), Gr/CC was hushed with MoS_2, resulting in a 3D sea-grape-like heterostructure. For 2 days at different potentials, MoS_2/Gr/CC disclosed extraordinary HER results, with low Tafel slope value (i.e., 48 mV dec^{-1}), low $\eta 10$ (91 mV, overpotential at 10 mA cm^{-2}), onset potential (50 mV), high stability in 0.5 M of H_2SO_4, and high EDLC, i.e., electrical double layer capacitance (Cdl, 239 mF cm^{-2}). The MoS_2/Gr/CC have a significantly better electrocatalytic activity than most electrocatalysts based on graphene and MoS_2 because of their 3D sea-grape-like morphology and their synergistic effects. For better HER and OER, a variety of graphene-abbreviated transition metals and nanoparticles of metal alloys are shown in Table 9.3.

9.5 Graphene in Combination with Other Nanostructures

The first-row transition elements (Fe, Mn, Ni, and Co) and metal oxides are currently receiving a lot of attention from researchers due to their abundance, superior electrochemical occurrence, high stability in basic media, low toxicity, and being environmentally benign [112, 113]. The transition metal-based spinel compounds (AB_2X_4; where X = chalcogens and A, B = metals) are highly regarded as promising catalysts for the electrolysis of water because they are affordable, involve a facile synthesis process, and occupy variable valence states during redox reactions. They also have high tuneability due to the inclusion of metallic materials and are extremely resistant to corrosion. As a result, these materials have received a lot of attention as reliable and common catalysts for the electrolysis of water [114].

9.5.1 Graphene–Bimetal

However, due to several intrinsic features, pure transition metals (M)/ metal oxides (MO_x) exhibit low thermodynamically electrocatalytic activity [115]. To mitigate this problem, carbon substrates can be coupled with (M)/(MO_x), such as activated carbon, conductive polymers, carbon nanosheets, carbon nanotubes, and graphene. These catalysts, particularly those with hierarchical nanosheet architectures, can provide a large specific surface area, high loading capacity, and high chemical/thermal stability [116].

A novel bimetallic graphene–phosphide hybrid-based catalyst was proposed by Lu *et al.* by converting a molecule with dual ligands into nanorod-like embellished (Fe-Co) P with nitrogen (N)-doped graphene nanosheets (i.e., DLC FCP@NG) [117]. This was made by phosphorization, which is a simple co-coordination reaction at room temperature. The exterior of the DLC FCP@NG has been wrapped with N-doped graphene, and the unwinding nanorod-like structure LC FCP@NG was obtained. The DLC FCP@NG composites have a higher overpotential of 189 mV and 269 mV at 10 mA cm^{-2} and a reduced Tafel slope of 66.7 mV dec^{-1} and 49.2 mV dec^{-1} for HER and OER, which results in exceptional electrocatalytic performance. Additionally, it demonstrates outstanding stability over 30 h and only requisites a relatively less voltage of about 1.63 V for achieving the current density of 10 mA cm^{-2} because it serves both as an anode and a cathode for water splitting [117]. Chaudhary *et al.* used a hydrothermal approach aided by wet chemicals to produce the multifunctional Ti_3C_2 MXene/graphene composite aerogel (abbreviated as NCSe@MGA) [118]. The hierarchical aerogel structure's 3D spatial microsphere arrangement of NCSe increased the exposed electroactive surface area. Rapid multidimensional ion-phase transport can be achieved through robust surface-to-volume ratios (S/V) and 3D porous frameworks in aerogel. In the case of electrocatalytic water splitting, 10 mA cm^{-2} of current density at overpotentials of 201 mV and 78 mV for OER and HER, respectively, can be achieved. The Tafel slope measurements as reduced to 79 mV dec^{-1} for OER and 55 mV dec^{-1} for HER indicated rapid kinetics. The electrolytic activity of NCSe@MGA was ultimately increased in comparison to NCSe@GA and NCSe due to the symbiotic effects resulting from close contact between NCSe, graphene, and MXene. A bifunctional electrocatalyst using GO doped with iron and nickel at MXene (GMX) was developed by Raj *et al.* [119]. By annealing at various temperatures in an inert environment, GMX-based electrocatalytic materials were prepared. A current density of 10 mA cm^{-2} is attained with the GMX-500, which is regarded as a competent bifunctional electrocatalyst, at reduced overpotentials of 370 mV for OER and 470 mV for HER and also low Tafel slopes of 70 and 110 mV dec^{-1}, respectively. The studies show that GMX-500 performs superbly as a bifunctional catalyst in water splitting. Sarkar *et al.* present a low-cost approach for fabricating a bifunctional electrocatalyst made of cobalt–chromium nanostructures enmeshed in an N-doped graphene tube (i.e., denoted as CoCr@NGT) [120]. As a result, CoCr@NGT can be utilized both as an anode and a cathode catalyst in a complete alkaline water electrolyzer for an extended period of time, and this delivers 10 mA cm^{-2} current density at roughly 1.58 V.

9.5.2 Graphene–Metal–Organic Frameworks

Metal–organic frameworks (MOFs) have a wide surface area and high porosity, with various applications like gas adsorption, drug delivery, sensing, and energy storage. Thus, researchers have recently focused on electrocatalytic water-splitting applications. MOF has poor catalytic activity and stability in general. In provision to enhance the catalytic activity and stability of MOFs, metal nanoparticle loading and heteroatom loading are effective methods for upgrading defects [121].

Palani *et al.* demonstrated a simple method for fabricating bifunctional electrocatalysts through the use of graphene nanosheets (i.e., GNS) sheathed with the highly refined NiCo-MOF nanocomposite, which has a cavity structure filled with N-doped defects (called NiCo-MOF@GNS) via one-pool usage of MOFs [122]. In comparison to other NiCoMOF@GNS (500, 600, 800, 900), the NiCo-MOF@GNS-700 demonstrated remarkable OER activity and less overpotential at 10 mA cm^{-2} of 420 mV and a reduced Tafel slope at 74.5 mV dec^{-1}. Gopi *et al.* testified vanadium-doped bimetallic nickel–iron nanoarray supported by a MOF@GO architecture (i.e., V-NiFe MOF/GO composite) for electrochemical water splitting via a carbonization process [123]. In both alkaline and acidic media, Ni/Fe catalysts with equal mole ratio (0.06:0.06) demonstrated high catalytic activity for OER and HER. Furthermore, the results showed that vanadium doping has a substantial synergetic effect with Ni/Fe, resulting in a tiny overpotential and high current density of 90 mV at 208 mA cm^{-2} and 210 mV at 579 mA cm^{-2} for HER at 0.5 M H_2SO_4 and OER at 1 M KOH solution, respectively. Jin *et al.* used hexahydroxytriphenylene (HHTP) ligand strategically to construct a remarkable Ni–Fe bimetallic conductivity MOF grown by the *in-situ* method on carbon cloth (i.e., $[Ni_{2.87}Fe_{0.13}(HHTP)_2]_n$/CC) by the traditional hydrothermal one-step process [124]. The ($[Ni_{2.87}Fe_{0.13}(HHTP)_2]_n$/CC) exhibits wide specific surface area, good electrical conductivity, and abundant nanopores, allowing the ($[Ni_{2.87}Fe_{0.13}(HHTP)_2]_n$/CC) to upgrade electron conductivity with additional active catalytic sites. The ($[Ni_{2.87}Fe_{0.13}(HHTP)_2]_n$/CC) catalyst exhibits considerably improved catalytic activity with less overpotential at 20 mA cm^{-2} of 310 mV on OER due to the symbiotic interaction of bimetallic Ni–Fe site and advantageous nanostructure.

9.5.3 Graphene–Spinel

Currently, several metal oxides (MO_x) [125], hydroxides [126], sulfides [127], and selenides [125, 128] have been used in electrocatalytic applications and supercapacitors. Metal selenides (MSe) such as $FeSe_2$, $NiSe_2$, $MoSe_2$, CuSe, and CoSe have emerged as sophisticated futuristic electrode materials due to their high electrical conductivity (for Se = 1×10^{-3} Sm^{-1}) and detailed electrochemical activity. Because the chemical bond energy of metal selenide is lower than oxides and selenides, sulfides can engage in their electrochemical reactions more actively than other counterparts. The electrochemical characteristics of MMSe, i.e., mixed-valent bimetallic selenides in particular, will be greatly improved by adding heterometallic ions for symbiotic effects [118].

Debata *et al.* described the production of $NiCo_2O_4$ nanostructure, a bifunctional coral sustained on rGO (reduced graphene oxide) sheets via a simple, controlled hydrothermal reaction technique [114]. This nanomaterial is a good candidate for electrochemical water splitting due to its substantial electroactive surface area, distinctive shape, and symbiotic influence of the metal ions on the electrocatalytic activity. The coral-like rGO-$NiCo_2O_4$ nanostructure has demonstrated high catalytic activity and revealed its low onset potential of 0.144 and 0.048 V, high current density of 1,550 mA cm^{-2} and 600 mA cm^{-2}, less Tafel slope of 90.63 and 46.74 mV dec^{-1}, and long-term stability in both OER as well as HER. Simultaneously using both hydrothermal and chemical reduction methods, Hanan *et al.* prepared cobalt ferrite oxide-reduced graphene oxide (Co_2FeO_4@rGO) electrocatalyst for water splitting [129]. Excellent multifunctional capabilities are displayed by the optimized Co_2FeO_4@rGO composite, which has an overpotential at 20 mA cm^{-2} of 240 mV for OER and at 10 mA cm^{-2} of 320 mV for HER, respectively. This results in a high density of active sites, high electrical conductivity, oxygen vacancies, crystal defects, and favorable electronic structure resulting through the substitution of Fe for Co atoms in binary spinel oxide phase associated with multifunctional properties of the optimized Co_2FeO_4@rGO composite. In addition, the surface features of Co_2FeO_4@rGO composites synergistically enhanced their electrocatalytic properties.

9.6 Conclusion

The method of hydrogen production via electrochemical water splitting presents a renewable approach for generating carbon-free hydrogen. To meet the future energy demand, the large-scale applicability of electrochemical water splitting became more promising and, hence, has been extensively researched with major attention on developing low-cost, highly active, and stable electrocatalytic material. While transition metal-based electrocatalysts presented great performance on water splitting, they suffered from problems like low stability, low electron transfer rate, and low conductivity, hindering their large-scale commercialization. There are many drawbacks of commercially used Pt/C (for HER) and Ru or Ir (for OER) as they are scarce and very expensive, making their large-scale application not viable, and also, they suffer from longtime stability which drove researchers to look for better alternatives. Recently, graphene-based metal and their nanohybrid structures employed in various configurations are being utilized widely. Therefore, this chapter covered electrochemical water splitting through graphene-based electrocatalytic materials in detail. These include heteroatom-doped graphene-based electrocatalysts, non-precious transition metal-doped graphene catalysts, functionalized graphene catalysts, electrocatalysts based on other carbon materials/graphene hybrids, and electrocatalysts based on graphene in combination with other nanostructures. The application of all these strategies has made considerable progress in the field with improved HER and OER activities. Despite the progress made in graphene-based electrocatalysts, there are some challenges to be addressed for their commercial viability. There is room for enhancement in the performance of graphene-based electrocatalysts. More robust synthesis processes and fabrication techniques need to be implemented to get the unique structural benefit of graphene mostly. Further simulation and theoretical studies are needed for designing new electrocatalysts based on graphene, understanding HER mechanism, and analyzing their stability and activity. To conclude, graphene-based electrocatalytic materials have presented enormous opportunities and certain challenges. The overall progress and the currently growing research interest in the topic hold a promising future in clean hydrogen production via electrochemical water splitting.

Acknowledgments

PP acknowledges the Science and Engineering Research Board for the NPDF fellowship (Sanction No. PDF/2021/001550). TS would also like to acknowledge SERB for the funding (Grant No. SRG/2019/1303).

References

1. Li, X., Xu, Y., Li, Y. *et al.*, Increasing the heteroatoms doping percentages of graphene by porous engineering for enhanced electrocatalytic activities. *J. Colloid Interface Sci.*, 577, 101–108, 2020, doi: 10.1016/J.JCIS.2020.05.089.
2. Tam, T.V., Kang, S.G., Kim, M.H. *et al.*, Novel graphene hydrogel/b-doped graphene quantum dots composites as trifunctional electrocatalysts for zn–air batteries and overall water Splitting. *Adv. Energy Mater.*, 9, 26, 1900945, 2019, doi: 10.1002/AENM.201900945.
3. Mettenbörger, A., Singh, T., Singh, A.P. *et al.*, Plasma-chemical reduction of iron oxide photoanodes for efficient solar hydrogen production. *Int. J. Hydrogen Energy*, 39, 10, 4828–4835, 2014, doi: 10.1016/J.IJHYDENE.2014.01.080.
4. Singh, T., Müller, R., Singh, J., Mathur, S., Tailoring surface states in WO3 photoanodes for efficient photoelectrochemical water splitting. *Appl. Surf. Sci.*, 347, 448–453, 2015, doi: 10.1016/J.APSUSC.2015.04.126.
5. Singh, T., Lehnen, T., Leuning, T., Mathur, S., Atomic layer deposition grown MOx thin films for solar water splitting: Prospects and challenges. *J. Vac. Sci. Technol. A: Vac. Surf. Films*, 33, 1, 010801, 2014, doi: 10.1116/1.4904729.
6. Sasinska, A., Bialuschewski, D., Islam, M.M., Singh, T., Deo, M., Mathur, S., Experimental and theoretical insights into influence of hydrogen and nitrogen plasma on the water splitting performance of ALD grown TiO_2 thin films. *J. Phys. Chem. C*, 121, 29, 15538–15548, 2017, doi: 10.1021/ACS.JPCC.7B03424/ASSET/IMAGES/LARGE/JP-2017-034244_0010.JPEG.
7. Pattanayak, P., Singh, P., Bansal, N.K. *et al.*, Recent progress in perovskite transition metal oxide-based photocatalyst and photoelectrode materials for solar- driven water splitting. *J. Environ. Chem. Eng.*, 10, 5, 108429, 2022, doi: 10.1016/J. JECE.2022.108429.
8. Yu, J., He, Q., Yang, G., Zhou, W., Shao, Z., Ni, M., Recent advances and prospective in ruthenium-based materials for electrochemical water splitting. *ACS Catal.*, 9, 11, 9973–10011, 2019, doi: 10.1021/ACSCATAL.9B02457/ASSET/ IMAGES/LARGE/CS9B02457_0022.JPEG.

9. Yue, X., Huang, S., Cai, J., Jin, Y., Shen, P.K., Heteroatoms dual doped porous graphene nanosheets as efficient bifunctional metal-free electrocatalysts for overall water-splitting. *J. Mater Chem. A*, 5, 17, 7784–7790, 2017, doi: 10.1039/ C7TA01957B.
10. Pandey, A., Mukherjee, A., Chakrabarty, S., Chanda, D., Basu, S., Interface engineering of an RGO/MoS2/Pd 2D heterostructure for electrocatalytic overall water splitting in alkaline medium. *ACS Appl. Mater Interfaces*, 11, 45, 42094–42103, 2019, doi: 10.1021/ACSAMI.9B13358/ASSET/IMAGES/LARGE/AM9B13358_0009.JPEG.
11. Stern, L.A., Feng, L., Song, F., Hu, X., Ni 2 P as a janus catalyst for water splitting: The oxygen evolution activity of Ni 2 P nanoparticles. *Energy Environ. Sci.*, 8, 8, 2347–2351, 2015, doi: 10.1039/C5EE01155H.
12. Shi, H., Liang, H., Ming, F., Wang, Z., Efficient overall water-splitting electrocatalysis using lepidocrocite VOOH hollow nanospheres. *Angew Chemie.*, 129, 2, 588–592, 2017, doi: 10.1002/ANGE.201610211.
13. Zhu, Y., Zhou, W., Zhong, Y. *et al.*, A perovskite nanorod as bifunctional electrocatalyst for overall water splitting. *Adv. Energy Mater.*, 7, 8, 1602122, 2017, doi: 10.1002/AENM.201602122.
14. Mishra, S., Ghosh, S., Singh, T., Progress in materials development for flexible perovskite solar cells and future prospects. *ChemSusChem.*, 14, 2, 512–538, 2021, doi: 10.1002/CSSC.202002095.
15. Pattanayak, P., Pramanik, N., Kumar, P., Kundu, P.P., Fabrication of cost-effective non-noble metal supported on conducting polymer composite such as copper/polypyrrole graphene oxide (Cu2O/PPy–GO) as an anode catalyst for methanol oxidation in DMFC. *Int. J. Hydrogen Energy*, 43, 25, 11505–11519, 2018, doi: 10.1016/J.IJHYDENE.2017.04.300.
16. Rudra, R. *et al.*, Conducting polymer-based microbial fuel cells. *Enzymatic Fuel Cells: Mater. Appl.*, 44, 173–187, 2019, Material Research Forum LLC, US, http://dx.doi.org/10.21741/9781644900079-8.
17. Kumar, V., Pattanayak, P., Hait, S., Applicability of emerging nanomaterials in microbial fuel cells as cathode catalysts, in: *Nanotechnol Life Sci*, pp. 643–664, 2022, Published online.
18. Du, Z., Shen, S.L., Tang, Z.H., Yang, J.H., Graphene quantum dots-based heterogeneous catalysts. *New Carbon Mater.*, 36, 3, 449–467, 2021, doi: 10.1016/S1872-5805(21)60036-7.
19. Tian, J., Chen, J., Liu, J., Tian, Q., Chen, P., Graphene quantum dot engineered nickel-cobalt phosphide as highly efficient bifunctional catalyst for overall water splitting. *Nano Energy*, 48, 284–291, 2018, doi: 10.1016/J.NANOEN.2018.03.063.
20. Kuang, P., Sayed, M., Fan, J., Cheng, B., Yu, J., 3D graphene based H2-production photocatalyst and electrocatalyst. *Adv. Energy Mater.*, 10, 14, 1903802, 2020, doi: 10.1002/AENM.201903802.
21. Guex, L.G., Sacchi, B., Peuvot, K.F. *et al.*, Experimental review: Chemical reduction of graphene oxide (GO) to reduced graphene oxide (rGO) by

aqueous chemistry. *Nanoscale.*, 9, 27, 9562–9571, 2017, doi: 10.1039/C7NR02943H.
22. Verdejo, R., Barroso-Bujans, F., Rodriguez-Perez, M.A., De Saja, J.A., Lopez-Manchado, M.A., Functionalized graphene sheet filled silicone foam nanocomposites. *J. Mater Chem.*, 18, 19, 2221–2226, 2008, doi: 10.1039/B718289A.
23. Fang, M., Wang, K., Lu, H., Yang, Y., Nutt, S., Covalent polymer functionalization of graphene nanosheets and mechanical properties of composites. *J. Mater Chem.*, 19, 38, 7098–7105, 2009, doi: 10.1039/B908220D.
24. Lee, C., Wei, X., Kysar, J.W., Hone, J., Measurement of the elastic properties and intrinsic strength of monolayer graphene. *Science (80-).*, 321, 5887, 385–388, 2008, doi: 10.1126/SCIENCE.1157996/SUPPL_FILE/LEE-SOM.PDF.
25. Shah, R., Datashvili, T., Cai, T. *et al.*, Effects of functionalised reduced graphene oxide on frictional and wear properties of epoxy resin. *Mater. Res. Innovations*, 19, 2, 97–106, 2014, https://doi.org/10.1179/1433075X14Y.0000000220.
26. Lerf, A., He, H., Forster, M., Klinowski, J., Structure of graphite oxide revisited. *J. Phys. Chem. B*, 102, 23, 4477–4482, 1998, doi: 10.1021/JP9731821/ASSET/ IMAGES/LARGE/JP9731821F00007.JPEG.
27. Wang, Z., Nelson, J.K., Hillborg, H., Zhao, S., Schadler, L.S., Graphene oxide filled nanocomposite with novel electrical and dielectric properties. *Adv. Mater.*, 24, 23, 3134–3137, 2012, doi: 10.1002/ADMA.201200827.
28. Pang, H., Chen, T., Zhang, G., Zeng, B., Li, Z.M., An electrically conducting polymer/ graphene composite with a very low percolation threshold. *Mater. Lett.*, 64, 20, 2226–2229, 2010, doi: 10.1016/J.MATLET.2010.07.001.
29. Bin, Z.H., WG, Zheng, Yan, Q. *et al.*, Electrically conductive polyethylene terephthalate/graphene nanocomposites prepared by melt compounding. *Polym. (Guildf)*, 51, 5, 1191–1196, 2010, doi: 10.1016/J.POLYMER.2010.01.027.
30. Gómez, H., Ram, M.K., Alvi, F., Villalba, P., Stefanakos, E., Kumar, A., Graphene- conducting polymer nanocomposite as novel electrode for supercapacitors. *J. Power Sources*, 196, 8, 4102–4108, 2011, doi: 10.1016/J.JPOWSOUR.2010.11.002.
31. Wei, C., Xu, Z.J., Wei, C., Xu, Z.J., The comprehensive understanding of as an evaluation parameter for electrochemical water splitting. *Small Methods*, 2, 11, 1800168, 2018, doi: 10.1002/SMTD.201800168.
32. Li, L., Wang, P., Shao, Q., Huang, X., Metallic nanostructures with low dimensionality for electrochemical water splitting. *Chem. Soc. Rev.*, 49, 10, 3072–3106, 2020, doi: 10.1039/D0CS00013B.
33. Zhu, J., Hu, L., Zhao, P., Lee, L.Y.S., Wong, K.Y., Recent advances in electrocatalytic hydrogen evolution using nanoparticles. *Chem. Rev.*, 120, 2, 851–918, 2020, doi: 10.1021/ACS.CHEMREV.9B00248/ASSET/IMAGES/MEDIUM/ CR9B00248_0039.GIF.
34. Zou, X. and Zhang, Y., Noble metal-free hydrogen evolution catalysts for water splitting. *Chem. Soc. Rev.*, 44, 15, 5148–5180, 2015, doi: 10.1039/C4CS00448E.

35. Mahmood, N., Yao, Y., Zhang, J.W., Pan, L., Zhang, X., Zou, J.J., Electrocatalysts for hydrogen evolution in alkaline electrolytes: Mechanisms, challenges, and prospective solutions. *Adv. Sci.*, 5, 2, 1700464, 2018, doi: 10.1002/ADVS.201700464.
36. Suen, N.T., Hung, S.F., Quan, Q., Zhang, N., Xu, Y.J., Chen, H.M., Electrocatalysis for the oxygen evolution reaction: Recent development and future perspectives. *Chem. Soc. Rev.*, 46, 2, 337–365, 2017, doi: 10.1039/C6CS00328A.
37. Li, S., Li, E., An, X., Hao, X., Jiang, Z., Guan, G., Transition metal-based catalysts for electrochemical water splitting at high current density: Current status and perspectives. *Nanoscale*, 13, 30, 12788–12817, 2021, doi: 10.1039/D1NR02592A.
38. Yan, Y., Xia, B.Y., Zhao, B., Wang, X., A review on noble-metal-free bifunctional heterogeneous catalysts for overall electrochemical water splitting. *J. Mater Chem. A*, 4, 45, 17587–17603, 2016, doi: 10.1039/C6TA08075H.
39. Wu, Z.P., Lu, F., Zang, S.Q. *et al.*, Non-noble-metal-based electrocatalysts toward the oxygen evolution reaction. *Adv. Funct. Mater.*, 30, 15, 1910274, 2020, doi: 10.1002/ADFM.201910274.
40. Zou, X. and Zhang, Y., Noble metal-free hydrogen evolution catalysts for water splitting. *Chem. Soc. Rev.*, 44, 15, 5148–5180, 2015, doi: 10.1039/C4CS00448E.
41. Gong, Y., Zhi, Y., Lin, Y. *et al.*, Controlled synthesis of bifunctional particle-like Mo/Mn-NixSy/NF electrocatalyst for highly efficient overall water splitting. *Dalt Trans.*, 48, 20, 6718–6729, 2019, doi: 10.1039/C9DT00957D.
42. Doyle, R.L. and Lyons, M.E.G., The oxygen evolution reaction: Mechanistic concepts and catalyst design, in: *Photoelectrochem Sol Fuel Prod From Basic Princ to Adv Devices*, pp. 41–104, January 1, 2016, Published online.
43. Anantharaj, S., Ede, S.R., Sakthikumar, K., Karthick, K., Mishra, S., Kundu, S., Recent trends and perspectives in electrochemical water splitting with an emphasis on sulfide, selenide, and phosphide catalysts of Fe, Co, and Ni: A review. *ACS Catal.*, 6, 12, 8069–8097, 2016, doi: 10.1021/ACSCATAL.6B02479/ASSET/IMAGES/LARGE/CS-2016-02479S_0016.JPEG.
44. Guo, Y., Park, T., Yi, J.W. *et al.*, Nanoarchitectonics for transition-metal-sulfide-based electrocatalysts for water splitting. *Adv. Mater.*, 31, 17, 1807134, 2019, doi: 10.1002/ADMA.201807134.
45. Pattanayak, P., Papiya, F., Kumar, V., Pramanik, N., Kundu, P.P., Deposition of Ni–NiO nanoparticles on the reduced graphene oxide filled polypyrrole: Evaluation as cathode catalyst in microbial fuel cells. *Sustain. Energy Fuels*, 3, 7, 1808–1826, 2019, doi: 10.1039/C9SE00055K.
46. Papiya, F., Pattanayak, P., Kumar, V., Das, S., Kundu, P.P., Sulfonated graphene oxide and titanium dioxide coated with nanostructured polyaniline nanocomposites as an efficient cathode catalyst in microbial fuel cells. *Mater. Sci. Eng. C*, 108, 110498, 2020, doi: 10.1016/J.MSEC.2019.110498.

47. Papiya, F., Das, S., Pattanayak, P., Kundu, P.P., The fabrication of silane modified graphene oxide supported Ni–Co bimetallic electrocatalysts: A catalytic system for superior oxygen reduction in microbial fuel cells. *Int. J. Hydrogen Energy*, 44, 47, 25874–25893, 2019, doi: 10.1016/J.IJHYDENE.2019.08.020.
48. Pattanayak, P., Pramanik, N., Papiya, F., Kumar, V., Kundu, P.P., Metal-free keratin modified poly(pyrrole-co-aniline)-reduced graphene oxide based nanocomposite materials: A promising cathode catalyst in microbial fuel cell application. *J. Environ. Chem. Eng.*, 8, 3, 103813, 2020, doi: 10.1016/J.JECE.2020.103813.
49. Papiya, F., Pattanayak, P., Biswas, A.K., Kundu, P.P., Polyaniline and sulfonated graphene oxide supported bimetallic manganese cobalt oxides as an effective and non-precious cathode catalyst in air-cathode microbial fuel cells. *J. Environ. Chem. Eng.*, 9, 5, 105992, 2021, doi: 10.1016/J.JECE.2021.105992.
50. Lee, X.J., Hiew, B.Y.Z., Lai, K.C. *et al.*, Review on graphene and its derivatives: Synthesis methods and potential industrial implementation. *J. Taiwan Inst Chem. Eng.*, 98, 163–180, 2019, doi: 10.1016/J.JTICE.2018.10.028.
51. Ambrosi, A., Chua, C.K., Bonanni, A., Pumera, M., Electrochemistry of graphene and related materials. *Chem. Rev.*, 114, 14, 7150–7188, 2014, doi: 10.1021/CR500023C/ASSET/IMAGES/LARGE/CR-2014-00023C_0020.JPEG.
52. Edwards, R.S. and Coleman, K.S., Graphene synthesis: Relationship to applications. *Nanoscale*, 5, 1, 38–51, 2012, doi: 10.1039/C2NR32629A.
53. Shams, S.S., Zhang, R., Zhu, J., Graphene synthesis: A Review. *Mater Sci. Pol.*, 33, 3, 566–578, 2015, doi: 10.1515/MSP-2015-0079.
54. Jacobberger, R.M., Machhi, R., Wroblewski, J., Taylor, B., Gillian-Daniel, A.L., Arnold, M.S., Simple graphene synthesis via chemical vapor deposition. *J. Chem. Educ.*, 92, 11, 1903–1907, 2015, doi: 10.1021/ACS.JCHEMED.5B00126/ SUPPL_FILE/ED5B00126_SI_002.DOCX.
55. Dato, A. and Frenklach, M., Substrate-free microwave synthesis of graphene: Experimental conditions and hydrocarbon precursors. *New J. Phys.*, 12, 12, 125013, 2010, doi: 10.1088/1367-2630/12/12/125013.
56. Yang, Y., Liu, R., Wu, J. *et al.*, Bottom-up fabrication of graphene on silicon/silica substrate via a facile soft-hard template approach. *Sci. Rep. 2015 51.*, 5, 1, 1–7, 2015, doi: 10.1038/srep13480.
57. Singh, V., Joung, D., Zhai, L., Das, S., Khondaker, S.I., Seal, S., Graphene based materials: Past, present and future. *Prog. Mater Sci.*, 56, 8, 1178–1271, 2011, doi: 10.1016/J.PMATSCI.2011.03.003.
58. Yazdi, G.R., Iakimov, T., Yakimova, R., Epitaxial graphene on SiC: A review of growth and characterization. *Cryst 2016*, 6, 5, 53, 2016, doi: 10.3390/CRYST6050053.
59. Hernaez, M., Zamarreño, C.R., Melendi-Espina, S., Bird, L.R., Mayes, A.G., Arregui, F.J., Optical fibre sensors using graphene based materials: A review. *Sensors 2017*, 17, 1, 155, 2017, doi: 10.3390/S17010155.

60. Yi, M. and Shen, Z., A review on mechanical exfoliation for the scalable production of graphene. *J. Mater Chem. A*, 3, 22, 11700–11715, 2015, doi: 10.1039/C5TA00252D.
61. Bastwros, M., Kim, G.Y., Zhu, C. *et al.*, Effect of ball milling on graphene reinforced Al6061 composite fabricated by semi-solid sintering. *Compos Part B Eng.*, 60, 111–118, 2014, doi: 10.1016/J.COMPOSITESB.2013.12.043.
62. Fan, X., Chang, D.W., Chen, X., Baek, J.B., Dai, L., Functionalized graphene nanoplatelets from ball milling for energy applications. *Curr. Opin. Chem. Eng.*, 11, 52–58, 2016, doi: 10.1016/J.COCHE.2016.01.003.
63. Zhuang, S., Nunna, B.B., Boscoboinik, J.A., Lee, E.S., Nitrogen-doped graphene catalysts: High energy wet ball milling synthesis and characterizations of functional groups and particle size variation with time and speed. *Int. J. Energy Res.*, 41, 15, 2535–2554, 2017, doi: 10.1002/ER.3821.
64. Kyesmen, P.I., Onoja, A., Amah, A.N., Fullerenes synthesis by combined resistive heating and arc discharge techniques. *Springerplus*, 5, 1, 1–7, 2016, doi: 10.1186/S40064-016-2994-7/FIGURES/5.
65. Li, Y., Chen, Q., Xu, K., Kaneko, T., Hatakeyama, R., Synthesis of graphene nanosheets from petroleum asphalt by pulsed arc discharge in water. *Chem. Eng J.*, 215-216, 45–49, 2013, doi: 10.1016/J.CEJ.2012.09.123.
66. Wang, Z., Li, N., Shi, Z., Gu, Z., Low-cost and large-scale synthesis of graphene nanosheets by arc discharge in air. *Nanotechnology*, 21, 17, 175602, 2010, doi: 10.1088/0957-4484/21/17/175602.
67. Shashurin, A. and Keidar, M., Synthesis of 2D materials in arc plasmas. *J. Phys. D Appl. Phys.*, 48, 31, 314007, 2015, doi: 10.1088/0022-3727/48/31/314007.
68. Chua, C.K. and Pumera, M., Chemical reduction of graphene oxide: A synthetic chemistry viewpoint. *Chem. Soc. Rev.*, 43, 1, 291–312, 2013, doi: 10.1039/ C3CS60303B.
69. Jankovský, O., Hrdličková Kučková, Š, Pumera, M., Šimek, P., Sedmidubský, D., Sofer, Z., Carbon fragments are ripped off from graphite oxide sheets during their thermal reduction. *New J. Chem.*, 38, 12, 5700–5705, 2014, doi: 10.1039/ C4NJ00871E.
70. Yu, H., Zhang, B., Bulin, C., Li, R., Xing, R., High-efficient synthesis of graphene oxide based on improved hummers method. *Sci. Rep. 2016 61*, 6, 1, 1–7, 2016, doi: 10.1038/srep36143.
71. Chen, J., Yao, B., Li, C., Shi, G., An improved hummers method for eco-friendly synthesis of graphene oxide. *Carbon N Y.*, 64, 225–229, 2013, doi: 10.1016/J.
72. Chen, J., Zhang, Y., Zhang, M. *et al.*, Water-enhanced oxidation of graphite to graphene oxide with controlled species of oxygenated groups. *Chem. Sci.*, 7, 3, 1874–1881, 2016, doi: 10.1039/C5SC03828F.
73. Hernandez, Y., Nicolosi, V., Lotya, M. *et al.*, High-yield production of graphene by liquid-phase exfoliation of graphite. *Nat. Nanotechnol. 2008 39*, 3, 9, 563– 568, 2008, doi: 10.1038/nnano.2008.215.

74. Ciesielski, A. and Samorì, P., Graphene via sonication assisted liquid-phase exfoliation. *Chem. Soc. Rev.*, 43, 1, 381–398, 2013, doi: 10.1039/C3CS60217F.
75. Xu, J., Dang, D.K., Tran, V.T. *et al.*, Liquid-phase exfoliation of graphene in organic solvents with addition of naphthalene. *J. Colloid Interface Sci.*, 418, 37–42, 2014, doi: 10.1016/J.JCIS.2013.12.009.
76. Alaferdov, A.V., Gholamipour-Shirazi, A., Canesqui, M.A., Danilov, Y.A., Moshkalev, S.A., Size-controlled synthesis of graphite nanoflakes and multi-layer graphene by liquid phase exfoliation of natural graphite. *Carbon N Y.*, 69, 525–535, 2014, doi: 10.1016/J.CARBON.2013.12.062.
77. Campos-Delgado, J., Botello-Méndez, A.R., Algara-Siller, G. *et al.*, CVD synthesis of mono- and few-layer graphene using alcohols at low hydrogen concentration and atmospheric pressure. *Chem. Phys. Lett.*, 584, 142–146, 2013, doi: 10.1016/J.CPLETT.2013.08.031.
78. Min, B.H., Kim, D.W., Kim, K.H., Choi, H.O., Jang, S.W., Jung, H.T., Bulk scale growth of CVD graphene on Ni nanowire foams for a highly dense and elastic 3D conducting electrode. *Carbon N Y.*, 80, 1, 446–452, 2014, doi: 10.1016/J. CARBON.2014.08.084.
79. Chen, X., Wu, B., Liu, Y., Direct preparation of high quality graphene on dielectric substrates. *Chem. Soc. Rev.*, 45, 8, 2057–2074, 2016, doi: 10.1039/ C5CS00542F.
80. Brownson, D.A.C. and Banks, C.E., The electrochemistry of CVD graphene: progress and prospects. *Phys. Chem. Chem. Phys.*, 14, 23, 8264–8281, 2012, doi: 10.1039/ C2CP40225D.
81. Lee, X.J., Lee, L.Y., Foo, L.P.Y., Tan, K.W., Hassell, D.G., Evaluation of carbon-based nanosorbents synthesised by ethylene decomposition on stainless steel substrates as potential sequestrating materials for nickel ions in aqueous solution. *J. Environ. Sci.*, 24, 9, 1559–1568, 2012, doi: 10.1016/ S1001-0742(11)60987-X.
82. Nemiwal, M., Zhang, T.C., Kumar, D., Graphene based electrocatalysts: Hydrogen evolution reactions and overall water splitting. *Int. J. Hydrogen Energy*, 46, 41, 21401–21418, 2021, doi: 10.1016/J.IJHYDENE.2021.04.008.
83. Deng, D., Pan, X., Yu, L. *et al.*, Toward N-doped graphene via solvothermal synthesis. *Chem. Mater.*, 23, 5, 1188–1193, 2011, doi: 10.1021/CM102666R/ SUPPL_FILE/CM102666R_SI_001.PDF.
84. Xu, H., Cheng, D., Cao, D., Zeng, X.C., A universal principle for a rational design of single-atom electrocatalysts. *Nat. Catal. 2018 15*, 1, 5, 339–348, 2018, doi: 10.1038/s41929-018-0063-z.
85. Wang, X.R., Liu, J.Y., Liu, Z.W. *et al.*, Identifying the key role of pyridinic-N–Co bonding in synergistic electrocatalysis for reversible ORR/OER. *Adv. Mater.*, 30, 23, 1800005, 2018, doi: 10.1002/ADMA.201800005.
86. Zhu, J., Hu, L., Zhao, P., Lee, L.Y.S., Wong, K.Y., Recent advances in electrocatalytic hydrogen evolution using nanoparticles. *Chem. Rev.*, 120, 2, 851–918, 2020, doi: 10.1021/ACS.CHEMREV.9B00248/ASSET/IMAGES/ MEDIUM/CR9B00248_0040.GIF.

87. Chen, Z., Wu, R., Liu, Y. *et al.*, Ultrafine Co nanoparticles encapsulated in carbon-nanotubes-grafted graphene sheets as advanced electrocatalysts for the hydrogen evolution reaction. *Adv. Mater.*, 30, 30, 1802011, 2018, doi: 10.1002/ADMA.201802011.
88. Jiang, K., Siahrostami, S., Zheng, T. *et al.*, Isolated Ni single atoms in graphene nanosheets for high-performance CO2 reduction. *Energy Environ. Sci.*, 11, 4, 893–903, 2018, doi: 10.1039/C7EE03245E.
89. Shen, Y., Zhou, Y., Wang, D., Wu, X., Li, J., Xi, J., Nickel–Copper alloy encapsulated in graphitic carbon shells as electrocatalysts for hydrogen evolution reaction. *Adv. Energy Mater.*, 8, 2, 1701759, 2018, doi: 10.1002/AENM.201701759.
90. Ma, R., Lin, G., Zhou, Y. *et al.*, A review of oxygen reduction mechanisms for metal-free carbon-based electrocatalysts. *NPJ Comput. Mater. 2019 51*, 5, 1, 1–15, 2019, doi: 10.1038/s41524-019-0210-3.
91. Lai, L., Potts, J.R., Zhan, D. *et al.*, Exploration of the active center structure of nitrogen-doped graphene based catalysts for oxygen reduction reaction. *Energy Environ. Sci.*, 5, 7, 7936–7942, 2012, doi: 10.1039/C2EE21802J.
92. Navadeepthy, D., Rebekah, A., Viswanthan, C., Ponpandian, N., Boosting the kinetics of oxygen and hydrogen evolution in alkaline water splitting using nickel ferrite/N-graphene nanocomposite as a bifunctional electrocatalyst. *Int. J. Hydrogen Energy*, 46, 41, 21512–21524, 2021, doi: 10.1016/J.
93. Li, Y., Ai, C., Deng, S. *et al.*, Nitrogen doped vertical graphene as metal-free electrocatalyst for hydrogen evolution reaction. *Mater. Res. Bull.*, 134, 111094, 2021, doi: 10.1016/J.MATERRESBULL.2020.111094.
94. Begum, H., Ahmed, M.S., Jung, S., Hollow carbon nanoballs on graphene as metal-free catalyst for overall electrochemical water splitting. *Adv. Mater. Interfaces*, 8, 19, 2101265, 2021, doi: 10.1002/ADMI.202101265.
95. Zhang, C., Mahmood, N., Yin, H., Liu, F., Hou, Y., Synthesis of phosphorus-doped graphene and its multifunctional applications for oxygen reduction reaction and lithium ion batteries. *Adv. Mater.*, 25, 35, 4932–4937, 2013, doi: 10.1002/ADMA.201301870.
96. Sheng, Z.H., Gao, H.L., Bao, W.J., Bin, W.F., Xia, X.H., Synthesis of boron doped graphene for oxygen reduction reaction in fuel cells. *J. Mater Chem.*, 22, 2, 390–395, 2011, doi: 10.1039/C1JM14694G.
97. Zheng, Y., Jiao, Y., Ge, L., Jaroniec, M., Qiao, S.Z., Two-step boron and nitrogen doping in graphene for enhanced synergistic catalysis. *Angew Chemie Int. Ed.*, 52, 11, 3110–3116, 2013, doi: 10.1002/ANIE.201209548.
98. Nivetha, R. and Grace, A.N., Manganese and zinc ferrite based graphene nanocomposites for electrochemical hydrogen evolution reaction. *J. Alloys Compd.*, 796, 185–195, 2019, doi: 10.1016/J.JALLCOM.2019.05.021.
99. Maio, A., Pibiri, I., Morreale, M., La Mantia, F.P., Scaffaro, R., An overview of functionalized graphene nanomaterials for advanced applications. *Nanomater 2021*, 11, 7, 1717, 2021, doi: 10.3390/NANO11071717.

100. Xia, B.Y., Yan, Y., Wang, X., Lou, X.W., Recent progress on graphene based hybrid electrocatalysts. *Mater Horizons*, 1, 4, 379–399, 2014, doi: 10.1039/C4MH00040D.
101. Abdolmaleki, A., Mohamadi, Z., Ensafi, A.A., Atashbar, N.Z., Rezaei, B., Efficient and stable HER electrocatalyst using Pt-nanoparticles@ poly(3,4–ethylene dioxythiophene) modified sulfonated graphene nanocomposite. *Int. J. Hydrogen Energy*, 43, 17, 8323–8332, 2018, doi: 10.1016/J.IJHYDENE.2018.03.142.
102. Yang, B., Huang, Z., Wu, H. *et al.*, Sea urchin-like CoSe2 nanoparticles modified graphene oxide as an efficient and stable hydrogen evolution catalyst. *J. Electroanal Chem.*, 907, 116037, 2022, doi: 10.1016/J.JELECHEM.2022.116037.
103. dos Santos, P.L., Rowley-Neale, S.J., Ferrari, A.G.M., Bonacin, J.A., Banks, C.E., Ni–Fe (Oxy)hydroxide modified graphene additive manufactured (3D-Printed) electrochemical platforms as an efficient electrocatalyst for the oxygen evolution reaction. *ChemElectroChem.*, 6, 22, 5633–5641, 2019, doi: 10.1002/ CELC.201901541.
104. Huang, Y., Tian, F., Liu, Y. *et al.*, Mesoporous cobalt ferrite phosphides/ reduced graphene oxide as highly effective electrocatalyst for overall water splitting. *J. Colloid Interface Sci.*, 605, 667–673, 2022, doi: 10.1016/J.JCIS.2021.07.117.
105. Perumal, S., Atchudan, R., Cheong, I.W., Poly[2-(methacryloyloxy)ethyl phosphorylcholine]-stabilized graphene-iron oxide composites for water splitting. *Int. J. Hydrogen Energy*, 46, 18, 10850–10861, 2021, doi: 10.1016/J.IJHY DENE.2020.12.156.
106. Tran, D.T., Hoa, V.H., Prabhakaran, S., Kim, D.H., Kim, N.H., Lee, J.H., Activated CuNi@Ni core@shell structures via oxygen and nitrogen dual coordination assembled on 3D CNTs-graphene hybrid for high-performance water splitting. *Appl. Catal. B Environ.*, 294, 120263, 2021, doi: 10.1016/J.
107. Rosalbino, F., Scavino, G., Actis Grande, M., Electrocatalytic activity of Ni–Fe–M (M = Cr, Mn, Cu) sintered electrodes for hydrogen evolution reaction in alkaline solution. *J. Electroanal Chem.*, 694, 114–121, 2013, doi: 10.1016/J.
108. Nivetha, R., Chella, S., Kollu, P., Jeong, S.K., Bhatnagar, A., Andrews, N.G., Cobalt and nickel ferrites based graphene nanocomposites for electrochemical hydrogen evolution. *J. Magn. Magn. Mater.*, 448, 165–171, 2018, doi: 10.1016/J.
109. Franceschini, E.A. and Lacconi, G.I., Synthesis and performance of nickel/reduced graphene oxide hybrid for hydrogen evolution reaction. *Electrocatalysis*, 9, 1, 47–58, 2018, doi: 10.1007/S12678-017-0415-5/FIGURES/9.
110. Suryawanshi, U.P., Ghorpade, U.V., Lee, D.M. *et al.*, Colloidal Ni2P nanocrystals encapsulated in heteroatom-doped graphene nanosheets: A synergy of 0D@2D heterostructure toward overall water splitting. *Chem. Mater.*,

33, 1, 234–245, 2021, doi: 10.1021/ACS.CHEMMATER.0C03543/ASSET/IMAGES/LARGE/CM0C03543_0007.JPEG.

111. Li, X., Duan, X., Han, C. *et al.*, Chemical activation of nitrogen and sulfur co-doped graphene as defect-rich carbocatalyst for electrochemical water splitting. *Carbon N Y.*, 148, 540–549, 2019, doi: 10.1016/J.CARBON.2019.04.021.
112. Trotochaud, L., Young, S.L., Ranney, J.K., Boettcher, S.W., Nickel-iron oxyhydroxide oxygen-evolution electrocatalysts: The role of intentional and incidental iron incorporation. *J. Am. Chem. Soc.*, 136, 18, 6744–6753, 2014, doi: 10.1021/ JA502379C/SUPPL_FILE/JA502379C_SI_001.PDF.
113. Singh, A. and Spiccia, L., Water oxidation catalysts based on abundant 1st row transition metals. *Coord Chem. Rev.*, 257, 17-18, 2607–2622, 2013, doi:10.1016/J. CCR.2013.02.027.
114. Debata, S., Patra, S., Banerjee, S., Madhuri, R., Sharma, P.K., Controlled hydrothermal synthesis of graphene supported NiCo2O4 coral-like nanostructures: An efficient electrocatalyst for overall water splitting. *Appl. Surf Sci.*, 449, 203–212, 2018, doi: 10.1016/J.APSUSC.2018.01.302.
115. Yang, L., Yu, J., Wei, Z. *et al.*, Co-N-doped MoO2 nanowires as efficient electrocatalysts for the oxygen reduction reaction and hydrogen evolution reaction. *Nano Energy*, 41, 772–779, 2017, doi: 10.1016/J.NANOEN.2017.03.032.
116. Yaseen, W., Xie, M., Yusuf, B.A. *et al.*, Hierarchical Co/MoO2@N-doped carbon nanosheets derived from waste lotus leaves for electrocatalytic water splitting. *Int. J. Hydrogen Energy*, 47, 35, 15673–15686, 2022, doi: 10.1016/J.
117. Lu, Y., Chen, Y., Srinivas, K. *et al.*, Employing dual-ligand co-coordination compound to construct nanorod-like Bi-metallic (Fe, Co)P decorated with nitrogen-doped graphene for electrocatalytic overall water splitting. *Electrochim Acta*, 350, 136338, 2020, doi: 10.1016/J.ELECTACTA.2020.136338.
118. Chaudhary, K., Zulfiqar, S., Somaily, H.H., Aadil, M., Warsi, M.F., Shahid, M., Rationally designed multifunctional Ti3C2 MXene@Graphene composite aerogel integrated with bimetallic selenides for enhanced supercapacitor performance and overall water splitting. *Electrochim Acta*, 431, 141103, 2022, doi: 10.1016/J.ELECTACTA.2022.141103.
119. Raj, S.K., Bhadu, G.R., Upadhyay, P., Kulshrestha, V., Three-dimensional Ni/Fe doped graphene oxide @ MXene architecture as an efficient water splitting electrocatalyst. *Int. J. Hydrogen Energy*, 47, 99, 41772–41782, June 11, 2022. Published online. https://doi.org/10.1016/j.ijhydene.2022.05.204
120. Sarkar, B., Barman, B.K., Nanda, K.K., Non-precious bimetallic CoCr nanostructures entrapped in bamboo-like nitrogen-doped graphene tube as a robust bifunctional electrocatalyst for total water splitting. *ACS Appl. Energy Mater.*, 1, 3, 1116–1126, 2018, doi: 10.1021/ACSAEM.7B00233/ASSET/IMAGES/LARGE/AE-2017-00233S_0006.JPEG.
121. Li, H., Li, L., Lin, R.B. *et al.*, Porous metal-organic frameworks for gas storage and separation: Status and challenges. *EnergyChem.*, 1, 1, 100006, 2019, doi: 10.1016/J.ENCHEM.2019.100006.

122. Palani, R., Anitha, V., Karuppiah, C. *et al.*, Imidazolatic-framework bimetal electrocatalysts with a mixed-valence surface anchored on an rGO matrix for oxygen reduction, water splitting, and dye degradation. *ACS Omega*, 6, 24, 16029–16042, 2021, doi: 10.1021/ACSOMEGA.1C01870/ASSET/IMAGES/MEDIUM/AO1C01870_M016.GIF.
123. Gopi, S., Panda, A., Ramu, A.G., Theerthagiri, J., Kim, H., Yun, K., Bifunctional electrocatalysts for water splitting from a bimetallic (V doped-NixFey) metal–organic framework MOF@graphene oxide composite. *Int. J. Hydrogen Energy*, 47, 100, 42122–42135, May 28, 2021. Published online. https://doi.org/10.1016/j.ijhydene.2021.05.028
124. Jin, Y., Zhang, T., Pan, N. *et al.*, Surface functionalization of carbon cloth with conductive Ni/Fe-MOFs for highly efficient oxygen evolution. *Surf. Interfaces*, 33, 102294, 2022, doi: 10.1016/J.SURFIN.2022.102294.
125. Wang, A.Y., Chaudhary, M., Lin, T.W., Enhancing the stability and capacitance of vanadium oxide nanoribbons/3D-graphene binder-free electrode by using VOSO4 as redox-active electrolyte. *Chem. Eng J.*, 355, 830–839, 2019, doi: 10.1016/J.CEJ.2018.08.214.
126. Liu, Y., Cao, X., Cui, L. *et al.*, Zn-Ni-Co trimetallic carbonate hydroxide nanothorns branched on Cu(OH)2 nanorods array based on Cu foam for high-performance asymmetric supercapacitors. *J. Power Sources*, 437, 226897, 2019, doi: 10.1016/J.JPOWSOUR.2019.226897.
127. Lin, T.W., Tsai, H.C., Chen, T.Y., Shao, L.D., Facile and controllable one-pot synthesis of hierarchical co9s8 hollow microspheres as high-performance electroactive materials for energy storage and conversion. *ChemElectroChem.*, 5, 1, 137–143, 2018, doi: 10.1002/CELC.201700886.
128. Vidhya, M.S., Yuvakkumar, R., Ravi, G., Pannipara, M., Al-Sehemi, A.G., Velauthapillai, D., Hydrothermal synthesis of Cu2Se–CoSe nanograin for electrochemical supercapacitor applications. *Appl. Nanosci.*, 11, 6, 1881–1888, 2021, doi: 10.1007/S13204-021-01890-9/TABLES/1.
129. Hanan, A., Shu, D., Aftab, U. *et al.*, Co2FeO4@rGO composite: Towards trifunctional water splitting in alkaline media. *Int. J. Hydrogen Energy*, 47, 80, 33919–33937, 2022, doi: 10.1016/J.IJHYDENE.2022.07.269.

10

Graphene Electrocatalysts: New Insights Into the Current State of Water Splitting

R. Rajalakshmi, A. Rebekah and N. Ponpandian*

Department of Nanoscience and Technology, Bharathiar University, Coimbatore, India

Abstract

Due to the high use of fossil fuels, environmental degradation, and effective use of excess energy from intermittent renewable energy sources, water splitting has emerged as the most promising hydrogen production method in the 21st century. Water electrolysis is a method for producing hydrogen fuel that is cleanly produced by electrochemical water splitting. With the help of the required capable electrocatalysts, two essential reactions—hydrogen evolution reaction (HER) and oxygen evolution reaction (OER)—complete the electrochemical water-splitting process. Even though noble metals dominate the field of electrocatalysts, their cost needs attention. Given its prominence as a carbon allotrope, graphene supports and improves the electrochemical performance of both its HER (anodic) and OER (cathodic) counterparts for electrochemical water splitting. A tremendous amount of work has been put forward in the past few decades by graphene-based electrocatalysts as support materials with noble metals, nanocomposites, and even hybrid architectures for efficient overall water splitting serving as a bifunctional electrocatalyst. These could be effectively attributed to the intrinsic property of the graphene components, their 2D structure, and excellent conducting behavior. This chapter focuses on the performance of graphene-based electrocatalysts employed for HER, OER, and overall water splitting.

Keywords: Electrochemical, water splitting, HER, OER, mechanism, graphene, carbon, electrocatalyst

**Corresponding author*: ponpandian@bud.edu.in

Sudheesh K. Shukla, Chaudhery Mustansar Hussain, Santanu Patra, Meenakshi Choudhary (eds.) Electrocatalytic Materials for Renewable Energy, (271–302) © 2024 Scrivener Publishing LLC

10.1 Introduction

The urgent need to replace traditional energy sources with eco-friendly renewable energy resources has become imperative due to severe environmental problems caused by global warming and the increasing global energy demand. The demand for energy is expected to reach 30–50 TW by 2050 [1, 2]. Although renewable energy resources like wind, solar, hydrothermal, geothermal, and tidal offer sustainable solutions, the fluctuations caused by seasonal changes necessitate the expansion of competent and cost-effective energy conversion and storage devices [3]. The world's energy situation is still worrisome, as fossil fuels account for 83% of the total energy resource. These fuels release carbon dioxide, which contributes 76% of yearly harmful gas releases. This increase in emissions has resulted in negative impacts on the surroundings and local health, including the degradation of marine life and a rise in environmental catastrophes. It is therefore crucial to immediately investigate alternate fuel sources that are both eco-friendly and economically feasible [4].

Solar energy, with a substantial magnitude available, is viewed as an alluring source of renewable power. Photovoltaic (PV) cells, also referred to as solar cells, offer a direct method of converting solar energy into electricity. As of 2018's end, the worldwide accumulated capacity of installed PV reached approximately 512 GW, showcasing an annual growth rate of 30% [5]. However, there is a drawback to solar cells because the electricity generated is highly influenced by the presence of adequate sunlight. It poses a challenge for power managers to distribute the inconsistency of solar energy. Therefore, the widespread utilization of solar energy necessitates an efficient energy storage solution. The storage of the sun's energy chemically through bonds is a desirable option for energy storage. The photons from the visible light carry an energy range of 1–3 eV, which is sufficient for various chemical synthesis procedures [6]. Chemical fuels are advantageous over other energy storage systems due to their great storage capabilities and ease of transport. Chemical fuels offer a convenient and efficient solution for energy storage and transportation needs [7, 8].

Hydrogen has emerged as a promising clean fuel option due to its extraordinary gravimetric density and lack of harmful by-products. However, the efficient operation of eco-friendly energy conversion technologies like metal–air batteries and water electrolyzers requires an effective electrocatalyst, while a clean energy source is necessary for powering these systems. Hydrogen's potential as a transportation fuel and its high gravimetric energy density make it an attractive option for meeting current and future energy needs [9]. Furthermore, hydrogen can generate electricity through OERs, producing only water as a by-product. Additionally, it plays a critical role in the production of various valuable chemicals. Unfortunately, hydrogen's low volumetric density and limited availability pose challenges to its production [10]. Currently, the vast majority (96%) of hydrogen production is reliant on fossil fuels, while the remaining 4% is from renewable sources. Nonetheless, research is ongoing to enhance the production of sustainable hydrogen economically, to become competitive with fossil fuel-based production [11, 12].

10.2 Overview of Electrochemical Water Splitting

Water decomposition into hydrogen and oxygen, known as water splitting, is a process that can be performed using various methods such as electrolysis, thermolysis, photoelectrolysis, solar light splitting, and photovoltaic and photochemical reactions [13]. Hydrogen production through water splitting is a sustainable and eco-friendly means of generating gas. The methods of water splitting can be classified into three systems:

(1) Electrolytic, where the water molecule is broken down into hydrogen and oxygen through an electrical current, with the release of greenhouse gases depending on the energy source used.
(2) Thermal, where energy from fossil fuels, biomass, and others are converted into hydrogen through a thermochemical reaction.
(3) Photonic, where the sun's energy is utilized to convert water into hydrogen and oxygen and is considered the most sustainable and advanced method for producing gas, with photoelectrochemical and photobiological processes as examples.

Hydrogen production by way of water splitting, either through the electrochemical or photoelectrochemical method, is deemed a desirable solution for retaining solar energy as it can be stored, transported, and utilized as needed. The reverse reaction, which results in the creation of only water, also makes this process eco-friendly. This reaction requires a minimum potential of 1.23 V to carry out the splitting of water molecules into hydrogen and oxygen [14].

$$2H_2O\ (l)\ O_2\ (g) + 2H_2\ (g)\ \Delta G° = +237.1\ \text{kJ mol}^{-1},\ E_o = 1.23\ \text{V} \tag{10.1}$$

To make the water-splitting process into hydrogen and oxygen happen at a significant speed, extra energy is needed. To overcome this, either light energy or electrical energy is utilized as a driving force, making the reaction occur at a noticeable pace. This simple process is one of the efficient methods of obtaining hydrogen as a fuel source.

Water splitting through electricity is made up of two half-cell reactions separated by a membrane. These reactions involve two electrochemical processes known as the hydrogen evolution reaction (HER) and oxygen evolution reaction (OER), occurring at the cathode and anode, respectively [15]. However, the slow rate of these two processes demands an increased overpotential to achieve a substantial catalytic current density, leading to reduced energy conversion efficiency (Figure 10.1). Thus, the potential for water splitting under standard conditions is considerably greater than 1.23 V represented as

$$E_{op} = 1.23\ \text{V} + +\eta_c + \eta_{other} \tag{10.2}$$

In the equation, "η_a" denotes the anode overpotential and "η_c" the cathode overpotential, while "η_{other}" is the summation of voltage drops across other resistances from the membrane, contact solution, etc. in the working cell. The overpotentials "η_a" and "η_c" arise from the high energies required to form reaction intermediates on the electrode surface. The two half-reactions can be expressed differently depending on the reaction conditions as follows:

(a) In acidic solution

$$\text{Cathode: } 2H^+ + 2e^- \rightarrow H_2 \tag{10.3}$$

$$\text{Anode: } H_2O \rightarrow 2H^+ + 1/2O_2 + 2e^- \tag{10.4}$$

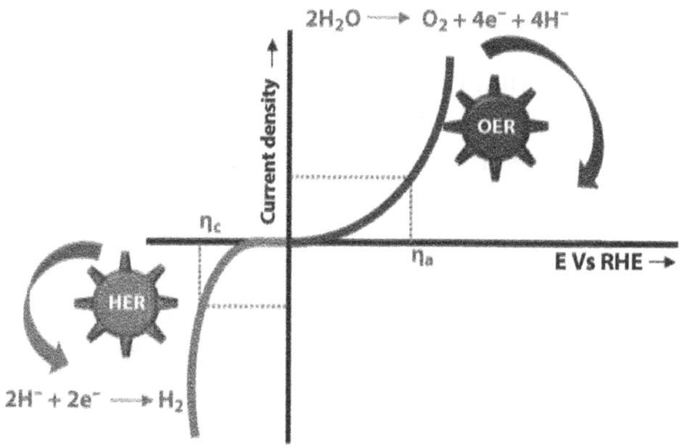

Figure 10.1 Hydrogen evolution and oxygen evolution reaction with their respective polarization curves.

(b) In neutral solution

$$\text{Cathode: } 2H_2O + 2e^- \rightarrow H_2 + 2OH^- \tag{10.5}$$

$$\text{Anode: } H_2O \rightarrow 2H^+ + 1/2O_2 + 2e^- \tag{10.6}$$

(c) In alkaline solution

$$\text{Cathode: } 2H_2O + 2e^- \rightarrow H_2 + 2OH^- \tag{10.7}$$

$$\text{Anode: } 2 \rightarrow OH^- \; H_2O + 1/2O_2 + 2e^- \tag{10.8}$$

10.2.1 Hydrogen Evolution Reaction (HER)

HER reactions are conducted in an acidic or alkaline medium, so accordingly, in different pH media, electron transfer steps are different as depicted in Figure 10.2. In an acidic medium, the HER reaction is facile because abundant protons are offered, and it proceeds through two steps. The primary phase is the Volmer phase, and it is also known as the electrochemical hydrogen adsorption step, where H^+ ions are reduced by receiving electrons from an electrode and adsorbed on the active catalytic sites

($H^+_{(aq)} + e^- = H_{ads}$), that is known as absorbed active species (H_{ads}). In the second step, H_2 desorption may occur following two different steps: one is the Heyrovsky step (electrochemical desorption) and the other one is the Tafel step (chemical desorption), as per the H_{ads} coverage on the catalyst surface. The electrochemical desorption step (Heyrovsky step) occurs only when the H_{ads} coverage is minimum, and the surface of the electrode has enough free active sites. In this case, the H_{ads} reacts with another H^+ and e^- (from the electrode) to evolve an H_2 molecule ($H_{ads} + H^+_{(aq)} + e^- = H_2$). The chemical desorption step occurs (Tafel step) when the coverage of H_{ads} on the catalyst surface is abundant. Then, two nearby absorbed active species (H_{ads}) easily react chemically to produce H_2 gas ($2H_{ads} = H_2$). The efficiency of the catalyst can be directly correlated with H_{ads} species on the catalyst surface. Under the alkaline condition, H_2O behaves as the proton donor. Table 10.1 shows the respective Tafel slope equations in different pH media [16]. A low concentration of proton in an alkaline medium increases the kinetic barrier. Hence, HER becomes more sluggish in basic pH.

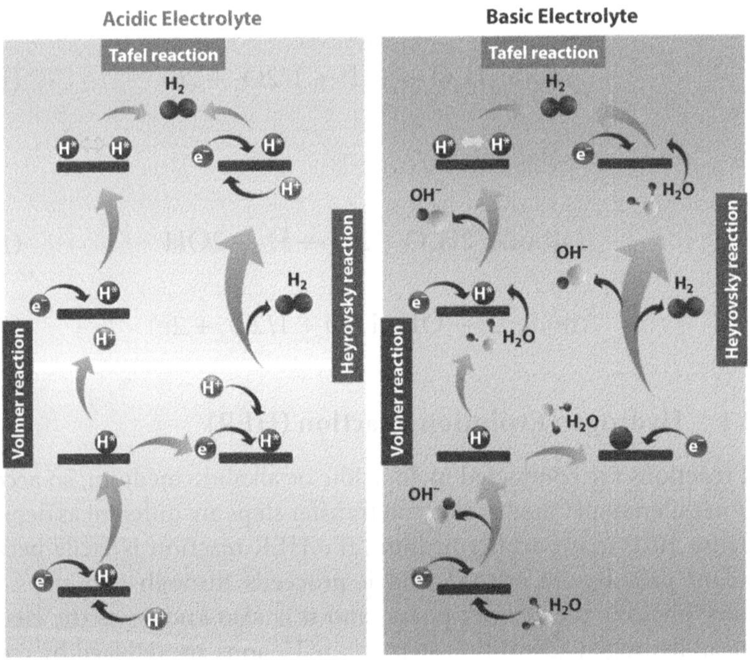

Figure 10.2 Reaction pathways of HER in acidic and alkaline electrolytes.

Table 10.1 Reaction mechanism of hydrogen evolution reaction with the respective Tafel slope equations.

	Acid	Alkaline	Tafel slope
Overall	$^* + 2H^+ + 2e^- \rightarrow H_2$	$^* + 2H_2O + 2e^- \rightarrow H_2 + 2OH^-$	$h = a + b \log(-j)$
Volmer	$^* + H^+ + e^- \rightarrow H^*$	$^* + H_2O + e^- \rightarrow H^* + OH^-$	$b = 2.3RT/aF$ ~120 mV/dec
Heyrovsky	$^* + H^+ + e^- + H^* \rightarrow H_2 + ^*$	$^* + H_2O + e^- + H^* \rightarrow H_2 + OH^- + ^*$	$B = 2.3RT/(1 + a)F$ ~40 mV/dec
Tafel	$2H^* \rightarrow H_2 + 2^*$	$2H^* \rightarrow H_2 + 2^*$	$b = 2.3RT/2F$ ~30 mV/dec
Where "*" denotes the active site on the surface of the electrocatalyst.			

10.2.2 Oxygen Evolution Reaction (OER)

The process of catalytic water oxidation involves four steps. The first step involves transferring an electron from the catalyst to the oxidizing agent, which can be facilitated by incorporating strong electron-donating ligands into the catalyst. In the second step, the catalyst must have a high oxidizing potential for water oxidation to occur. The third step is the formation of the O–O bond, which can occur through either a water nucleophilic attack on the oxo group of the metal complex or through the interaction of two M–O species to form a peroxo-intermediate. The final step is the release of molecular oxygen, which is a slow reaction [17]. The reaction mechanism for OER in both acidic and alkaline conditions is shown in Figure 10.3.

Under acidic conditions, the overall water oxidation reaction is as follows:

$$H_2O\ (l) \rightarrow O_2\ (g) + 4H^+ + 4e^-\ \Delta G_o = 4.92\ \text{eV} \quad (10.9)$$

The reaction kinetics in an acidic environment for oxygen evolution reaction takes place in four steps:

$$H_2O\ (l) + {*} \rightarrow HO{*} + H^+ + e^- \quad (10.10)$$

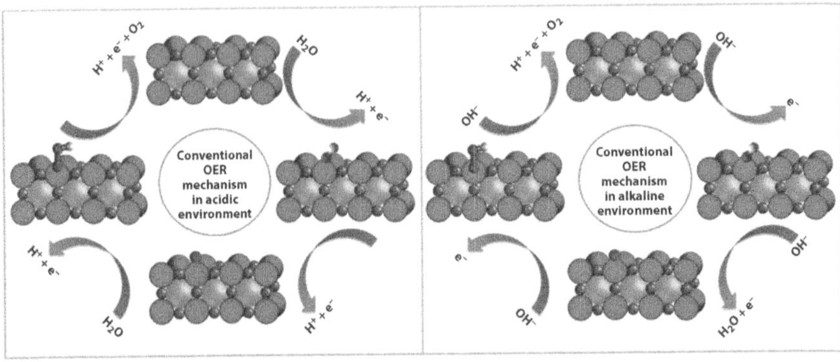

Figure 10.3 Conventional OER mechanism involving four proton–electron transfers on the surface metal centers in an acidic and alkaline environment.

$$HO^* \rightarrow O^* + H^+ \tag{10.11}$$

$$H_2O + O^* \rightarrow OOH^* + H^+ \tag{10.12}$$

$$OOH^* \rightarrow O_2\ (g) + {}^* + H^+ \tag{10.13}$$

The water oxidation under alkaline conditions is given by

$$4OH^- \rightarrow O_2\ (g) + 2H_2O\ (l) + 4e^- \tag{10.14}$$

The reaction kinetics in the basic environment is assumed to follow the four elementary steps through Equations 10.15 to 10.18.

$$^* + OH^- \rightarrow OH^* + e^- \tag{10.15}$$

$$OH^* + OH^- \rightarrow O^* + H_2O\ (l) + e^- \tag{10.16}$$

$$O^* + OH^- \rightarrow OOH^* + e^- \tag{10.17}$$

$$OOH^* + OH^- \rightarrow {}^* + O_2\ (g) + 2H_2O\ (l) + e^- \tag{10.18}$$

Thus, the transfer of four electrons is necessary for water oxidation in contrast to the two-electron transfer for HER, which results in higher overpotentials for OER. The process of hydrogen evolution reaction is relatively straightforward to understand, whereas the oxygen evolution reaction is a complex reaction from a kinetic perspective. It involves the formation of several adsorbed intermediates and various reactions, resulting in several possible pathways [17, 18].

10.3 Electrocatalyst Selection Criteria for Electrochemical Water Splitting

Efficient electrocatalysts made of earth-abundant materials are crucially important for the progress and advancement of solar water-splitting devices, particularly for oxygen evolution reaction (OER) and hydrogen evolution reaction (HER). The role of a catalyst in water splitting is to reduce the overpotential (activation energy) for OER and HER. The fundamental requirements for the electrocatalyst are as follows:

- Catalytically active–good electrical conductivity, surface energetics
- Chemical compatibility
- Scalable preparation process
- Earth-abundant
- Environmentally benign
- Durability

The use of noble metals in the production of hydrogen through water splitting is limited in terms of cost and sustainability. Despite efforts to find more efficient alternatives, noble metals like platinum (Pt), ruthenium (Ru), and iridium (Ir) continue to be the most efficient catalysts for hydrogen evolution reaction (HER). Numerous studies have been conducted to investigate the oxygen evolution reaction (OER), leading to the development of several catalysts that enhance electrode kinetics and stability in diverse electrolyte environments. Rutile-type RuO_2 has shown excellent OER catalytic activity, but it is unstable and easily oxidized in both acidic and alkaline electrolytes. IrO_2 is also an effective OER catalyst, but it also dissolves during the reaction and is made from precious metal. Thus, both RuO_2 and IrO_2 are not practical for large-scale production [19].

10.4 Significance of Graphene as an Electrocatalyst

Graphene and its derivatives, due to their structural and conductive properties, show excellent electrochemical properties for water electrolysis and are cost-effective and environmentally friendly. Doping or encapsulating graphene with other active electrocatalysts enhances the electrochemical properties of HER and OER for water splitting. Doping with nitrogen and other heteroatoms, or encapsulation with carbon nanotubes, increases the electronegativity and the mechanical strength, resulting in better electronic properties for water splitting. The formation of binary, ternary, and hybrid nanocomposites between graphene layers and the reactive active components further improves the electrochemical properties.

10.4.1 Graphene: Synthesis and Its Derivatives

Graphene is a 2D material with a range of properties that have been the subject of extensive research for its large-scale industrial production. The synthesis of graphene can be done through solid, liquid, or gas phases of graphite and is usually done through chemical vapor deposition, exfoliation, or organic synthesis methods. Top-down and bottom-up approaches can be used for synthesis, with the top-down approach using graphite as a starting material and the bottom-up approach using different carbon sources. There are also other methods such as epitaxial growth and direct growth from allotropes of carbon [20]. The exfoliation process involves splitting graphite into single- or few-layered graphene using electromagnetic, electrostatic, or mechanical forces (Figure 10.4). The epitaxial growth process produces graphene on a semiconductor substrate and direct growth from carbon allotropes involves low-temperature methodology using precursors such as C60 fullerene and pyrolytic graphite. Graphene-based materials can be modified by covalent and non-covalent bonding, with non-covalent bonding preserving the natural structure of graphene but having low interaction with fillers, while covalent bonding results in strong mutual interaction but loss of natural graphene properties.

10.5 Graphene-Based HER Electrocatalyst

Graphene is needed for hydrogen evolution reaction (HER) because of its unique electronic and structural properties that make it an effective catalyst

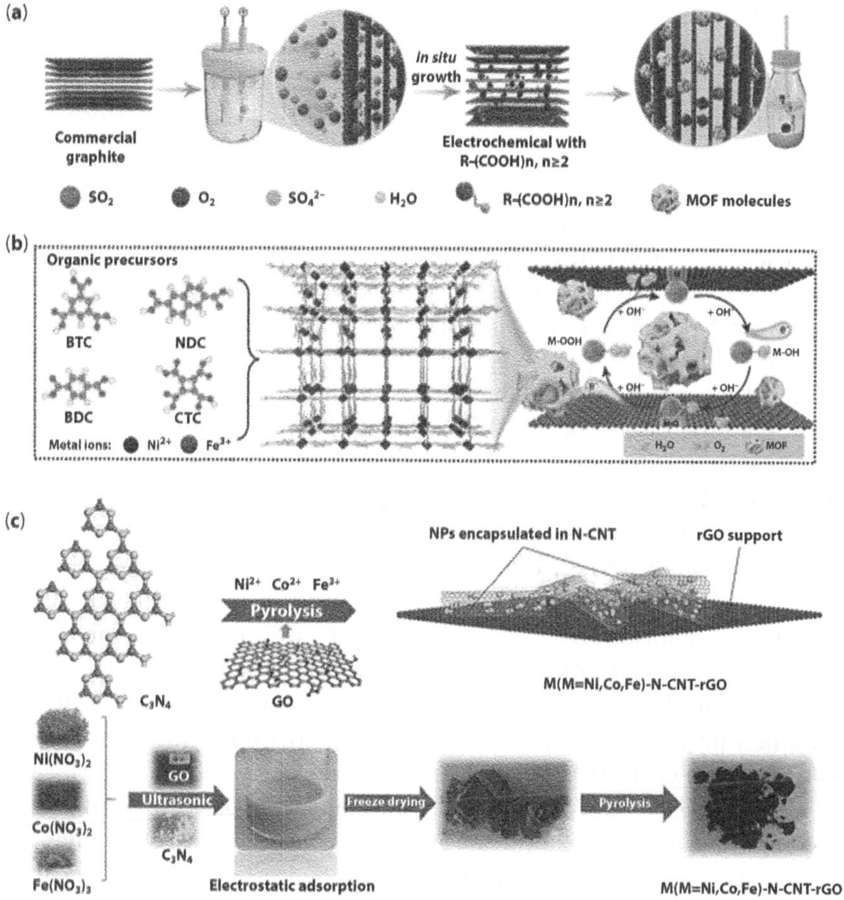

Figure 10.4 (a) Schematic illustration representing the synthesis workflow of MOF molecules through the electrochemical process and (b) The resultant NiFe-MOF//G. Adapted with permission from ref. [21], Springer Nature under a Creative Commons CC BY License. (c) Illustration of the synthesis procedure for the transition metals Ni, Fe, and Co nanoparticles encapsulated in the carbon nanotube and embedded on the rGO support through ultrasonication, electrostatic adsorption, and pyrolysis. Adapted with permission from ref. [22], Royal Society of Chemistry under a Creative Commons Attribution-NonCommercial 3.0 Unported License.

for the production of hydrogen gas. The high conductivity and large surface area of graphene enable fast electron transfer and provide ample active sites for the adsorption of reactant molecules, respectively. Additionally, graphene has high chemical stability, which allows it to resist corrosion in harsh electrolyte environments, making it more durable and efficient compared with traditional catalysts. Furthermore, graphene-based HER catalysts can be synthesized in various forms, such as graphene nanoflakes or graphene quantum dots, to further tailor their catalytic performance [23, 24].

Though the performance of the noble metal ruthenium is unparalleled with other non-noble metal-based electrocatalysts, it is not cost-friendly. To tackle this problem, a support material such as conductive reduced graphene oxide could be inserted with Ru taking into consideration the least weight percentage of Ru, hence overcoming the cost issue. Recent work on sulfur-doped reduced graphene oxide was composited with ruthenium sulfide nanospheres delivering excellent catalyzing behavior toward HER in a broad range of pH. An extremely small overpotential at 10 mA cm^{-2} (η@10 mA cm^{-2}) was obtained (25 mV) with a Tafel slope of 29 mV dec^{-1} in an alkaline solution and the η touched 56 mV at 50 mA cm^{-2} delivering outstanding stability for 100 h [25].

The co-doping strategy of nitrogen and sulfur into the graphene structure subsequently increased the porous structure through chemical activation by zinc chloride [26]. The η@10 mA cm^{-2} was only 310 mV and 290 mV vs. RHE in extremely acidic and basic pH, respectively. The robust electrochemical activity toward catalyzing HER could be attributed to the extremely precise surface area available for kinetics due to the alternations in the surface chemistry of graphene by co-doping with N and S (Figure 10.5). A hybrid architecture of an ultrastable alloy of cobalt and nickel was reported to showcase extraordinary HER activity in harsh environments. The effectiveness of the alloy toward hydrogen evolution reaction was accredited to the unique flower-like morphology obtained through a facile chemical method and also the encapsulation of the metal alloy with N-doped graphene to boost its existing HER performance. This nanoflower framework along with N-doped graphene was composited with a porous network of N-doped g-C_3N_4. This led to an exceptional display in HER characteristics with 122 mV and 95 mV (η@10 mA cm^{-2}) in basic and acidic electrolytes, respectively. The Tafel slopes in the respective pH environments were 84 and 77 mV dec^{-1} [24].

A group of researchers developed an affordable electrocatalyst by combining a nanosheet-like morphology of cobalt phosphide with a few layers of N-doped graphene nanosheets (CoP@NG). The synthesis technique was a facile approach adopting hydrothermal and phosphorization procedures. The HER onset potential stood at 87 mV vs. RHE for acidic and 115 mV

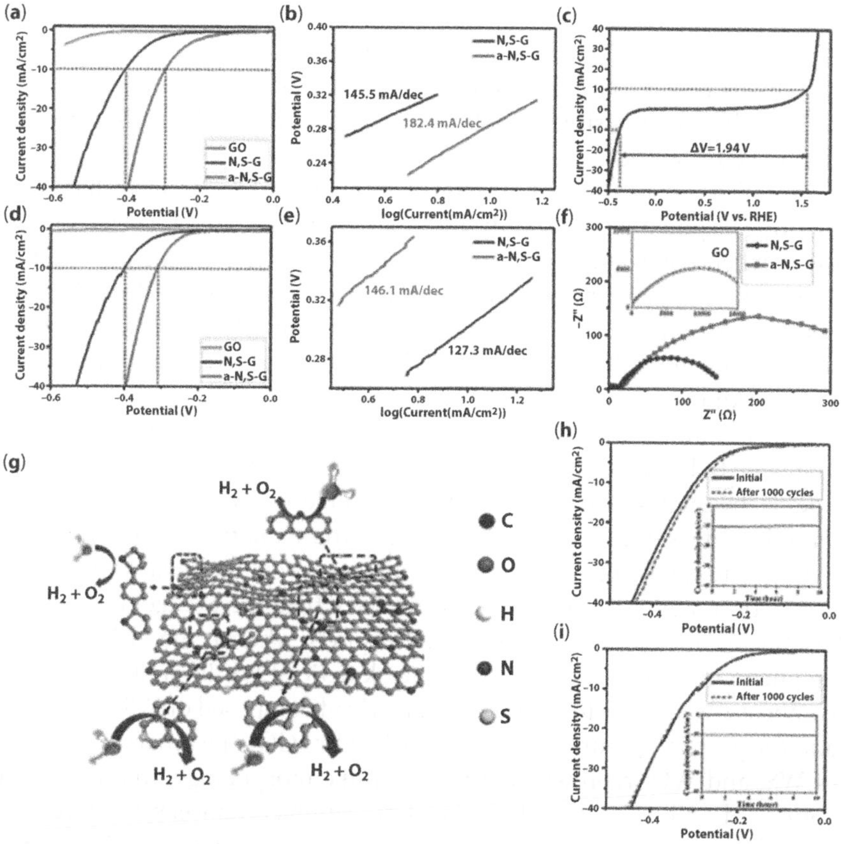

Figure 10.5 (a) The HER polarization curves of GO, N,S-G, and a-N,S-G co-doped GO in 1 M KOH. (b) The corresponding Tafel slopes of the electrocatalysts in 1 M KOH. (c) Water-splitting polarization of a-N,S-G. (d) The HER polarization curves of GO, N,S-G, and a-N,S-G co-doped GO in 0.5 M H_2SO_4. (e) The corresponding Tafel slopes of the electrocatalysts in 0.5 M H_2SO_4. (f) The EIS spectra of GO, N,S-G, and a-N,S-G in 0.5 M H_2SO_4. (g) Illustration of active sites and mechanism of overall water splitting in the electrocatalyst a-N,S-G. (h) The before and after LSV curves of a-N,S-G in 0.5 M H_2SO_4. (i) The before and after LSV curves of a-N,S-G in 1 M KOH. Adapted with permission from ref. [26].

vs. RHE for alkaline electrolytes, respectively, with the Tafel slope of 64 and 60 mV dec^{-1} in 0.5M H_2SO_4 and 1 M KOH electrolytes, respectively [27]. A similar work where CoPx was uniformly distributed on two-dimensional N-doped rGO revealed a superior catalytic performance of η = 104 mV and a calculated Tafel slope of 65 mV dec^{-1} in a basic electrolyte medium.

The following HER performances could be attributed to the excellent porosity developed by doping the graphene-based rGO with nitrogen thereby increasing the active surface area and also the nanodimension of cobalt sulfide demonstrating only 4 nm size for the excellent adsorption of the active H^+ species necessary for HER [28].

An allotrope of the carbon family graphene in its 2D form has been marking its potential as a significant catalyst for HER. A 2D-layered graphdiyne when conjugated with WS_2 showed a reduction in the potential to as low as 140 mV with the least Tafel slope of 54 mV dec^{-1}. The enhanced performance was ascribed to the inherent electric field which was built up between the graphdiyne and the WS_2 layers that led to the efficient charge transfer kinetics and heightened activity in the WS_2 basal plane sites [23]. Another sulfide-based catalyst (MoS_2) was combined with hydrogenated graphene to heighten the electronic and structural modulations of the sulfide electrocatalyst. The overpotential accessible at 10 mA cm^{-2} was 124 mV with the Tafel slope as low as 41 mV dec^{-1}. The stability was also incredible with the durability of a whole day of continuous operation at 30 mA cm^{-2}. Simulation studies also confirmed the fact that the S-edge atoms were responsible for the active hydrogen adsorption, whereas the hydrogenated graphene increased the charge transfer kinetics [29]. In another work, a composite was in the form of a ternary hybrid composed of WS_2 and WO_3, and the graphene counterpart was satisfied by rGO. The binary composite of WS_2 and WO_3 represented a nanorod-like morphology, whereas when combined with rGO, the electrocatalyst crumbled between the layers of the reduced graphene oxide and showed excellent catalytic activity toward HER. The η@10 mA cm^{-2} displayed 96 mV, and the Tafel slope calculated using the Tafel equation was 37 mV dec^{-1} [30].

Under peculiar circumstances, biological cells of *Saccharomycetes* yeasts were developed as a template to prevent aggregation by adsorbing the metal ions from synthesized nanoparticles such as Co_2P. This aided in the robust stability and the least inset potential of an electrocatalyst composed of Co_2P nanoparticles, N- and P-doped carbon, and reduced graphene oxide. This ternary composite (Co-Co_2P@NPC/rGO) developed excellent stability (20 h at a current density of 1,000 mA cm^{-2}) due to the *Saccharomycetes* cells as the template and even withstood tough pH conditions (0.5 M H_2SO_4) demonstrating a low overpotential of 61.5 mV [31].

Single-atom dispersion into carbon species has attracted electrocatalysts for an effective increment in the electrocatalytic HER. The noble metal Pt was dispersed into N-doped porous carbon, developed rich defect sites, and delivered an exuberant η@10 mA cm^{-2} of 13 mV way lesser than the benchmark Pt/C commercial electrocatalyst for HER. The restructuring of the porous carbon structure with optimized Pt atom dispersion (Pt0.8) paved the way for the electronic states to be coordinated well for the enhanced catalytic activity revealing a mass activity of 11,284 mA mg^{-1} at 0.1 V for effective hydrogen evolution reaction [32]. In similar research involving single-atom dispersion for increased HER kinetics, Pt atoms were anchored with graphene stacked with aniline. The HER activity demonstrated a mass current density of 22,400 Ag_{Pt}^{-1} at a current density of 50 mV and an outstanding η@10 mA cm^{-2} of 12 mV. This mass current density surpassed the existing value of Pt/C at a rate of 46% higher and the excellent kinetics could be ascribed to the organization of Pt atoms with the aniline structure which is composed of nitrogen atoms to enhance the existing hydrogen adsorption energy of the Pt atoms [33].

10.6 Graphene-Based OER Electrocatalyst

Graphene-based oxygen evolution reaction (OER) electrocatalysts have garnered interest in recent years due to their high efficiency and stability in producing oxygen gas through an electrochemical process. Graphene, with its large surface area and good conductivity, provides an ideal platform for catalytic activity in the OER process. These materials have shown promising results in laboratory experiments and hold great potential for practical applications in water splitting, which is a crucial step in the production of hydrogen fuel and renewable energy technologies. The development of efficient and stable graphene-based OER electrocatalysts will play a key role in promoting the widespread adoption of these technologies.

The OER performance of N, S co-doped graphene structures delivered an η@10 mA cm^{-2} of 0.33 V in an extreme basic pH =14 in KOH electrolyte. The results exceeded the recently reported carbon-based electrocatalyst ruling the OER counterpart [26]. Also, the encapsulation of graphene on an alloy made up of Ni and Mo was utilized for OER which displayed excellent catalytic activity with an ultrastable condition in harsh alkaline (KOH) electrolyte. The DFT studies revealed that the graphene counterpart assisted in the reduction of the adsorption energy of the H^+ and OH^- intermediates thus improving the OER kinetics on the alloy surface [20]. The flexible layers of both cobalt phosphide and graphene layers when doped with nitrogen delivered an OER onset potential of 1.48 V vs. RHE, though the η@10 mA cm^{-2} crossed ~350 mV vs. RHE. The Tafel slope was as low as ~64 mV dec^{-1}. The mechanical stability of the nanosheet-like morphology along with the excellent electron transfer between the transition metal phosphide and the conductive and porous graphene layers due to nitrogen doping has provided exceptional reaction sites for electrochemical oxygen evolution reaction [27].

The integration of photocatalysis and the working of the OER combine to form a photocatalytic OER (PFC). The OER provides the necessary electrical input through its working principle as the photocatalytic electrocatalyst does the function of absorbing the photons from the light source to degrade or remove a component usually from water. The generated photoelectrons from the photoanode travel to the photocathode to generate electricity, and the holes produced are used to oxidize the organic substances or reduce H^+ and generate H_2 [34]. This integrated technology is usually used for waste water management technologies by utilizing solar light to degrade the harmful degradants (organic materials) from water. They could also be combined with other energy devices such as supercapacitors to power portable electronic devices. The development of PFCs in a positive direction could only be accomplished with the material aspect of the photocatalytic electrocatalysts and the MEA assembly of OERs both falling in place to ensure maximum efficacy, mechanical strength, and durability. In the case of flexible PFCs, the photocathode and the photoanode should possess a certain Fermi level of energy that hinders the recombination of the photogenerated ions [35, 36].

A chip consisting of zeolite architecture-based carbon (ZIF-C) coupled with NiCoP@NiOOH as a supercapacitor and TiO_2 as a photocatalytic OER was fabricated to power small portable electronics. The peak density was 3.04 μW cm^{-2} under 30 mW cm^{-2} solar intensity. Restocking the PFC with urea delivered continuous energy outputs for a longer duration of time. The micro supercapacitor exhibited a 54.7 mF cm^{-2} areal capacitance and 270.5 μW cm^{-2} of power density with exceptional retention of no decay after 8,000 charge–discharge cycles and 500 bending cycles [37]. On the contrary, a PFC, which was flexible and worked on feeding biowastes (ethanol, glycerol, lactic acid, and glucose), was successfully implemented by a research group. The flexible PFC worked as a sweatband generating a peak power of 4 mW cm^{-2} g^{-1} under both the flat and bending conditions. The fPFC with a 3-cm radius surpassed most of the photoelectrochemical tests as a potential candidate for wearable power-generating devices and also waste water remediation technologies surpassing the other categories of OERs [38]. Recyclable catalysts using rGO and graphitic nitride carbon dots (CNQDs) were implemented as active catalytic layers in a flexible PFC. The physical characterizations such as FESEM and HRTEM confirmed the presence of quantum dots wrapped into reduced graphene layers (Figure 10.6), and the aerogel hybrid effectively removed ciprofloxacin at a rate six times more advanced than the bulk g-C_3N_4 [39].

A carbon fiber cloth served the purpose of a flexible electrode and a support base for a highly active Fe-based nanocomposite comprising graphene oxide and titanium phosphate in different Fe ratios to serve the purpose of a photoanode. The photocatalytic cell degraded rhodamine (RhB) with the optimized weight percentage of Fe doping (5 wt%). The cell voltage progressed to 0.4 V as the concentration of RhB piped up to 100 mg/L. Though the cyclic stability was admirable when carbon cloth fiber was substituted with stainless steel, the adaptation in terms of flexibility was delivered only by the CF. The electrodes were tested in a single-chamber photocatalytic OER using a 50-W input supply of light source, and almost 90% of the RhB was degraded even in acidic electrolyte (pH = 1) when the resistance was 100 Ω. When the external resistance increased, the degradation was affected without affecting the cell voltage. In the absence of an external electric supply, almost 80% of the dye was degraded by photocatalytic activity [40]. Graphitic carbon nitride along with Ag_3PO_4 was synthesized to develop a z-scheme photocatalytic OER by implementing it as the anode. The degradation of a harmful compound called tetracycline was easily degraded by the nanocomposite along with copper oxide as the photocathode revealing a Pmax of 6.06 μW cm^{-2} when compared with the individual Ag_3PO_4 photoanode (4.16 μW cm^{-2}) and g-C_3N_4 (0.067 μW cm^{-2}), respectively. The photogenerated carriers developed excellent redox activity which was ~90% more than the pristine graphitic electrocatalyst. The oxidants used for the study were nitrogen and oxygen for electrical input and pollutant degradation, respectively [35].

A flexible electrode for use in photocatalytic OERs was developed with organic semiconductors like epindolidione (EPI) along with graphene and decorated with Ni nanoparticles for hydrogen generation. The method developed was coated with carbon-based graphene oxide on either side of EPI to reduce the recombination of excitons. The photo response was not satisfactory when only the organic semiconductor was employed as the flexible electrode. The addition of graphene along with the non-noble metal as a co-catalyst amplified the hydrogen generation with excellent durability without wearing off the active catalyst material. The photo response of the flexible PFC electrode was 4 μA cm^{-2} at 100 mW cm^{-2} yielding solar to

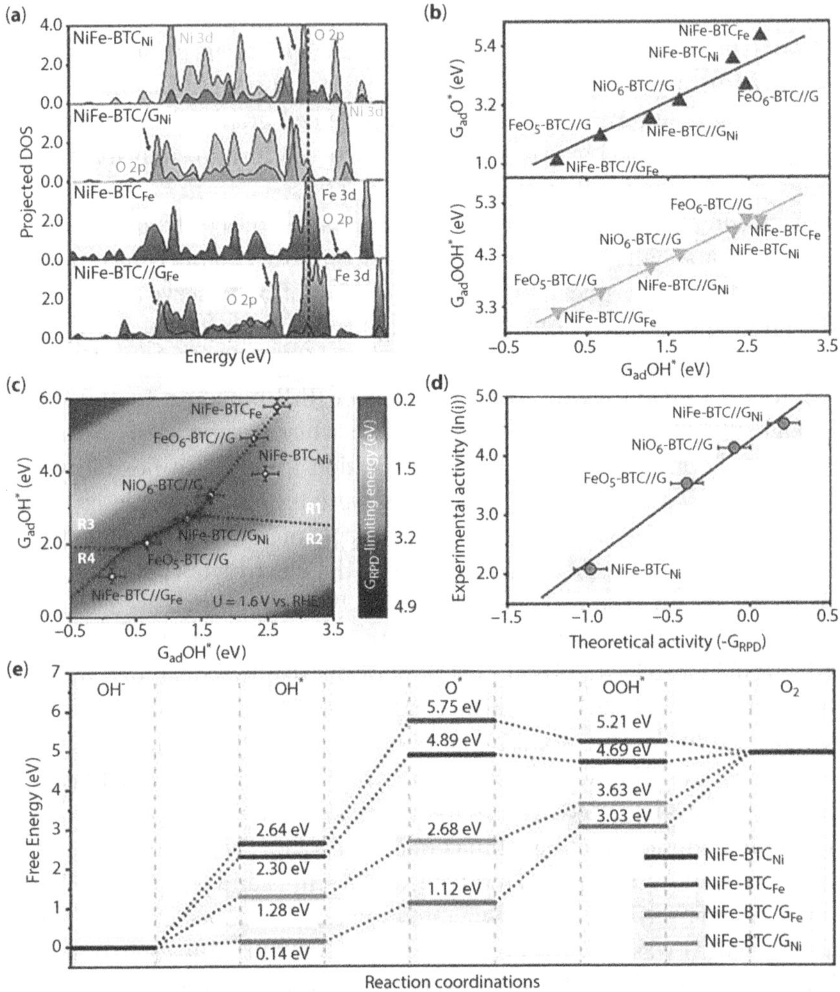

Figure 10.6 (a) The projected DOS file of different NiFe-based MOF electrocatalysts confined in graphene layers. (b) The linear relationship between adsorption energies of O*/OOH* vs. OH*. (c) The activity map of O* and OH* as two independent descriptors. (d) The relation between the theoretical values (–GRPD) and the experimental values (ln(*j*)). (e) The free energy diagram of the NiFe-based MOF electrocatalyst confined in graphene layers. Adapted from ref. [21], Springer Nature under a Creative Commons CC BY license.

hydrogen yield of $0.6 \times 10^{-2}\%$ [36]. Though the use of carbon-based materials such as nitrogen-doped graphene oxide coupled with cobalt nanocomposite (Co-N-RGO) only performed satisfactorily toward ORR in a basic electrolyte, the same material displayed excellent cathodic properties in a photocatalytic OER revealing an alternative approach to the platinum-free OER. The anodic cell was a titania/CdS and the Co-N-RGO catalyst served as the photocathode. The mechanism was emphasized on the active dopant centers of Co and nitrogen with intercalation between the enriched reduced graphene layers developed through pyrolysis [41].

The simple architecture including the fuel cartridge and instantaneous recharging options makes direct liquid OERs (DLFCs) stand out from other types of OERs [42]. Also, the liquid in these OERs depicts different alcohols such as methanol, ethanol, and even oxides such as hydrogen peroxide [43, 44]. The least developed DLFCs include formate, formic acid, hydrazine, borohydride, and dimethyl ether liquid OERs [45]. The foremost hindrance in this liquid type of direct OER is the catalyst loading and the leakage of the liquid fuel which might be a major hazardous issue. The flexible DLFCs are still an emerging technology with a research emphasis mainly on the long-term durability of these bendable components. The commercialization of these OERs is still in the pipeline [42]. Though Pt and other noble metal alloys are significantly used in the fabrication of other components in a flexible DLFC, carbon-based materials are also gaining interest with different nanocomposites to showcase their working ability in both acidic and alkaline conditions [46].

In a recent study, a T-shaped flexible methanol OER was developed where the active catalytic layer consisted of N-doped graphene oxide (GO) composited with $NiCo_2O_4$ microspheres which depicted a flower-like morphology. Along with methanol as an oxidant, potassium permanganate was also utilized as an added advantage. This nanocomposite displayed five times more Pmax performance when compared with the catalyst without N-doped GO (0.93 mW cm^{-2}). The catalyst coated on paper cells forming two cells (anode and cathode, respectively) demonstrated a Pmax of 4.67 mW cm^{-2} which could easily light an LED bulb for 30 mins [46].

Meanwhile, though the Nafion® membrane is used as a penetration medium in methanol OERs, its costs are a concern to be discussed. To find an alternative, a polymer electrolyte membrane without Nafion but delivering high conductivity of protons with low permeability of methanol was derived. It was a polymer matrix consisting of sulfonated graphene oxide (SGO) and sodium alginate (SA). This combination when combined under appropriate proportions (0.2 wt% SGO) displayed excellent proton conductivity (13.2×10^{-3} S cm^{-1}), while the methanol permeability was very low (1.535×10^{-7} cm^2 s^{-1}). Such incomparable performance was attributed to the existence of SGO in optimized proportion along with SGO considered on par with the commercially available Nafion membranes [47].

In another study, graphene flakes were synthesized through bath sonication coupled with spray pyrolysis to be implemented for a methanol flexible OER and deposited above the anode catalytic layer just adjacent to the MEA in different weight proportions. The layers deposited were on a nanometer scale and 5 wt% of nanographene layers displayed a Pmax of 91 mW·cm^{-2}. The methanol permeation was also reduced from ~150 to 110 mA·cm^{-2} with only a negligible decrease in the proton conductivity. At very high concentrations, the methanol OER performance even surpassed 75% of its initial performance demonstrating marginal voltage losses. This 38% increase in the Pmax and the sturdiness of voltage operation up to 20 h was associated with the effectiveness of the synthesis approach and the complete use of nanographene layers as an ion-transport membrane [48].

Flexible ethanol OERs in an alkaline medium were investigated with the aid of a novel catalyst made up of graphene oxide combined with quaternized polyvinyl alcohol (QPVA/GO). When the loading percentage of GO was optimized to 15 wt% in an alkaline solution of 1 M KOH, the ethanol permeation was reduced to 1.49×10^{-7} cm^2 s^{-1} at room temperature. This combination especially assisted in the reduction of the ethanol permeation and enhanced the maximum power density by 11.5 mW cm^{-2} when the commercial Nafion and rigid alkaline DEFCs displayed only 7.68 mW cm^{-2} and 9.1 mW cm^{-2} at room temperature, respectively. This QPVA/GO composite could act as a replacement for Pt-based catalyst membranes to enhance the existing performance of flexible ethanol OERs [49]. A catalyst based on the combination of two different sulfides (MoS_2 and Ni_3S_2) when composited with reduced graphene oxide delivered exceptional performance in both methanol and ethanol OERs in basic media. Even after repeated CV cycles (200 cycles), the cyclic stability increased to 105% for the methanol oxidation reaction (MOR) and 104% hike for the ethanol oxidation reaction (EOR), respectively. The reason was accredited to the morphology of Ni_3S_2 which represented a starfish-like appearance along with the porous structure of MoS_2 and the excellent conductivity and active surface area of rGO. The synergy between their combined effects facilitated an open circuit voltage of 0.59 V and a Pmax of 4.42 mW cm^{-2} for flexible DMFC, whereas the open circuit voltage was reduced to 0.48 V and a Pmax of 3.5 mW cm^{-2} for flexible DEFC [43].

10.7 Graphene-Based Electrocatalyst for Overall Water Splitting

Graphene-based electrocatalysts are materials that use graphene as a substrate to support catalytic nanoparticles, which are responsible for promoting the electrolysis of water into hydrogen and oxygen. Graphene's unique physical and chemical properties, such as high electrical conductivity, high stability in harsh environments, and large surface area, make it a suitable material for water electrolysis. The combination of graphene with catalytic nanoparticles, such as metal nanoparticles or transition metal dichalcogenides, results in a highly efficient and stable electrocatalyst for overall water splitting. This technology has the potential to provide a sustainable and environmentally friendly solution for hydrogen production, which can be used as a clean energy source in various applications.

An overall water splitting was achieved when the noble metal ruthenium was utilized to synthesize ruthenium sulfide sphere-like morphologies along with S-doped rGO to match the existing Pt/C benchmark HER electrocatalyst. When inserted into a homemade electrolyzer with the synthesized RuS_2/S-rGO as the cathode and the RuO_2/C as the anode, a potential of 1.54 V at 20 mA cm^{-2} was delivered with stability for 24 h without decay. The superior electrochemical characteristics could be due to the Ru sites in RuS_2 responsible for initiating the HER activity through adsorption of the active species with the conductivity of S-doped rGO for the diffusion of electrons to evolve the H_2 gas [25]. When nitrogen doping and the concentration of ruthenium nanoparticles were in the ratio of 2.4:1.2 at wt%, a 3D architecture of Ru-based electrocatalyst encapsulated in the graphene aerogel exhibited a leastη@10 mA cm^{-2} of 140 mV at a pH = 1 in the 0.5-M H_2SO_4 electrolyte toward catalyzing OER. The same electrocatalyst even delivered a Tafel slope of 78 mV dec^{-1} in the same electrolyte conditions toward HER catalyzation. The 3D electrocatalyst with the size of only 2–4 nm surpassed the commercially available electrocatalysts (Pt/C and RuO_2) to reach a potential of 1.62 V vs. RHE for overall water splitting [50].

In a similar trend, instead of Ru nanoparticles, Ir nanoparticles were incorporated with N-doped graphene nanosheets to deliver an η@10 mA cm^{-2} of 270 mV and 303 mV in 1 M KOH and 0.5 M H_2SO_4, respectively, with Tafel slope being the least for OER catalysis to date −60 and 50 mV dec^{-1} in basic and acidic solutions, respectively. The electrocatalyst also performed well in the acidic electrolyte when it came to hydrogen evolution reaction with an η@10 mA cm^{-2} of 19 mV in pH = 0 when compared with 43 mV at pH = 14. The respective Tafel slopes also altered accordingly, with the least value of 26 mV dec^{-1} (pH = 0) and around 35 mV dec^{-1} (pH = 14), respectively. The Ir/N-G boosted an overall water-splitting voltage of 1.6 V at a current density of 20 mA cm^{-2} when the loading of the electrocatalyst was only 23 μg cm^{-2} [51]. In a similar work, three different electrolyte media (1.0 M KOH, 0.1 M KOH, and 0.5 M H_2SO_4) were taken and studied with the same catalyst where Ir was embedded on N-doped graphene.

The HER performance outperformed in the acidic electrolyte with an η@10 mA cm^{-2} of 25 mV compared with a large onset potential of 114 mV in an alkaline electrolyte. The Tafel slope values were also least in the 0.5 M H_2SO_4 electrolyte medium with 15 mV dec^{-1} than in 0.1 M KOH with a Tafel slope of 113 mV dec^{-1}. In the case of OER, the 1.0-M KOH demonstrated an η@10 mA cm^{-2} of 273 mV with a Tafel slope of 64 mV dec^{-1}. The overpotential values were considerably larger for pH = 14 (340 mV) and pH = 0 (371 mV), respectively (Figure 10.7). Owing to their excellent stability in the acidic electrolyte exhibiting very low onset and Tafel slope values, the overall water-splitting cell voltage using the Ir@NG as a bifunctional electrocatalyst was determined to be 1.7 V at a current density of 10 mA cm^{-2} [52].

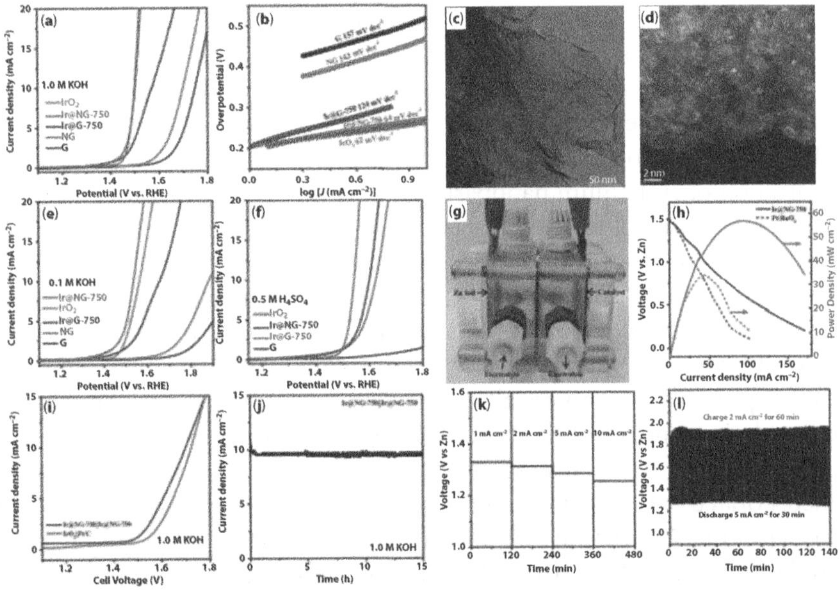

Figure 10.7 (a) The LSV polarization curves of the Ir@NG-based electrocatalysts in 1.0 M KOH. (b) Tafel plots derived from the LSV curves of panel (a). (c) TEM images of Ir@NG electrocatalyst. (d) HAADF-STEM image of Ir@NG. (e) The LSV polarization curves of the Ir@NG-based electrocatalysts in 0.1 M KOH. (f) The LSV polarization curves of the Ir@NG-based electrocatalysts in 0.5 M H_2SO_4. (g) Device setup. (h) The polarization curve vs. the plot of power density. (i) The LSV polarization curves of the bifunctional electrocatalyst Ir@NG comparing it with IrO_2||Pt/C. (j) Chronopotentiometry analysis (at 1.7 V) for the bifunctional electrocatalyst in 1.0 M KOH electrolyte. (k) Chronoamperometric analysis at different current densities (1, 2, 5, and 10 mA cm^{-2}) in a time interval of 2 h. (l) The cyclic performance of the Ir@NG electrocatalyst for a period of 140 h. Adapted with permission from ref. [52], Royal Society of Chemistry.

On the contrary, noble metal-free electrocatalysts are considered a hot topic taking into consideration the cost parameter. A complicated yet ultra-efficient electrocatalyst for overall water splitting was synthesized incorporating transition metals such as NiFe encapsulated with N-doped CNTs and finally dispersed into reduced graphene oxide layers. The encapsulated electrocatalyst delivered the least Tafel slope for OER (42 mV dec^{-1}) and 84 mV dec^{-1} for HER. The cell voltage demonstrated by the NiFe-N-CNT-rGO electrocatalyst was 1.5 V at 10 mA cm^{-2}. The least cell voltage and excellent long-term durability could be attributed to the N entries and the intermediates of carbon content in both carbon nanotubes and reduced graphene oxide fastening the two-electron transfer process for HER and the four-electron transfer process for OER, respectively [22].

10.8 Graphene in Combination with Other Nanostructures for Overall Water Splitting

Graphene is an effective material that is used as a catalyst in water splitting as it provides a large surface area for the catalytic reaction to occur, and its high electrical conductivity allows for the efficient transfer of electrons. Though the material is found to possess great catalytic activity as a support material, its combination with other nanostructures such as bimetal nanomaterials, metal–organic frameworks (MOFs), and spinel ferrites has also shown potential in the overall water-splitting process.

1. Bimetal nanomaterials: Bimetal nanomaterials are made up of two different metals and are often used in catalytic applications. In water splitting, bimetal nanomaterials can be used as efficient catalysts to accelerate the reaction and improve the overall efficiency of the process.

In a recent work, the authors present a cost-effective and efficient electrocatalyst for hydrogen evolution reaction (HER) and oxygen evolution reaction (OER) that are crucial for the design of water electrolyzers. The proposed catalyst was a cobalt–chromium nanostructure that was contained within a nitrogen-doped graphene tube (CoCr@NGT). The large-sized graphene in the form of a tube, with a diameter of around 100 nm, has greatly improved the performance of both the anodic and the cathodic counterparts with exuberant electrochemical performances. The CoCr@NGT catalyst could serve for both HER and OER in an alkaline water electrolyzer delivering a current density of 10 mA cm^{-2} at 1.58 V for an extended period of time. The excellent electrochemical performance of this catalyst is attributed to the presence of chromium, which enhances stability and activity due to its deterioration resilience properties [53].

2. Metal–organic frameworks (MOFs): MOFs are a type of nanomaterial composed of metal ions linked together by organic ligands. They have a high surface area and porosity, which make them ideal for catalytic applications in water splitting. Another study using a metal–organic framework of vanadium-doped bimetallic nickel–iron nanoarray was made using a supported architecture of carbon through a pyrolysis process. Different molar ratios of Ni/Fe were developed, and the results showed that the catalyst with an equal mole ratio of Ni and Fe had pronounced activity in both acidic and alkaline media. Vanadium doping had a strong synergistic effect, leading to $\eta = 90$ mV (HER) and $\eta = 210$ mV (OER) at 10 mA cm^{-2}, respectively. The catalysts also showed excellent durability of 20 h of unfailing performance at a fixed overpotential of 250 mV. This ratio also had a better corrosion potential of 0.209 V vs. Ag/AgCl and at $A = 0.594 \times 10^{-12}$ A cm^{-2} at an extreme pH of 14 [54].

3. Spinel ferrites: Spinel ferrites are a type of iron oxide mineral with a unique crystal structure that gives them unique magnetic and catalytic properties. In water splitting, spinel ferrites can be used as catalysts to improve the efficiency and stability of the process. In an investigation involving the spinel composition of $NiCo_2O_4$, rGO was incorporated to achieve a coral-like nanostructure through a time-controlled hydrothermal method for superior electrocatalytic activity catalyzing both HER and OER. The material's electrochemical performance was increased due to its large electroactive surface area, good connection between the $NiCo_2O_4$ and highly conductive graphene sheets, and the synergistic effect of the material's electroactive sites. The rGO-$NiCo_2O_4$ catalyst also has excellent water-splitting ability in an alkaline environment with a low onset potential of 1.5 V, making it suitable for use in a highly efficient and low-cost electrolyzer [55].

Overall, the combination of graphene with these nanostructures has the potential to improve the efficiency, stability, and scalability of the water-splitting process, making it a promising solution for the production of clean hydrogen fuel.

10.9 Conclusion and Future Perspectives

The progress of efficient catalyst materials for electrochemical water splitting is very well the need of the hour, acknowledging the future energy demands. Graphene-based electrocatalysts in their allotropic forms are widely used as support materials with other existing nanocatalysts to enhance the existing electrochemical performance of the water-splitting components—the HER and OER. The synthesis of these graphene-based materials is tailored with high precision through quality-controlled chemical and physical synthesis techniques. The applications implementing these graphene structures with oxides, sulfides, etc. are innumerable, and the emerging field is a shot to be taken in the future where energy-related devices have reigned in almost all sectors of human life. However, there are also some cumbersome challenges to be handled with carbon structures in this field, where the cost of manufacturing and commercialization should be moderate to increase the supply. Investments in research in the field of electrolyzers should also be appreciated by different organizations both public and private to cut down the rate of raw materials (carbon sources) and reduce the complexity of compositions without compromising on the efficacy and long-term durability of the graphene-supported competent electrocatalysts. Taking care of these opinions, the graphene-based electrocatalysts for electrochemical water splitting are the apt option for minimizing the present energy-related climatic challenges. The future views taking into account the green energy perspectives are also discussed below:

i. Though graphene-based materials are already in line for water electrolysis, improving catalytic activity is crucial for efficient water splitting. Research is underway to modify graphene with other materials to enhance its catalytic performance.
ii. Another important area of research is to improve the stability of graphene-based catalysts under long-term exposure to water and other conditions. Integrating graphene-based water-splitting systems with renewable energy sources such as solar or wind power could provide a sustainable and eco-friendly solution for hydrogen production.
iii. Scaling up the production of graphene-based materials to meet the demand for large-scale water-splitting systems is a challenge that must be addressed in the future.

iv. From a green energy perspective, graphene-based materials and water-splitting systems are expensive, which limits their widespread adoption. Reducing the cost of production and improving efficiency are key to making these systems more accessible.
v. Safety is also a measure to be taken into account. Hydrogen is highly flammable, so the storage and transportation of hydrogen produced through water splitting must be safe and secure.
vi. Governments must provide clear regulations and incentives for the production and use of hydrogen produced through water splitting to reduce the regulatory barriers.

References

1. van Ruijven, B.J., de Cian, E., Sue Wing, I., Amplification of future energy demand growth due to climate change. *Nat. Commun.*, 10, 2762, 2019, https:// doi.org/10.1038/s41467-019-10399-3.
2. Romitti, Y. and Sue Wing, I., Heterogeneous climate change impacts on electricity demand in world cities circa mid-century. *Sci. Rep.*, 12, 4280, 2022, https:// doi.org/10.1038/s41598-022-07922-w.
3. Landrigan, P.J., Fuller, R., Acosta, N.J.R., Adeyi, O., Arnold, R., Basu, N., Baldé, A.B., Bertollini, R., Bose-O'Reilly, S., Boufford, J.I., Breysse, P.N., Chiles, T., Mahidol, C., Coll-Seck, A.M., Cropper, M.L., Fobil, J., Fuster, V., Greenstone, M., Haines, A., Hanrahan, D., Hunter, D., Khare, M., Krupnick, A., Lanphear, B., Lohani, B., Martin, K., v Mathiasen, K., McTeer, M.A., Murray, C.J.L., Ndahimananjara, J.D., Perera, F., Potočnik, J., Preker, A.S., Ramesh, J., Rockström, J., Salinas, C., Samson, L.D., Sandilya, K., Sly, P.D., Smith, K.R., Steiner, A., Stewart, R.B., Suk, W.A., van Schayck, O.C.P., Yadama, G.N., Yumkella, K., Zhong, M., The commission on pollution and health. *Lancet.*, 391, 462–512, 2018, https://doi.org/10.1016/S0140-6736(17)32345-0.
4. Lelieveld, J., Klingmüller, K., Pozzer, A., Burnett, R.T., Haines, A., Ramanathan, V., Effects of fossil fuel and total anthropogenic emission removal on public health and climate. *Proc. Natl. Acad. Sci.*, 116, 7192–7197, 2019, https://doi.org/10.1073/pnas.1819989116.
5. Capellán-Pérez, I., de Castro, C., Arto, I., Assessing vulnerabilities and limits in the transition to renewable energies: Land requirements under 100% solar energy scenarios. *Renewable Sustain. Energy Rev.*, 77, 760–782, 2017, https:// doi.org/https://doi.org/10.1016/j.rser.2017.03.137.
6. Impram, S., Varbak Nese, S., Oral, B., Challenges of renewable energy penetration on power system flexibility: A survey. *Energy Strategy Rev.*, 31, 100539, 2020, https://doi.org/https://doi.org/10.1016/j.esr.2020.100539.

7. Mitali, J., Dhinakaran, S., Mohamad, A.A., Energy storage systems: A review. *Energy Storage Sav.*, 1, 166–216, 2022, https://doi.org/https://doi.org/10.1016/j.enss.2022.07.002.
8. Rahman, M.M., Oni, A.O., Gemechu, E., Kumar, A., Assessment of energy storage technologies: A review. *Energy Convers Manage.*, 223, 113295, 2020, https://doi.org/https://doi.org/10.1016/j.enconman.2020.113295.
9. Andersson, J. and Grönkvist, S., Large-scale storage of hydrogen. *Int. J. Hydrogen Energy.*, 44, 11901–11919, 2019, https://doi.org/https://doi.org/10.1016/j. ijhydene.2019.03.063.
10. Gür, T.M., Review of electrical energy storage technologies, materials and systems: Challenges and prospects for large-scale grid storage. *Energy Environ. Sci.*, 11, 2696–2767, 2018, https://doi.org/10.1039/C8EE01419A.
11. Mazloomi, K. and Gomes, C., Hydrogen as an energy carrier: Prospects and challenges. *Renewable Sustain. Energy Rev.*, 16, 3024–3033, 2012, https://doi.org/https://doi.org/10.1016/j.rser.2012.02.028.
12. Tashie-Lewis, B.C., Nnabuife, S.G., Hydrogen production distribution, storage and power conversion in a hydrogen economy - A technology review. *Chem. Eng. J. Adv.*, 8, 100172, 2021, https:// doi.org/https://doi.org/10.1016/j.ceja.2021.100172.
13. Song, H., Luo, S., Huang, H., Deng, B., Ye, J., Solar-driven hydrogen production: Recent advances challenges, and future perspectives. *ACS Energy Lett.*, 7, 1043–1065, 2022, https://doi.org/10.1021/acsenergylett.1c02591.
14. Yilmaz, F., Balta, M.T., Selbaş, R., Retracted: A review of solar based hydrogen production methods. *Renewable Sustain. Energy Rev.*, 56, 171–178, 2016, https://doi.org/https://doi.org/10.1016/j.rser.2015.11.060.
15. Wang, S., Lu, A., Zhong, C.-J., Hydrogen production from water electrolysis: Role of catalysts. *Nano Converg.*, 8, 4, 2021, https://doi.org/10.1186/s40580-021-00254-x.
16. Lasia, A., Mechanism and kinetics of the hydrogen evolution reaction. *Int. J. Hydrogen Energy*, 44, 19484–19518, 2019, https://doi.org/10.1016/j.ijhydene.2019.05.183.
17. Liang, Q., Brocks, G., Bieberle-Hütter, A., Oxygen evolution reaction (OER) mechanism under alkaline and acidic conditions. *J. Phys.: Energy.*, 3, 026001, 2021, https://doi.org/10.1088/2515-7655/abdc85.
18. Alobaid, A., Wang, C., Adomaitis, R.A., Mechanism and kinetics of HER and OER on NiFe LDH films in an alkaline electrolyte. *J. Electrochem. Soc.*, 165, J3395–J3404, 2018, https://doi.org/10.1149/2.0481815jes.
19. Li, C. and Baek, J.-B., Recent advances in noble metal (Pt Ru, and Ir)-based electrocatalysts for efficient hydrogen evolution reaction. *ACS Omega.*, 5, 31–40, 2020, https://doi.org/10.1021/acsomega.9b03550.
20. Jeong, S., Hu, K., Ohto, T., Nagata, Y., Masuda, H., Fujita, J., Ito, Y., Effect of graphene encapsulation of nimo alloys on oxygen evolution reaction. *ACS Catal.*, 10, 792–799, 2020, https://doi.org/10.1021/acscatal.9b04134.

21. Lyu, S., Guo, C., Wang, J., Li, Z., Yang, B., Lei, L., Wang, L., Xiao, J., Zhang, T., Hou, Y., Exceptional catalytic activity of oxygen evolution reaction via two-dimensional graphene multilayer confined metal-organic frameworks. *Nat. Commun.*, 13, 6171, 2022, https://doi.org/10.1038/s41467-022-33847-z.
22. Wan, W., Wei, S., Li, J., Triana, C.A., Zhou, Y., Patzke, G.R., Transition metal electrocatalysts encapsulated into N-doped carbon nanotubes on reduced graphene oxide nanosheets: Efficient water splitting through synergistic effects. *J. Mater Chem. A Mater.*, 7, 15145–15155, 2019, https://doi.org/10.1039/C9TA03213D.
23. Yao, Y., Jin, Z., Chen, Y., Gao, Z., Yan, J., Liu, H., Wang, J., Li, Y., Liu, S., Graphdiyne-WS2 2D-nanohybrid electrocatalysts for high-performance hydrogen evolution reaction. *Carbon N Y.*, 129, 228–235, 2018, https://doi.org/https://doi.org/10.1016/j.carbon.2017.12.024.
24. Liu, B., Peng, H.-Q., Cheng, J., Zhang, K., Chen, D., Shen, D., Wu, S., Jiao, T., Kong, X., Gao, Q., Bu, S., Lee, C.-S., Zhang, W., Nitrogen-doped graphene-encapsulated nickel–copper alloy nanoflower for highly efficient electrochemical hydrogen evolution reaction. *Small.*, 15, 1901545, 2019, https://doi.org/https://doi.org/10.1002/smll.201901545.
25. Yu, J., Guo, Y., Miao, S., Ni, M., Zhou, W., Shao, Z., Spherical ruthenium disulfide-sulfur-doped graphene composite as an efficient hydrogen evolution electrocatalyst. *ACS Appl. Mater. Interfaces*, 10, 34098– 34107, 2018, https://doi.org/10.1021/acsami.8b08239.
26. Li, X., Duan, X., Han, C., Fan, X., Li, Y., Zhang, F., Zhang, G., Peng, W., Wang, S., Chemical activation of nitrogen and sulfur co-doped graphene as defect-rich carbocatalyst for electrochemical water splitting. *Carbon N Y.*, 148, 540–549, 2019, https://doi.org/https://doi.org/10.1016/j.carbon.2019.04.021.
27. Lu, Y., Hou, W., Yang, D., Chen, Y., CoP nanosheets *in-situ* grown on N-doped graphene as an efficient and stable bifunctional electrocatalyst for hydrogen and oxygen evolution reactions. *Electrochim Acta*, 307, 543–552, 2019, https://doi.org/https://doi.org/10.1016/j.electacta.2019.03.208.
28. Zheng, L., Hu, W., Shu, X., Zheng, H., Fang, X., Ultrafine CoPx nanoparticles anchored on nitrogen doped reduced graphene oxides for superior hydrogenation in alkaline media. *Adv. Mater. Interfaces*, 5, 1800515, 2018, https://doi.org/https://doi.org/10.1002/admi.201800515.
29. Han, X., Tong, X., Liu, X., Chen, A., Wen, X., Yang, N., Guo, X.-Y., Hydrogen evolution reaction on hybrid catalysts of vertical MoS2 nanosheets and hydrogenated graphene. *ACS Catal.*, 8, 1828–1836, 2018, https://doi. org/10.1021/acscatal.7b03316.
30. Chen, Y., Ren, R., Wen, Z., Ci, S., Chang, J., Mao, S., Chen, J., Superior electrocatalysis for hydrogen evolution with crumpled graphene/tungsten disulfide/tungsten trioxide ternary nanohybrids. *Nano Energy*, 47, 66–73, 2018, https://doi.org/https://doi.org/10.1016/j.nanoen.2018.02.023.
31. Li, G., Yu, J., Jia, J., Yang, L., Zhao, L., Zhou, W., Liu, H., Cobalt–cobalt phosphide nanoparticles@nitrogen-phosphorus doped carbon/graphene derived

from cobalt ions adsorbed saccharomycete yeasts as an efficient, stable, and large-current-density electrode for hydrogen evolution reactions. *Adv. Funct. Mater.*, 28, 1801332, 2018, https://doi.org/https://doi.org/10.1002/adfm.201801332.

32. Chen, S., Lv, C., Liu, L., Li, M., Liu, J., Ma, J., Hao, P., Wang, X., Ding, W., Xie, M., Guo, X., High-temperature treatment to engineer the single-atom Pt coordination environment towards highly efficient hydrogen evolution. *J. Energy Chem.*, 59, 212–219, 2021, https://doi.org/https://doi. org/10.1016/j.jechem.2020.11.004.
33. Ye, S., Luo, F., Zhang, Q., Zhang, P., Xu, T., Wang, Q., He, D., Guo, L., Zhang, Y., He, C., Ouyang, X., Gu, M., Liu, J., Sun, X., Highly stable single Pt atomic sites anchored on aniline-stacked graphene for hydrogen evolution reaction. *Energy Environ. Sci.*, 12, 1000–1007, 2019, https://doi.org/10.1039/C8EE02888E.
34. Jin, Y., Luan, Y., Wu, Z., Wen, W., Zhang, X., Wang, S., Photocatalytic fuel cell-assisted molecularly imprinted self-powered sensor: A flexible and sensitive tool for detecting aflatoxin B1. *Anal. Chem.*, 93, 13204– 13211, 2021, https://doi.org/10.1021/acs.analchem.1c02074.
35. Yu, H., Xue, Y., Lu, Y., Wang, X., Zhu, S., Qin, W., Huo, M., Novel application of a Z-scheme photocatalyst of Ag3PO4@g-C3N4 for photocatalytic fuel cells. *J. Environ. Manage.*, 254, 109738, 2020, https://doi.org/https://doi.org/10.1016/j.jenvman.2019.109738.
36. Kondo, K., Watanabe, Y., Kuno, J., Ishii, Y., Kawasaki, S., Kato, M., Kalita, G., Hattori, Y., Mashkov, O., Sytnyk, M., Heiss, W., Flexible photo-catalytic electrode using graphene, non-noble metal, and organic semiconductors for hydrogen evolution reaction. *Energy Technol.*, 9, 2100123, 2021, https://doi.org/10.1002/ente.202100123.
37. Qiu, M., Sun, P., Cui, G., Tong, Y., Mai, W., A flexible microsupercapacitor with integral photocatalytic fuel cell for self-charging. *ACS Nano*, 13, 8246–8255, 2019, https://doi.org/10.1021/acsnano.9b03603.
38. Lui, G., Jiang, G., Lenos, J., Lin, E., Fowler, M., Yu, A., Chen, Z., Advanced biowaste-based flexible photocatalytic fuel cell as a green wearable power generator. *Adv. Mater. Technol.*, 2, 1600191, 2017, https://doi.org/https://doi.org/10.1002/admt.201600191.
39. Zhao, P., Jin, B., Yan, J., Peng, R., Fabrication of recyclable reduced graphene oxide/graphitic carbon nitride quantum dot aerogel hybrids with enhanced photocatalytic activity. *RSC Adv.*, 11, 35147–35155, 2021, https://doi.org/10.1039/D1RA06347B.
40. Nahyoon, N.A., Liu, L., Rabe, K., Thebo, K.H., Yuan, L., Sun, J., Yang, F., Significant photocatalytic degradation and electricity generation in the photocatalytic fuel cell (PFC) using novel anodic nanocomposite of Fe, graphene oxide, and titanium phosphate. *Electrochim Acta*, 271, 41–48, 2018, https://doi.org/https://doi.org/10.1016/j.electacta.2018.03.109.

41. Raptis, D., Ploumistos, A., Zagoraiou, E., Thomou, E., Daletou, M., Sygellou, L., Tasis, D., Lianos, P., Co-N doped reduced graphene oxide as oxygen reduction electrocatalyst applied to photocatalytic fuel cells. *Catal Today*, 315, 31–35, 2018, https://doi.org/https://doi.org/10.1016/j.cattod.2018.02.047.
42. Ong, B.C., Kamarudin, S.K., Basri, S., Direct liquid fuel cells: A review. *Int. J. Hydrogen Energy*, 42, 10142–10157, 2017, https://doi.org/https://doi.org/10.1016/j.ijhydene.2017.01.117.
43. Salarizadeh, P., Askari, M.B., di Bartolomeo, A., MoS2/Ni3S2/reduced graphene oxide nanostructure as an electrocatalyst for alcohol fuel cells. *ACS Appl. Nano Mater.*, 5, 3361–3373, 2022, https://doi.org/10.1021/acsanm.1c03946.
44. Li, B., Song, C., Huang, X., Ye, K., Cheng, K., Zhu, K., Yan, J., Cao, D., Wang, G., A novel anode for direct borohydride-hydrogen peroxide fuel cell: Au nanoparticles decorated 3D self-supported reduced graphene oxide foam. *ACS Sustain Chem. Eng.*, 7, 11129–11137, 2019. https://doi.org/10.1021/acssuschemeng.9b00192.
45. Luo, S., Wang, Y., Kong, T.C., Pan, W., Zhao, X., Leung, D.Y.C., Flexible direct formate paper fuel cells with high performance and great durability. *J. Power Sources*, 490, 229526, 2021, https://doi.org/https://doi.org/10.1016/j.jpowsour.2021.229526.
46. Lal, S., Deepa, M., Sahu, K.C., Janardhanan, V.M., Methanol-based fuel cell on paper support with N-doped graphene oxide/nickel cobaltite composite catalyst. *J. Electrochem. Soc.*, 166, F190, 2019, https://doi.org/10.1149/2.0301904jes.
47. Shaari, N., Kamarudin, S.K., Basri, S., Shyuan, L.K., Masdar, M.S., Nordin, D., Enhanced proton conductivity and methanol permeability reduction via sodium alginate electrolyte-sulfonated graphene oxide bio-membrane. *Nanoscale Res. Lett.*, 13, 82, 2018, https://doi.org/10.1186/s11671-018-2493-6.
48. Balan, A.E., Bita, B.I., Vizireanu, S., Dinescu, G., Stamatin, I., Trefilov, A.M.I., Carbon-nanowall microporous layers for proton exchange membrane fuel cell. *Membranes (Basel)*, 12, 1064, 2022, https://doi.org/10.3390/membranes12111064.
49. Zakaria, Z., Kamarudin, S.K., Timmiati, S.N., Influence of graphene oxide on the ethanol permeability and ionic conductivity of QPVA-based membrane in passive alkaline direct ethanol fuel cells. *Nanoscale Res. Lett.*, 14, 28, 2019, https://doi.org/10.1186/s11671-018-2836-3.
50. Zhu, B., Qu, C., Gao, S., Liang, Z., Zhang, H., Zou, R., Ultralow loading ruthenium nanoparticles on nitrogen-doped graphene aerogel for trifunctional electrocatalysis. *ChemCatChem.*, 10, 1113–1121, 2018, https://doi.org/https://doi.org/10.1002/cctc.201701691.
51. Wu, X., Feng, B., Li, W., Niu, Y., Yu, Y., Lu, S., Zhong, C., Liu, P., Tian, Z., Chen, L., Hu, W., Li, C.M., Metal-support interaction boosted electrocatalysis of ultrasmall iridium nanoparticles supported on nitrogen doped graphene for highly efficient water electrolysis in acidic and alkaline media.

Nano Energy, 62, 117–126, 2019, https://doi.org/https://doi.org/10.1016/j.nanoen.2019.05.034.

52. Zhang, Q., Duan, Z., Wang, Y., Li, L., Nan, B., Guan, J., Atomically dispersed iridium catalysts for multifunctional electrocatalysis. *J. Mater. Chem. A Mater.*, 8, 19665–19673, 2020, https://doi.org/10.1039/D0TA05750A.
53. Sarkar, B., Barman, B.K., Nanda, K.K., Non-precious bimetallic CoCr nanostructures entrapped in bamboo-like nitrogen-doped graphene tube as a robust bifunctional electrocatalyst for total water splitting. *ACS Appl. Energy Mater.*, 1, 1116–1126, 2018, https://doi.org/10.1021/acsaem.7b00233.
54. Gopi, S., Panda, A., Ramu, A.G., Theerthagiri, J., Kim, H., Yun, K., Bifunctional electrocatalysts for water splitting from a bimetallic (V doped-NixFey) metal–organic framework MOF@graphene oxide composite. *Int. J. Hydrogen Energy*, 47, 42122–42135, 2022, https://doi.org/https://doi.org/10.1016/j. ijhydene.2021.05.028.
55. Debata, S., Patra, S., Banerjee, S., Madhuri, R., Sharma, P.K., Controlled hydrothermal synthesis of graphene supported NiCo2O4 coral-like nanostructures: An efficient electrocatalyst for overall water splitting. *Appl. Surf. Sci.*, 449, 203–212, 2018, https://doi.org/https://doi.org/10.1016/j.apsusc.2018.01.302.

11

Environmental Electrocatalysis for Air Pollution Applications

Anupama M. Pillai and Tanvir Arfin*

Air Pollution Control Division, CSIR-National Environmental Engineering Research Institute (CSIR-NEERI), Nehru Marg, Nagpur, India

Abstract

Air pollution is a noteworthy threat to the rising population of the world. According to the data, 90% of the world's population lives in areas where pollution levels go above and beyond the World Health Organization (WHO) guidelines. Immediate measures and stringent regulatory policies should accomplish a sustainable and habitable environment. Environmental catalysis has called attention to air pollutant mitigation, comprising nitrogen oxides (NOx), volatile organic compounds (VOCs), and formaldehyde, because of its favorable process circumstances and being environment-friendly. We refer to catalysis and room-temperature catalysis methods by environmental catalysis. By utilizing heat or light energy as the primary source, we can excite the catalysts and instigate O_2 activation to generate numerous reactive oxygen species liable for adsorbed gas pollutant oxidation, leading to ambient air purification. In this aspect, transitional metal oxides and their assisted noble metal nanoparticles are the most familiar and assuring catalysts notable of study because of their outstanding performance regarding O_2 activation. Various methods have been established to mold the particle size, morphology, and exposed surface to exploit the electronic structure and create additional active sites for better reaction and adsorption. Supremely, environmental electrocatalysis uses additive source control approaches for ambient air purification, putting forward higher efficacy and low energy consumption. Electrocatalyst strategy, synthesis, description, and optimization are the keys to this technology and are of chief reputation to this chapter.

Keywords: Environmental electrocatalysis, air purification, VOC removal, CO_2 reduction and utilization, degradation

**Corresponding author*: tanvirarfin@ymail.com

Sudheesh K. Shukla, Chaudhery Mustansar Hussain, Santanu Patra, Meenakshi Choudhary (eds.) Electrocatalytic Materials for Renewable Energy, (303–332) © 2024 Scrivener Publishing LLC

11.1 General Introduction

The advancement of materials for environmental presentations is based chiefly on two vital aims: (i) diminishment of circuits and (ii) reduction of cost [1–3]. Nanotechnology is the latest tool to draw up high-level materials that are highly efficient and smaller in size [4–6]. Advanced materials based on nanotechnology play a necessary part in environmental implementation. Nanotechnology has both direct and indirect applications for the production of eco-friendly products [7–9]. Direct applications are the identification and elimination of pollutants, and indirect applications are the designing or planning of engineering and industrial procedures. Molecular-level approaches are advantageous as many benefits are attained using size imprisonment, quantum effects, and interfacial interaction [10–12].

Nanotechnology techniques for environmental recovery are deterioration and segregation, which reduce the threats to humans and the environment [13–15]. Using nanomaterials for environmental recovery is expeditious, cost-effective, and safe. Specific modulation and designing of the materials at the molecular level may offer increased pollutants' affinity, capacity, and selectivity [16–18]. The utilization of nanotechnology is highly assured in the field of pollution [19]. Nanomaterials are developed for the purification of indoor air volumes [20, 21]. Nanofilters of the next generation are being used in automotive tailpipes and manufacturing smokestacks to remove particulate matter pollutants [22]. Because of their good theoretical energy density and unlimited oxygen fuel from the air, rechargeable metal–air batteries offer considerable potential for replacing traditional batteries [23].

11.2 Introduction of Air Pollution

Since ancient times, air pollution has been a severe issue faced worldwide. In the earlier periods, air pollution issues were not considered, but as humans progressed, air pollution issues also increased, worsening the condition. Several scientists defined air pollution differently, but as humans moved, the definition of air pollution also changed accordingly; for instance, according to Vallero *et al.* [24], air pollution is defined as the presence of pollutants in quantities that are harmful to human health, as well as damage to ecosystems as well as other socially valuable items such as structure and materials. According to Azam *et al.* [25], air pollution is all the harmful impacts of any foundations that subsidize atmospheric pollution and ecosystem degradation. According to Clark *et al.* [26], air pollution occurs when humans release solid and gaseous contaminants into the environment in quantities high enough to harm plants, other life forms, animals, works of art, and ecosystems.

Air plays an inevitable part in the existence of the biological scheme. All organisms that live on water, air, and earth require air or oxygen to survive. Air is essential for respiration, in which living beings inhale oxygen and exhale carbon dioxide (CO_2). Respiration provides energy to all physical coordination to eat, live life, and grow. This released CO_2 in the air lengthways with sunlight is used by plants to make food and then releases oxygen (O_2) into the atmosphere, which helps preserve the air quality. This superiority has been affected in the last few decades because of natural and human activities that pollute the air. Polluted air is destructive to human well-being and other biological organizations. This is accountable for acid rain and smog, cancer and respiratory diseases, ozone-depleting layer from the atmosphere, and global warming subsidies.

Air pollution issues were encountered for the first time in London during the 13th century due to coal burning. England's King Edward took a measure to decrease air pollution by raising consequences against the Londoners who burned coal. Again, the issue of air pollution arose in the 16th century. Due to the industrial revolution during the 18th and 19th centuries, coal feasting became extensive. This caused soot and smog in the atmosphere, which caused severe health issues for the residents progressing in urban areas. Kelly and Fussell [27] and Ritchie [28] deliberate the history of London air pollution, providing an image of the upcoming megacities' cloud. The historical appearance of pollution stages has been found in London.

11.3 Global Scenario of Air Pollution

The air pollution problems were undoubtedly from the past, but as technology develops and *the population rises, the issues were not confined to some countries. In London and other countries, air pollution-related problems were multiplying.* The original reason for air pollution was only coal burning because of the industrial revolution, but as technology improved and the population rose, the logic of air pollution increased in several ways; for instance, Romieu *et al.* [29] studied *air pollution in the Caribbean* and Latin America at the end of 19th century. According to their studies, air pollution is a chief *threat in* the Caribbean and Latin American nations, and industrialization and urbanization are the prime causes.

The chief cause of air pollutants is the smoke released from vehicles and domestic fuel burning that discharges lead (Pb), sulfur dioxide (SO_2), particulate matter (PM), etc. All these pollutants surpass the limit of air quality standards that have been recognized in Latin American countries. A significant part of these urban zones is pretentious by anthropogenic pollutant emissions, which includes the territory of Brazil and the city of Chile. Air pollution, climate fluctuation, and human health in the various zones of Brazil have been studied by Menezes *et al.* [30]. According to their findings, the main reason for air pollution is biomass burning and transportation which releases pollutants like SO_2, aldehydes, nitrogen dioxide (NO_2), hydrocarbons, ozone (O_3), and PM. All these pollutants accrue in the stratosphere layer of the atmosphere and affect climate and human health directly or indirectly.

African nations' emissions of aerosol particles have been estimated through satellite observation. A group of researchers have also forecasted the degree and transport of PM, O_3, carbon monoxide (CO), and SO_2 with the help of climate and atmospheric modeling using epidemiological prototypes, and they also predicted the health issues caused by pollution in 2016. The observations from an examination surprised Bauer's team, which shows the significant impact of dust storms on public health, and exposure to these natural emissions led to 556,475 premature deaths in African nations in 2016, and around 43,489 deaths typically happened in West African countries. According to the studies, 182,398 deaths were caused due to industrial emissions and 43,374 deaths due to biomass burning. According to the findings of Bauer's team, poor air quality is similar in scale to the human immunodeficiency virus (HIV) or acquired immunodeficiency syndrome (AIDS) humanity rate in the zone.

Generally, it is observed that seven million people die because of exposure to air pollutants, as per the WHO report, which produced respiratory diseases, heart diseases, lung cancer, and other various health issues; more than two-thirds of the total deaths have occurred in Asia. One out of 10 children younger than 5 years old died due to air pollution. It is observed that every day 99% of children in Southeast Asia inhale polluted air. In several countries like South Korea, Thailand, and India, approximately the main cities face problems due to poor quality. This caused authorities to take measures to reduce smog and toxic particles that enter the human body. The primary sources of air contaminants are heavy vehicular traffic, burning agricultural leftovers, coal burning for electricity, industrial processes, and construction work. Toxic air $PM_{2.5}$ in cities is the most significant protuberant pollutant, which affects the lungs and is also the reason for respiratory problems and heart disease.

In the last few decades, the air quality has been worsening substantially in several cities in emerging countries, and nowadays, several efforts are being made to enforce this problem. The leading cause for taking the initiative to control air pollutants such as silicon dioxide (SiO_2) or PM is the harmful effects they have on social fitness. These harmful effects comprise premature mortality and a rise in cases of chronic heart and lung diseases. Evaluation of the health issues caused by air pollution is significant because it provides an image of both the effect of environmental control measures and an assessment of the changes brought by implementing specific pollution control policies.

Electrocatalysis is a vital area in electrochemistry that developed and flourished after the late 1980s because of comprehensive requests for novel hybrid procedures. However, the majority of the recommendations have been used for academic commitments and not for technology usage in businesses. Implementing original ideas of electrocatalysis for manufacturing electrochemical systems is inevitable, requiring immediate consideration from chemists and planners.

Consequently, manufacturing electrocatalytic progressions have only been depicted primarily in the works from the chemical manufacturing aspects with short supplementary electrochemical descriptions throughout the 1970s. The modeling and production of an electrocatalyst are purely created on innovative impressions, including a measured external unevenness, atomic topographic outlines, and definite catalytic center locations, in the sequence of the electrochemical responses.

Electrocatalysis is developed as an excellent worldwide platform purely enthusiastic about discussing new thoughts in electrocatalysis for educational, administrative, and manufacturing researchers.

Electrocatalysis promotes logical inventions and innovations that stimulate precise engineering theories by rapidly publishing actual outcomes, perceptions, and originations, which is vital in expanding new electrochemical skills.

The ideas and boundaries of electrocatalysis in the workshop, which included its theoretic practice, also provide the readers with a stronger opinion of the difficulties related to electrocatalytic responses. This particular matter has gained the interest of several writers who studied the ideologies of electrocatalysis with applicability to untried circumstances of special priority in electrochemical manufacturing.

11.4 Halogenated Organic Compounds (HOPs)

Halogen-containing organic compounds, such as organic compounds comprising one or more halogens like fluorine (F), chlorine (Cl), bromine (Br), or iodine (I), have caused several menaces to the ecosystem, mainly to the aquatic ecosystem and to human beings, thus arousing attention nowadays. These organic compounds are primarily used in several manufacturing industries and factories, including the manufacturing of flame extinguishers, insecticides, pesticides, and fungicides [31]. These compounds are identified in water, air, and soil as they have been widely used for the last few years, leading to harmful effects on the environment due to their wide distribution, consistency, toxicity, and bioaccumulation [32].

Several degradation processes treat such compounds containing halogens that cause contamination. In the previous few decades, tremendous efforts have been taken to treat and degrade such halogen-containing compounds. The bond between carbon and halogens (C–X bonds, X can be considered as F, Cl, Br, and I) in halogenated organic compounds are solid; thus, several methods like electrocatalysis are developed to break those bonds [33].

It is challenging to correlate the efficiency of several degradation processes because of several target contaminants, the concentration at initials, causes of pollution, arrangements of the experiment, and several physicochemical criteria of the method used [34]. Electrocatalysis differentiates itself from other forms as it is accessible to implement, is highly energy-efficient, has manageable automation and safety, is unique, has high efficiency and unique deterioration ability, and is environmentally friendly [35].

Nowadays, electrocatalysis has gained more attention as it can be used as a deterioration method for hydrogenated organic compounds, specifically for perfluorinated compounds (PFCs) [36]. For instance, Rondinini *et al.* [37] discussed electrode-related reaction mechanisms and processes for the electroreduction of halogenated organic compounds. Martin *et al.* [31] considered several essential features of electroreduction methods of halogenated organic compounds, including experimental techniques, designing of the cell, and degradation procedures.

11.4.1 Electrooxidative Dehalogenation (EOD)

Electrooxidation is a successful technique to mineralize halogen-containing organic contaminants. Several catalysts have been used in electrooxidative dehalogenation (EOD) methods. Several adjustment methods have been used to improve the presentation of those electrodes. The important electrooxidative dehalogenation methods are discussed below.

11.4.1.1 PbO_2-Based Catalysts

As PbO_2-based catalysts are cost-effective, highly conductive, and chemically resistant and have high oxygen overvoltage [38], these materials are widely used in electrooxidative dehalogenation processes. Table 11.1 contains the latest illustrative studies of electrooxidative dehalogenation using several PbO_2-based catalytic materials. From these consequences, it is observed that the components and structure of the catalysts have an essential role in their decomposing presentation. The successful methods that can increase the potential of PbO_2-based catalysts are explained in this chapter.

11.4.1.2 SnO_2-Based Catalysts

Although SnO_2 belongs to an n-type semiconductor and its bandgap value is 3.6 eV, SnO_2-based electrocatalysts are more often used as catalysts to mineralize halogenated organic compounds [44]. To advance its application, the intrinsic properties of SnO_2 have to be changed, which can be achieved by the method of doping and, thus, make it a demandable electrode material. Table 11.2 represents the latest illustrative studies of SnO_2-based catalytic anodes, which are widely applied in the forms of HOP oxidation.

Table 11.1 Conclusion of illustration HOP-EOD studies applying PbO_2-based anodes.

S. no.	Anode	HOPs	Current density (mA cm^{-2})	HOP removal (%)	Time (h)	Reference
1.	GNS-PbO_2	2-Chlorophenol	50	95.42	2	[39]
2.	3 DEM-PbO_2	Flutriafol	5	73.8	3	[40]
3.	PbO_2-CeO_2	Malachite green	30	95.4	1.5	[41]
4.	Yb-PbO_2	Acetamiprid	250	98.96	2	[42]
5.	PTFE-doped b-PbO_2	Norfloxacin	10	100	12	[43]

Table 11.2 Conclusion of illustration HOP-EOD studies applying SnO_2-based anodes.

S. no.	Anode	HOPs	Current density (mA cm^{-2})	HOP removal (%)	Time (h)	Reference
1.	MI-meso SnO_2	2,4-Dichlorophenoxyacetic acid	6.34	99.15	3	[45]
2.	SnO_2-Sb_2O_5	2,2-Dichlorovinyl dimethyl phosphate	–	100	100 min	[46]
3.	Ti/SnO_2-F	Perfluorooctanoic acid	20	23.8	3.5	[47]
4.	SnO_2-Sb/CA	Perfluorooctanoate	20	91	5	[48]
5.	Mp-SnO_2/BDD	Clofibric acid	30	95	240 min	[49]

11.4.1.3 Other Metal Oxide-Based Catalysts

In recent years, various metal oxides like ruthenium (IV) oxide (RuO_2), bismuth (III) oxide (Bi_2O_3), iridium (IV) oxide (IrO_2), and titanium oxide (4/7) dititanium (III) oxide (Ti_4O_7) have also been used as electrodes as they are

Table 11.3 Conclusion of illustration HOP-EOD studies applying metal oxide (except PbO_2 and SnO_2)-based anodes.

S. no.	Anode	HOPs	Current density (mA cm^{-2})	HOP removal (%)	Time (h)	Reference
1.	Ti/RuO_2	Ofloxacin	–	80	30 min	[51]
2.	Ti/RuO_2-IrO_2	4-Chlorophenol	25	57.61	89 min	[52]
3.	RuO_2-coated Ti	Rhodamine 6G	16.52	99	–	[53]
4.	Ti-Bi_2O_3	Reactive Orange 4	40	100	40 min	[54]
5.	Porous Ti_4O_7 ceramic	Perfluorooctanesulfonate	5	93.1	3.4 min	[55]

comparatively of minimum cost, have high chemical stability, and higher electrocatalytic properties [50]. Table 11.3 represents the latest illustrative studies, including the electrochemical oxidation of HOPs by these metal oxides.

11.4.2 Electrooxidative Dehalogenation Mechanisms

By using electrochemical oxidation in the presence of catalysts resulting from labile species or electrons, the breakage of strong C–X bonds can be achieved successfully. Therefore, the EOD processes are categorized into straight indirect oxidation and anodic oxidation. Most of the degradation procedures include both paths. However, the direct oxidation reaction plays a significant role as it is the rate-determining step in the degradation procedures of many pollutants which is inactive with active oxygen species, for example, fluorinated compounds [56].

Due to the high stability of the C–F bond and the harmfulness of PFCs, it is essential to put forward the methods to fossilize PFCs. Indeed, the various PFCs and degradation system processes may differ significantly. Previous review efforts [57] provide more extensive information, and Figure 11.1 depicts a typical PFC breakdown pathway.

11.5 Perfluorohexane Sulfonate (PFHxS)

Perfluorohexane sulfonate (PFHxS) is a member of the chemical group called perfluorinated alkylated substances (PFASs). The structure of

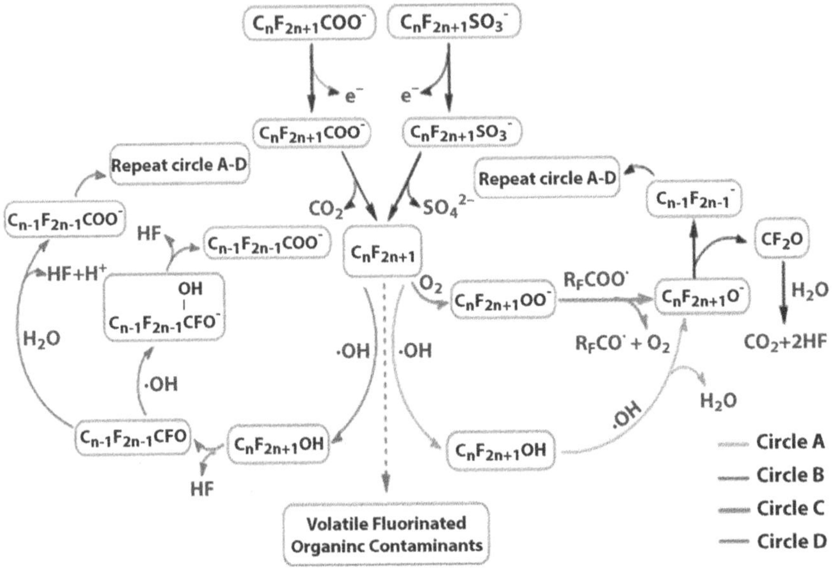

Figure 11.1 The electrochemical oxidation mechanism of PFCs. Reprinted from ref. [58], copyright (2016), with permission from Elsevier.

PFHxS is given in Figure 11.2. It is commonly applied in many industrial procedures and the production of many consumer goods. PFHxS can be recycled in firefighting foams, anti-dirt materials, food wraps, and surfactants. In 2002, a significant producer of PFHxS got ceased [59]. The manufacturing of PFHxS was discontinued in the whole country but continued in a few states. The crops wrapped with PFHxS are allowed to be imported to the USA.

Another PFAS, called perfluorohexane sulfonic corrosive (PFHS), is widely used to produce surfactants and fluoropolymers and as water- and tinge-defensive coverings of floor, materials, and paper. It is expected to be found on the earth and even in drinking groundwater. PFHxS and its salts are the raw materials used for the synthesis and fabrication of PFAS,

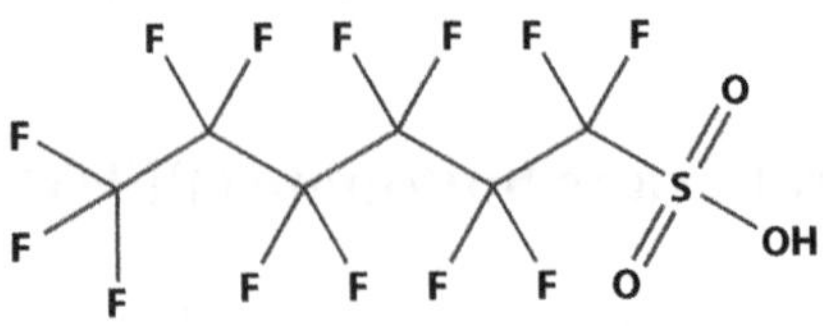

Figure 11.2 Structure of PFHxS.

which is used as a surfactant and surface, and sometimes it is accidentally produced during the industrial processes. PFHxS also belongs to a similar chemical class as perfluorooctane sulfonic corrosive (PFOS). PFAS is added to the Stockholm Convention (SC). Perfluorohexane sulfonate is one of the PFASs used as a mechanical surfactant. It is produced along with the production of fluorotelomer alcohols, perfluorosulfinic acids, and perfluorooctane and during the electrochemical fluorination methods [60].

These reports highlighted the importance of the risk management evaluation (RME) of PFHxS and its related salts, which are under the Annex F of the SC. Support data and research based on various dispersed operations and efforts were deliberately undertaken to add to the RME's data. This report gives an idea about the related four subjects: separation and segregation of PFHxS is a complicated process that is constant in soil and water. How PFHxS moves through the complete soil is related to the texture of dirt, and it is the science behind it. PFHxS has accumulated in the waterbodies of Minnesota for several days.

PFHS may remain as fumes in enclosed air [61] which has an expected weight of 0.0046 mm Hg at 25°C (SRC), which is predicted from a part of a steady technique using prototypes of gas or molecule and wrapping of semivolatile natural combinations in the air [62]. Further investigation reveals that these compounds persist in particulate and fume levels in the atmosphere [63]. The fume level PFHS is contaminated in the air along with photochemically delivered hydroxyl radicals (SRC). The rate reliability of 1.4×10^{-13} Cu cm/atom s at 25°C (SRC) has been disturbed through a structure approximation process, and thus, the half-life of this retort is found to be 115 days (SRC) [62]. The report may throw out particulate level PFHS from the air through wet and dry declaration (SRC). Perfluoroalkyl sulfonic acids may not go through straight photolysis [64]. It is observed that PFHS can preach over a long distance worldwide as it is tested and identified in the remote Arctic and marine air.

PFASs are not exposed to air-sparging methods as they have low Henry's law constants. Similarly, the treatment of PFAS with activated carbon gives no results because it has a low affinity for activated carbon adsorbents, and its adsorption rate is very low. Moreover, due to the occurrence of large amounts of co-contaminants, the effect of activated carbon declined as it competed for the available adsorption sites. In 2017, Bruton and Sedlak using small, obviously occurring reduction mechanisms, exhibited that PFASs abide and relocate on a large scale in the subsurface leading to the formation of huge dissolved plumes, and they also demonstrated that perfluorinated carboxylates are exposed to persulfate oxidation at high temperature and low pH circumstances and not perfluorinated sulfonates.

Sonication has also been used as a method to defluorinate PFAS, and the latest studies showed the usage of plasma for the destruction of PFAS [65].

The latest studies illustrated the potentiality of electrochemical (EC) oxidation of PFAS-wedged H_2O. These studies put forward the possibilities of titanium-based anodes [66] or boron-doped diamond (BDD) anodes. Due to their high oxidizing potential and durability, BDD anodes attained fascination [67]. Another merit is that BDD is commercially made in large surface area (>1 m^2) schemes. It is observed from the studies using BDD that the PFAS oxidation mechanism includes an initial step of controlling direct electron transfer at the anode [68] and then trailed by an "unzipping" mechanism of the perfluorinated chain that leads to the production of perfluorinated carboxylic acids (PFCAs) which is shortly chained, after the completion of defluorination. Earlier, much importance is not given to PFAA precursor oxidation, processing in groundwater systems at environment-friendly concentrations, and durability of PFAS treatment.

11.6 Methoxychlor (MXC)

Pure methoxychlor (MXC) appears as a less white contrast milky concentrated substance, which has a sweet and putrid odor. It will not quickly disappear in water and wind. The breakdown of MXC is done by a fungicide laborer using an oil improper fluid and introducing this as a rinse, and those mixtures formed with different artificial compounds are applied as a small stain. MXC is primarily used as a pesticide to kill creepy crawlies. Since MXC was used to control pests, its release into the earth in small amounts can cause serious environmental issues. As illustrated in Figure 11.3, MXC was released into the environment through herbal substance manufacturing. What type of MXC has been terminated due to assembly destinations? What figure items understand MXC? Likewise, a small amount may be released from harmful waste settings where it was discharged. An investigation was conducted in the Orontes River, where MXC, hexachlorobenzene (HCB), dichlorodiphenyltrichloroethane (DDT), and tetradifon were illustrious in the bottommost dimension (LOQ), and other organochlorine pesticide (OCP) analytes could not be distinguished [69]. No additional investigations were made in the Orontes because of its vicinity to Syria where a battle took place between 2012 and 2017.

MXC is a manufactured natural halogen-based compound that has been used as an insecticide. Though similar to DDT, MXC will not accumulate in the slippery tissue of organisms. The commercial names of MXC are Matos, Margate, Moxie, etc. Its molecular formula is $C_{16}H_{15}C_{13}O_2$. MXC is

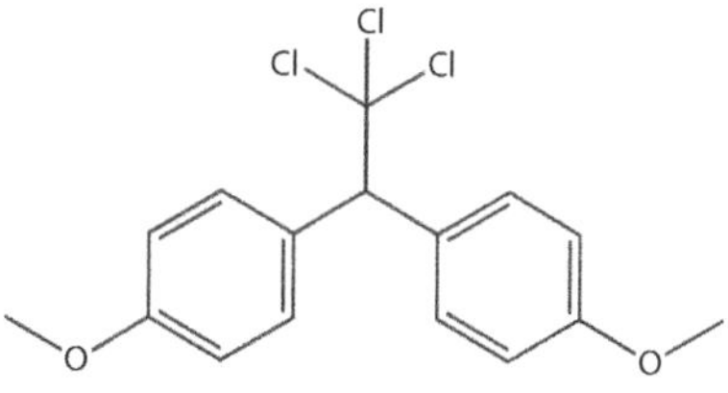

Figure 11.3 Structure of MXC.

yellowish dusty in appearance and possesses a sweetened aroma. MXC is used as a pesticide to control flies, cockroaches, mosquitoes, and vermin. It kills bugs through its usage or interaction. MXC is applied in different fields to keep away pests that attack numerous yields, plants, and trees.

Moreover, it was utilized to control parasites in cows. Attempts were made to replace DDT with MXC; however, it was found that MXC is highly toxic, has a short lifetime, and has a high bioaccumulation rate [70]. Later, MXC got boycotted due to its drawbacks related to metabolism [70]. An experiment using beetles gives proof of the breakdown of MXC, which is hydroxychlor, p,p′-hydroxy-DDT (HPTE), and it helps in the advancement of estronol-subordinate excellence, which can be done by treating it with antiestrogen [71]. It has been observed that male rodents exposed to MXC show a little load around the testicles and a small cell size as expected [72]. In a study involving rats and hamsters, MXC in combination with antiestrogen and estronol influenced the complications of the reproductive tract and deterioration of the ovary [73].

Furthermore, MXC incites an estronol influence on oophorectomy in female rodents, which is observed to affect weight in the womb [74]. An example of MXC in the Arctic shows a good lifespan and can spread through water and wind [75]. Research studies found that MXC has an estrogenic action, which may lead to adverse effects on reproduction and early pregnancy [76]. It appears to enhance abnormal cell growth [77], and its metabolite was observed to decrease testosterone production in rodents [78]. However, daylight can reduce its effect [79], but the outcome (MXC olefin) is constant and eliminated more gradually [80]. Biodegradation of MXC brings about damage through a small amount of halogen together with demethylate [81]. The ability of catalytic reduction of MXC in bis(-methoxybenzene) is completed using electrogenerated copper lessening together with ultraviolet light-activated charcoal [82]. Experiments were conducted using several kinds of TiO_2 below UV/O_2, among which 60%–90% in 30 min was produced by risks confirmed in crystal unfilled nanoparticles. Special permission has been granted to discard completely

99% of MXC. MXC disposal is carried out through dust covering with dehydrogenation, dechlorination, and replacement. In Argentina, a supplementary operation involving actinomycete was implemented [83].

MXC at a concentration of about 254 ng/m^3 in air samples was detected in an insecticide formulation plant in the southern part of California, USA. The other essential property that makes MXC familiar is that it stays in the air when used as an insecticide. Its fume mass is less or it mixes with dirt. The partial degradation MXC present in the air has not been found. The experiment was evaluated for 1 to 11 h [84]. High-separation techniques to detect MXC in the air during cold days were used [85]. Even in minor amounts, the blending of MXC with other organochlorinated insecticides may create global environmental issues. In two U.S. urban societies, a study related to the presence of insecticides in the air reported poor degrees of MXC of 0 to 100 pg/m^3 in the open air [86].

Furthermore, from the Canadian statement, the annual harmful amount of MXC can be attributed to the wind. The situation was crucial even after MXC was phased out as it continues to be detected [87]. The concentration of MXC in the air in the United States is minimal.

An organic solvent or water can be used as the reaction media of the electroreductive dehalogenation process. Utilizing organic solvent accompanies several qualities, such as high organohalide solubility, an increase in mass transfer, an extensive potential window, etc. The drawbacks include higher costs, consumption of energy, and the possibility of secondary pollution. Utilizing water for electroreductive dehalogenation is an eco-friendly and sustainable procedure. The problem is that the dehalogenation processes are highly repressed due to lateral reactions. The stable mass transfer limitation and C–Cl bonds occur due to organohalides' non-polar feature, stability, and catalytic activity. Moreover, because of the weak catalytic characteristic of electrocatalysts, electrocatalytic dehalogenation frequently cascades over the hydrogen evolution reaction (HER) method [88]. Thus, the benefit from electrocatalytic dehalogenation is poor. These demerits significantly affect the comprehensive utilization of electroreductive dehalogenation. Therefore, to address these problems, planning an original, multipurpose electrocatalytic scheme using well-organized catalysts for stimulating the C–Cl bond and with the capability to produce the H* species is necessary, which enables the transfer of electrons and hydrogen. In addition, capitalizing on their merits could both directly and indirectly decrease and increase their catalytic activity, and this could expose the chance of advancement of electroreductive dehalogenation (ERD) in H_2O and opens a window for upcoming comprehensive utilization in synthesis, catalysis, and environment.

Countless attempts have been made to improve the structure of electrocatalysts, such as their surface structure including size, morphology, plane, and electronic system; hence, the movement of the catalyst used in the ERD procedure can be improved [89]. Studies have been done to improve the dehalogenation action, which shows improvement by producing bimetallic or trimetallic nanocatalysts by alloying Pd with other transition metals [90]. The catalysts with measured imperfections have conquered the lure of booming attention and proved to be the electrocatalysts with greatly improved electrochemical activity for ORR and dehalogenation methods [91]. For this reason, we foresee that integrating the qualities of dissimilar transition metals in straight and subsidiary electrocatalytic processes to fabricate a faulty alloy will increase their catalytic dehalogenation ability because of the synergistic consequence of metals, which makes it available for the straight and subsidiary track to happen at the same time, and this two-fold practical electrocatalyst would surely improve the dehalogenation processes in H_2O and thus opens a path to overcome previous crucial scientific and practical experiments.

11.7 Dioxin and Furan

According to the SC, PCDD/Fs are categorized into dioxin and furan compounds. Polychlorinated dibenzodioxins (PCDDs) and polychlorinated dibenzofurans (PCDFs) are the unintended derivatives of the combustion of chlorinated products partially and are present globally [92], as shown in Figure 11.4. The quantity and location of chlorine atoms in the two aromatic rings determine the degree of harmfulness and chemical properties of these compounds. PCDDs and PCDFs constitute 75 and 135 homologs, respectively. Specific isomers of PCDD/Fs are acknowledged as harmful as they contain carcinogens.

Further studies clarified that dioxins and furan-like compounds may intervene in thyroid hormone levels, leading to development retardation, neurocognitive deficiencies, and reproductive impairment [93].

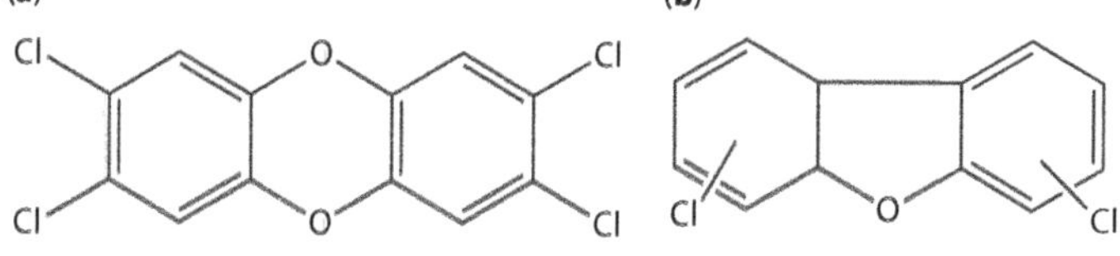

Figure 11.4 Structure of PCDD (a) and PCFD (b).

Exposure of infants to dioxin through breast milk and during pregnancy causes several health problems. Guo *et al.* [93] observed that exposure to these compounds might lead to a declining semen superiority, a rise in the probability of irregular sperm, and lessened daily sperm production in sexually active men. However, human exposure to PBDD/F is immense. The path of exposure to dioxin and furan varies, as shown in Figure 11.5. The studies on these rising pollutants are limited, and most of them focus on occupational liability [94].

The discarding of PCDD/F during incineration procedures initially came into notice in 1977. The researchers have calculated the discharge of those compounds during a sequence of thermal systems, including gasification and combined combustion [95]. The primary cause of the negative environmental fame of waste incineration is the release of PCDD/F and supplementary contaminants throughout the process, particularly for MSW incineration.

The amount of PCDD/Fs released is directly proportional to the quantity of carbon applied. Some of the essential components that aid in the formation of PCDD/Fs are carbon black (CB), compost (CP), residual carbon, and polycyclic aromatic hydrocarbons (PAHs). Additionally, other factors that contribute to the creation of PCDD/Fs include carbon black (CB), compost (CP), residual carbon, and polycyclic aromatic hydrocarbons (PAHs). These include PCDD, persistence, precursors, ignition temperature, chlorine in the maintenance process, and the ability to get more fuel and oxygen [96]. PCDFs are released between the cooling cycles of flue gas in the combustion process. The development procedure occurs through the two appliances discussed above [95]. The focal appliance for dioxin preparation in combustion schemes is the advanced protocol,

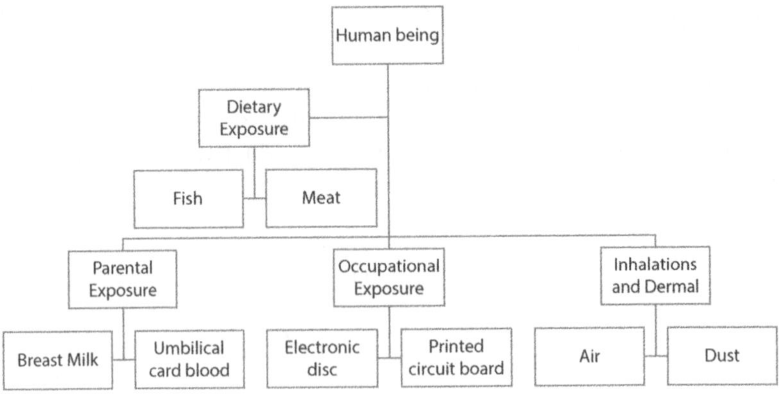

Figure 11.5 Summary of human exposure to POPs.

where carbon morphology from the decayed graphical outline is vital for dioxin preparation. Hence, such carbon morphologies were discovered. It was observed that the soot particles produced from the gas-phase combustion process involved deteriorated graphical protocol, and it has recently served as a foundation of the dioxin protocol.

Nowadays, dioxin shows an unspecified electrocatalytic attitude. However, electrocatalytic oxidation appears to show extraordinary competence and selectivity in the decomposition of toxicants. Thus, the gained outcomes made us form an outline for the electrocatalytic oxidation of dioxin above Pt and adsorbed oxygen to produce mineral yields. The electrocatalytic oxidation of dioxin can be explained according to the scheme in Figure 11.6. The characteristics of dioxin in electrocatalytic systems were

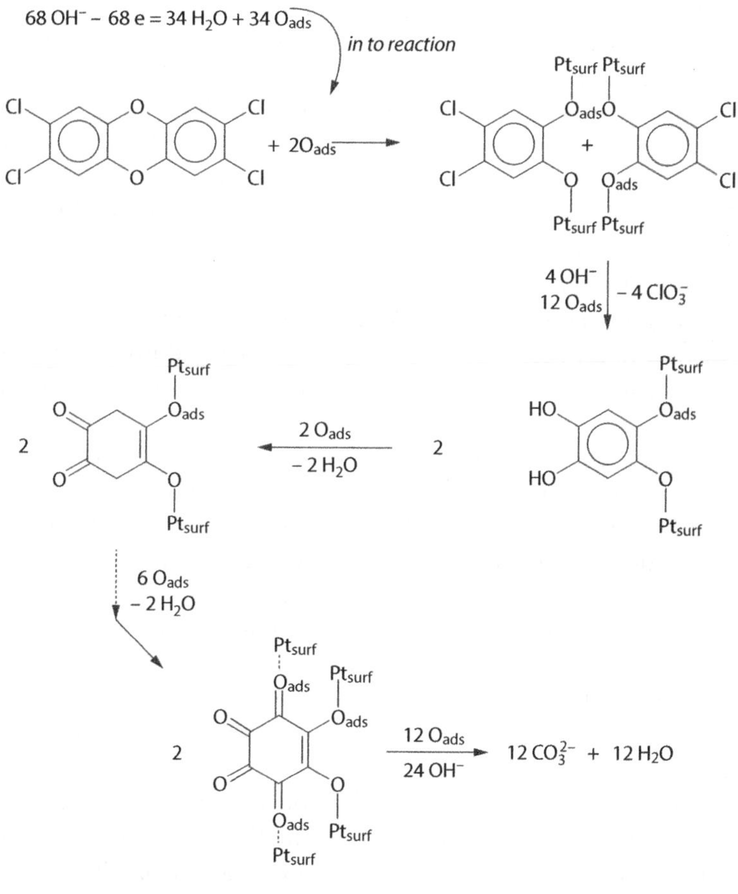

Figure 11.6 Dioxin reaction with adsorbed oxygen. Reprinted from ref. [97], copyright (2002), with permission from Elsevier.

put forward based on the analogy in arrangement and kinetic association in the oxidation of dioxin.

11.8 Volatile Organic Compounds (VOCs)

Volatile organic compounds (VOCs) are a combination of gaseous classes having extraordinary vapor pressures in the atmosphere. Typical organic compounds (OCs) include alkane, alkene, aromatic, halocarbon, and carbonyl [98]. Over the years, VOC emission from transportation and industries has increased and has excessively worsened, which is very harmful to the ecosystem. Most VOCs, such as benzene, formaldehyde, vinyl chloride, and polychlorinated biphenyls, are highly toxic and hazardous to human beings and are categorized as group 1 carcinogen protoplasmic poison by the International Agency for Research on Cancer (IARC) [99]. For example, benzene is ultimately the reason for low birth weight [100].

Moreover, VOCs are the main precursors for developing O_3 and secondary organic aerosols (SOAs). The rise in VOCs leads to growth in O_3 contamination and PM [99]. The decrease in extremely reactive VOCs controlled notable total radical manufacture and thus reduced the tropospheric O_3 level [101].

According to EPA's Total Exposure Assessment Methodology (TEAM) reports, it is observed that around 12 organic contaminants are found to be present 2 to 5 times higher in advanced privileged homes than in the external environment, irrespective of the location—rural areas or manufacturing zones. The TEAM reports also found that using products containing organic chemicals indoors leads to extreme contamination and the released concentrations can remain in the air for an extended time even after the activity is finished [102]. People spend more than 8% of their activities indoors. Therefore, it is necessary to eradicate indoor air VOCs to achieve environmental regulations and well-being.

Various observations and reports have improved the actual and suitable approaches to eradicate indoor air contamination. The most excellent operative method to decrease industrial VOC emissions is catalytic combustion since VOCs can be oxidized to CO_2 using the abovementioned catalysts.

Inappropriately, VOC's catalytic oxidation at room temperature was not attained up to the present time, and this method is not used for indoor VOC elimination. Recently, adsorption, photocatalysis, biofilters, and nonthermal plasma approaches have been utilized in ambient VOC elimination. However, there are limitations such as short light competence, by-product development, and hazards involved in the action. Photoelectrocatalysis (PEC) is an electrically improved photocatalysis technique conducted in ambient temperature air management nowadays [103]. The mineralization of typical obstinate indoor air pollutants such as xylene, benzene, and toluene cannot be done by applying an altogether solid PEC. Therefore, developing a new and well-organized technique suitable for VOC removal below indoor air circumstances is necessary. Correspondingly, no O_3 was identified in the study scheme, perhaps because of the absence of the inclusion of a secondary electrolyte. Seeing all the scenarios, due to the paucity of by-product development, the gas-phase electrochemical mineralization of VOCs is higher in grade than other present air pollution action skills.

The increased oxygen generation overpotential aids in the formation of •OH on the electrode as well as in the "electrochemical incineration" of organic contaminants. The increase in current caused by the accumulation of benzene was verified in 0.1 M H_2SO_4 after the presence of benzene. The cyclic voltammetry (CV) current density achieved in the H_2SO_4 solution was more significant than in pure water. The recent increase caused by benzene injection in the pure water system in the oxygen evolution reaction (OER) area is identical to that seen in 0.1 M H_2SO_4, indicating that benzene oxidation took place in the OER region with the polymer electrolytes [104].

An observable current rise was shown at 2.0 V practical potential, which was not relatable to the benzene concentration, indicating that the •OH formed on the electrode surface facilitated the benzene oxidation [105], and it mainly monitors a •OH-induced appliance [104].

11.9 Future Research Direction

Atmospheric contaminants, produced mainly by industrial processes and motor vehicles, are negatively impacting human health. One of the most efficient and cost-efficient methods for addressing significant air pollution issues is electrocatalysis; nevertheless, the main challenge is the lack of high-performance catalysis. Much work has gone into creating catalytic compounds that can remove atmospheric pollutants. The research during the last few decades has concentrated on the production and environmental uses of porous or nanosized catalysts. The enormous surface areas, rich surface states, and varied morphologies that electrocatalysis typically offers as a new form of material make it an up-and-coming candidate for use in environmental preservation and management. Improved electrocatalysis surface area, absorption of electrocatalysis, intrinsic conductivity, operational performance and stability, and porous engineering to better tolerate electrocatalysis conditions, such as pH, temperature, and toxic by-products, are among the research directions for the future.

On the other hand, it is advisable to look at how the degrading process might be combined with environmental applications in a future study to reduce operational costs, particularly the use of power.

11.10 Conclusions and Prospects

The growing concentration of contaminations in the atmosphere has driven extensive study toward implementing novel ecologically friendly and cost-effective debasing processes. Human health is harmed by air pollution, such as NOx, VOCs, and CO, which are produced mainly by industrial operations and conveyance vehicles. Electrocatalysis is a highly active and cost-effective solution for addressing significant air pollution issues; nevertheless, the presence of great-acting catalysts is critical. Considerable attempts have been done to produce catalytic materials that may be utilized to remove contaminants from the atmosphere. The production of homogeneous facilitators in aqueous environments is critical in the mediated electrocatalytic oxidation (MEO) of air contaminations. The development of low- and high-oxidation state metal complexes as facilitators for gas pollutant elimination has been achieved by uniting the potential spreading attained by a pH alteration in both edges of an electrolytic cell. To advance the correctness of VOC emissions in the future, it is important to develop foundation outlines covering oxygenated VOCs appropriately, which take an entire part in the control of events of an industrial method, rebuild extended term emissions through top-down approaches, and utilize satellite observations and surface monitoring to conduct additional emission verification effort.

In some border areas that are routinely exposed to elevated amounts of particulate matter PM_{10} and $PM_{2.5}$, ozone, and hazardous air pollutants, air pollution poses a significant environmental danger today. Significant factors in the poor air quality around the border include emissions from electricity generation and other industrial emissions, unpaved roads, diesel vehicles, buses, and cars, particularly those that were idle for extended periods of time at ports of entry. Urban and rural air quality has been significantly impacted by the region's economic and demographic growth.

Cities in the border region share airsheds; thus, decisions made in one might have an immediate impact on another, whether they are in the same nation or not. The public, as well as regional, state, congressional, and tribal authorities, must be actively involved in the development and implementation of global plans and solutions to combat pollution levels along the border.

Acknowledgments

The authors acknowledge the Knowledge Resource Centre, CSIR-NEERI (CSIR-NEERI/KRC/2022/JUNE/APC/1) for the support.

References

1. Arfin, T. and Mohammad, F., Synthesis, characterization and influence of electrolyte solutions towards the electric properties of nylon-6,6 nickel carbonate membrane: Test for the theory of uni-ionic potential based on the thermodynamics of irreversible processes, in: *Halides: Chemistry, Physical Properties and Structural Effects*, J. Lefebure (Ed.), pp. 39–66, Nova Science Publisher, New York, 2013.
2. Arfin, T. and Fatma, S., Anticipating behaviour of advanced materials in healthcare, in: *Advanced Biomaterials and Biodevices*, A. Tiwari and A.N. Nordin (Eds.), pp. 243–287, Scrivener Publishing LLC, USA, 2014.
3. Waghmare, S.S. and Arfin, T., Fluoride removal from water by calcium materials: A state-of-the-art: Review. *Int. J. Innov. Res. Sci. Eng. Technol.*, 4, 8090, 2015.
4. Arfin, T., Singh, B., Varshney, N., Biological adhesion behaviour of superhydrophobic polymer coating, in: *Superhydrophobic Polymer Coatings: Fundamentals, Design, Fabrication, and Applications*, S.K. Samal, S. Mohanty, S.K. Nayak (Eds.), pp. 161–177, Elsevier, Switzerland, 2019.
5. Arfin, T., Bhaisare, D.A., Waghmare, S.S., Development of a PANI/$Fe(NO_3)_2$ nanomaterials for reactive orange 16 (RO16) dye removal. *Anal. Methods*, 13, 5309, 2021.
6. Obulapuram, P.K., Arfin, T., Mohammad, F., Kumari, K., Khiste, S.K., Al-Lohedan, H.A., Chavali, M., Surface-enhanced biocompatibility and adsorption capacity of a zirconium phosphate-coated polyaniline composite. *ACS Omega*, 6, 33614, 2021.
7. Obulapuram, P.K., Arfin, T., Mohamamd, F., Khiste, S.K., Chavali, M., Albalawi, A.N., Al-Lohedan, H.A., Adsorption, equilibrium isotherm, and thermodynamic studies towards the removal of reactive orange 16 dye using Cu(I)-polyaniline composite. *Polymers*, 13, 3490, 2021.
8. Mohammad, F., Arfin, T., Al-Lohedan, H.A., Development of graphene-based nanocomposites as potential materials for supercapacitors and electrochemical cells, in: *Graphene-Based Nanotechnologies for Energy*

and Environmental Applications, M. Jawaid, A. Ahmad, D. Lokhat (Eds.), pp. 145–154, Elsevier, Switzerland, 2019.
9. Mohammad, F., Arfin, T., Al-Lohedan, H.A., Synthesis, characterization and applications of ethyl cellulose-based polymeric calcium (II) hydrogen phosphate composite. *J. Electron. Mater.*, 47, 2954, 2018.
10. Bushra, R., Arfin, T., Oves, M., Raza, W., Mohammad, F., Khan, M.A., Ahmad, A., Azam, A., Muneer, M., Development of PANI/MWCNTs decorated with cobalt oxide nanoparticles towards multiple electrochemical, photocatalytic and biomedical application site. *New J. Chem.*, 40, 9448, 2016.
11. Onwudiwe, D.C., Arfin, T., Strydom, C.A., Surfactant mediated synthesis of ZnO nanospheres at elevated temperature, and their dielectric properties. *Superlattices Microstruct.*, 81, 215, 2015.
12. Mohammad, F. and Arfin, T., Thermodynamics and electrochemical characterization of core-shell type gold-coated superparamagnetic iron oxide nanoparticles. *Adv. Mater. Lett.*, 5, 315, 2014.
13. Arfin, T. and Mohammad, F., DC electrical conductivity of nano-composite polystyrene-titanium-arsenate membrane. *J. Ind. Eng. Chem.*, 19, 2046, 2013.
14. Arfin, T. and Rangari, S.N., Graphene-ZnO nanocomposite modified electrode for the detection of phenol. *Anal. Methods*, 10, 347, 2018.
15. Arfin, T., Falch, A., Kriek, R.J., Evaluation of charge density and the theory for calculating membrane potential for a nano-composite nylon-6,6 nickel phosphate membrane. *Phys. Chem. Chem. Phys.*, 4, 16760, 2012.
16. Waghmare, S.S. and Arfin, T., Fluoride induced water pollution issue and its health efficacy in India: A review. *Int. J. Eng. Res. Gen. Sci.*, 3, 345, 2015.
17. Arfin, T. and Rafiuddin, Thermodynamics of ion conductivity of alkali halides across a polystyrene-based titanium arsenate membrane. *Electrochim. Acta*, 55, 8628, 2010.
18. Onwudiwe, D.C., Arfin, T., Styydom, C.A., Kriek, R.J., A study of the thermal and AC impedance properties of N-phenydithiocarbamate complexes of Zn(II). *Electrochim. Acta*, 109, 809, 2013.
19. Ahmad, S., Singh, R., Arfin, T., Neeti, K., Fluoride contamination, consequences and removal techniques in water: A review. *Environ. Sci. Adv.*, 1, 620, 2022.
20. Arfin, T., Pillai, A.M., Mathew, N., Turpude, A., Bang, R., Mondal, P., An overview of atmospheric aerosol and their effects on human health. *Environ. Sci. Pollut. Res.*, 30, 125347, 2023.
21. Mathew, N., Tirpude, A., Pillai, A.M., Mondal, P., Arfin, T., Emerging contaminants in air pollution and their sources, consequences, and future challenges, in: *Bioremediation technologies: For wastewater and sustainable circular bioeconomy*, R.T. Kapoor and M. Rafatullah, (Eds.), pp. 235–274, De Gruyter, Germany, 2023.
22. Fulekar, M.H., Singh, A., Dutta, D.P., Roy, M., Ballal, A., Tyagi, A.K., Ag incorporated Nano $BiPO_4$: Sonochemical synthesis, characterization and improved visible light photocatalytic properties. *RSC Adv.*, 4, 10097, 2014.

23. Gu, P., Xu, Y., Zhao, Y., Liu, W., Xue, H., Pang, H., Electrocatalysis of rechargeable non-lithium metal-air batteries. *Adv. Mater. Interfaces*, 4, 1700589, 2017.
24. Vallero, D., Air pollution, in: *Kirk-Othmer Encyclopedia of Chemical Technology*, pp. 1–48, Wiley, USA, 2015.
25. Azam, A.G., Zanjani, B.R., Mood, M.B., Effects of air pollution on human health and practical measures for prevention in Iran. *J. Res. Med. Sci.*, 21, 65, 2016.
26. Clark, N.A., Demers, P.A., Karr, C.J., Koehoorn, M., Lencar, C., Tamburic, L., Brauer, M., Effect of early life exposure to air pollution on development of childhood asthma. *Environ. Health Perspect.*, 118, 284, 2010.
27. Kelly, F.J. and Fussell, J.C., Air pollution and public health: emerging hazards and improved understanding of risk. *Environ. Geochem. Health*, 37, 631, 2015.
28. Ritchie, H., What the history of London's air pollution can tell us about the future of today's growing megacities, https://ourworldindata.org/london-air-pollution, 2017.
29. Romieu, I., Weitzenfeld, H., Finkelman, Urban air pollution in Latin America and the Caribbean: Health perspectives. *World Health Stat. Q.*, 43, 153, 1990.
30. Menezes, J.A., Margonari, C., Santos, R.B., Confalonieri, U., Air pollution, climate change, and human health in Brazil, pp. 375–403, Springer, Cham, 2017.
31. Martin, E.T., McGuire, C.M., Mubarak, M.S., Peters, D.G., Electroreductive remediation of halogenated environmental pollutants. *Chem. Rev.*, 116, 15198, 2016.
32. da Silva-Rackov, C.K.O., Lawal, W.A., Nfodzo, P.A., Vianna, M.M.G.R., do Nascimento, C.A.O., Choi, H., Degradation of PFOA by hydrogen peroxide and persulfate activated by iron-modified diatomite. *Appl. Catal. B.*, 192, 253, 2016.
33. Zhu, K., Wang, X., Ma, X., Sun, Z., Hu, X., Comparative degradation of atrazine by anodic oxidation at graphite and platinum electrodes and insights into electrochemical behavior of graphite anode. *Electrocatalysis*, 10, 35, 2019.
34. Trojanowicz, M., Bojanowska-Czajka, A., Bartosiewicz, I., Kulisa, K., Advanced oxidation/reduction processes treatment for aqueous perfluorooctanoate (PFOA) and perfluorooctanesulfonate (PFOS) - A review of recent advances. *Chem. Eng. J.*, 336, 170, 2018.
35. Fang, C., Megharaj, M., Naidu, R., Electrochemical advanced oxidation processes (EAOP) to degrade per- and polyfluoroalkyl substances (PFASs). *J. Adv. Oxid. Technol.*, 20, 20170014, 2017.
36. Ross, I., McDonough, J., Miles, J., Storch, P., Kochunarayanan, P.T., Kalve, E., Hurst, J., Dasgupta, S., Burdick, J., A review of emerging technologies for remediation of PFASs. *Remediation*, 28, 101, 2018.

37. Rondinini, S., Locatelli, C., Minguzzi, A., Vertova, A., Electroreduction, in: *Electrochemical Water and Wastewater Treatment*, C.A. Martinez-Huitle, M.A. Rodrigi, O. Scialdone (Eds.), pp. 3–28, Elsevier, Switzerland, 2018.
38. Zhang, Q., Guo, X., Cao, X., Wang, D., Wei, J., Facile preparation of a Ti/α-PbO_2/β-PbO_2 electrode for the electrochemical degradation of 2-chlorophenol. *Chin. J. Catal.*, 36, 975, 2015.
39. Duan, X., Zhao, C., Liu, W., Zhao, X., Chang, L., Fabrication of a novel PbO_2 electrode with a graphene nanosheet interlayer for electrochemical oxidation of 2-chlorophenol. *Electrochim. Acta*, 240, 424, 2017.
40. Liu, S., Wang, Y., Zhou, X., Han, W., Li, J., Sun, X., Shen, J., Wang, L., Improved degradation of the aqueous flutriafol using a nanostructure macroporous PbO_2 as reactive electrochemical membrane. *Electrochim. Acta*, 253, 357, 2017.
41. Yao, Y., Jiao, L., Cui, L., Yu, N., Preparation and Characterization of PbO_2-CeO_2 nanocomposite electrode with high cerium content and its application in the electrocatalytic degradation of malachite green. *J. Electrochem. Soc.*, 162, H693, 2015.
42. Yao, Y., Teng, G., Yang, Y., Huang, C., Liu, B., Guo, L., Electrochemical oxidation of acetamiprid using Yb-doped PbO_2 electrodes: Electrode characterization, influencing factors and degradation pathways. *Sep. Purif. Technol.*, 211, 456, 2019.
43. Sánchez-Montes, I., Fuzer Neto, J.R., Silva, B.F., Silva, A.J., Aquino, J.M., Rocha-Filho, R.C., Evolution of the antibacterial activity and oxidation intermediates during the electrochemical degradation of norfloxacin in a flow cell with a PTFE-doped β-PbO_2 anode: Critical comparison to a BDD anode. *Electrochim. Acta*, 284, 260, 2018.
44. Yang, S.Y., Choo, Y.S., Kim, S., Lim, S.K., Lee, J., Park, H., Boosting the electrocatalytic activities of SnO_2 electrodes for remediation of aqueous pollutants by doping with various metals. *Appl. Catal., B*, 317, 111– 112, 2012.
45. Tang, B., Shi, H., Fan, Z., Zhao, G., Preferential electrocatalytic degradation of 2,4-dichlorophenoxyacetic acid on molecular imprinted mesoporous SnO_2 surface. *Chem. Eng. J.*, 334, 882, 2018.
46. Vargas, R., Diaz, S., Viele, L., Nunez, O., Borras, C., Mostany, J., Scharifker, B.R., Electrochemical oxidation of dichlorvos on SnO_2-Sb_2O_5 electrodes. *Appl. Catal., B*, 144, 107, 2014.
47. Zhuo, Q., Li, X., Yan, F., Yang, B., Deng, S., Huang, J., Yu, G., Electrochemical oxidation of $_1H,_1H,_2H,_2H$-perfluorooctane sulfonic acid (6:2 FTS) on DSA electrode: Operating parameters and mechanism. *J. Environ. Sci.*, 26, 1733, 2014.
48. Zhao, H., Gao, J., Zhao, G., Fan, J., Wang, Y., Wang, Y., Fabrication of novel SnO_2-Sb/carbon aerogel electrode for ultrasonic electrochemical oxidation of perfluorooctanoate with high catalytic efficiency. *Appl. Catal. B*, 278, 136–137, 2014.

49. Chai, S., Wang, Y., Zhang, Y., Zhao, H., Liu, M., Zhao, G., Construction of a bifunctional electrode interface for efficient electrochemical mineralization of recalcitrant pollutants. *Appl. Catal. B*, 237, 473, 2018.
50. Petrović, M.M., Mitrović, J.Z., Radović, M.D., Kostić, M.M., Bojić, A.L.J., Preparation and characterisation of a new stainless steel/Bi_2O_3 anode and its dyes degradation ability. *Can. J. Chem. Eng.*, 92, 1000, 2014.
51. Kaur, R., Kushwaha, J.P., Singh, N., Electro-oxidation of ofloxacin antibiotic by dimensionally stable Ti/RuO_2 anode: Evaluation and mechanistic approach. *Chemosphere*, 193, 685, 2018.
52. De Coster, J., Vanherck, W., Appels, L., Dewil, R., Selective electrochemical degradation of 4-chlorophenol at a Ti/RuO_2-IrO_2 anode in chloride rich wastewater. *J. Environ. Manage.*, 190, 61, 2017.
53. Yunus, R.F., Zheng, Y., Nanayakkara, K.G.N., Chen, J.P., Electrochemical removal of rhodamine 6G by using RuO_2 coated Ti DSA. *Ind. Eng. Chem. Res.*, 48, 7466, 2009.
54. Petrović, M.M., Slipper, I.J., Antonijević, M.D., Nikolić, G.S., Mitrović, J.Z., Bojić, D.V., Bojić, A.L., Characterization of a Bi_2O_3 coat based anode prepared by galvanostatic electrodeposition and its use for the electrochemical degradation of Reactive Orange 4. *J. Taiwan Inst. Chem. Eng.*, 50, 282, 2015.
55. Lin, H., Niu, J., Liang, S., Wang, C., Wang, Y., Jin, F., Luo, Q., Huang, Q., Development of macroporous magnéli phase Ti_4O_7 ceramic materials: As an efficient anode for mineralization of poly- and perfluoroalkyl substances. *Chem. Eng. J.*, 354, 1058, 2018.
56. Huang, B., Li, J., Cao, X., Zhu, Y., Chen, W., Lei, C., Elevtrochemical reduction of p-chloronitrobenzene (p-CNB) are silver cathode in dimethyformamide. *Electrochim. Acta*, 196, 980, 2019.
57. Chaplin, B.P., Critical review of electrochemical advanced oxidation processes for water treatment applications. *Environ. Sci.: Processes Impacts, 16*, 1182, 2014.
58. Niu, J., Li, Y., Shang, E., Xu, Z., Liu, J., Electrochemical oxidation of perfluorinated compounds in water. *Chemosphere*, 146, 526, 2016.
59. US Environmental Protection Agency (EPA), Fact sheet: 2010/2015 PFOA stewardship program, https://www.epa.gov/assessmanaging-chemicalsunder-tsca/fact-sheet-20102015-ing-and-pfoa-stewardship-program#mfg, 2022.
60. Grandjean, P. and Landrigan, P.J., Neurobehavioural effects of developmental toxicity. *Lancet Neurol.*, 13, 330, 2014.
61. Bidleman, T.F., Atmospheric processes. *Environ. Sci. Technol.*, 22, 361, 1988.
62. US EPA, Estimation program interface (EPI) suite. Ver. 4.11, http://www.epa.gov/oppt/exposure/pubs/episuitedl.htm, 2022.
63. Kim, S.K. and Kannan, K., Perfluorinated acids in air, rain, snow, surface runoff, and lakes: Relative importance of pathways to contaminations of urban lakes. *Environ. Sci. Technol.*, 41, 8328, 2007.

64. *Draft toxicological profile for perfluoroalkyls*, Agency for toxic substances and disease registry, US Public Health Service, Atlanta, GA, http://www.atsdr.cdc.gov/toxprofiles/index.asp, 2022.
65. Bruton, T.A. and Sedlak, D.L., Treatment of aqueous film-forming foam by heat-activated persulfate under conditions representative of *in situ* chemical oxidation. *Environ. Sci. Technol.*, 51, 13878, 2017.
66. Yang, B., Jiang, C., Yu, G., Zhou, Q., Deng, S., Wu, J., Zhang, H., Highly efficient electrochemical degradation of perfluorooctanoic acid (PFOA) by F-doped Ti/SnO_2 electrode. *J. Hazard. Mater.*, 299, 417, 2015.
67. Urtiaga, A., Fernandez-Gonzalez, C., Gomez-Lavin, S., Ortiz, I., Kinetics of the electrochemical mineralization of perfluorooctanoic acid on ultrananocrystalline boron doped conductive diamond electrodes. *Chemosphere*, 129, 20, 2015.
68. Zhuo, Q., Deng, S., Yang, B., Huang, J., Wang, B., Zhang, T., Yu, G., Degradation of perfluorinated compounds on a boron-doped diamond electrode. *Electrochim. Acta*, 77, 17, 2012.
69. Kouzayha, A., Al Ashi, A., Al Akoum, R., Al Iskandarani, M., Budzinski, H., Jaber, F., Occurrence of pesticide residues in Lebanon's water resources. *Bull. Environ. Contam. Toxicol.*, 91, 503, 2013.
70. Wilson, V.S., Bobseine, K., Gray Jr., L.E., Development and characterization of a cell line that stably expresses an estrogen-responsive luciferase reporter for the detection of estrogen receptor agonist and antagonists. *Toxicol. Sci.*, 81, 69, 2004.
71. Chapin, R.E., Harris, M.W., Davis, B.J., Ward, S.M., Wilson, R.E., Mauney, M.A., Lockhart, A.C., Smialowicz, R.J., Moser, V.C., Burka, L.T., Collins, B.J., The effects of perinatal/juvenile methoxychlor exposure on adult rat nervous, immune, and reproductive system function. *Fundam. Appl. Toxicol.*, 40, 138, 1997.
72. Gray Jr., L.E., Ostby, J., Ferrell, J., Rehnberg, G., Linder, R., Cooper, R., Goldman, J., Slott, V., Laskey, J., A dose-response analysis of methoxychlor-induced alterations of reproductive development and function in the rat. *Fundam. Appl. Toxicol.*, 12, 92, 1989.
73. Gray Jr., L.E., Ostby, J.S., Ferrell, J.M., Sigmon, E.R., Goldman, J.M., Methoxychlor induces estrogen-like alterations of behavior and the reproductive tract in the femal rat and hamster: Effects on sex behavior, running wheel activity, and uterine morphology. *Toxicol. Appl. Pharmacol.*, 96, 525, 1988.
74. Hoferkamp, L., Hermanson, M.H., Muir, D.C.G., Current use pesticides in arctic media. *Sci. Total Environ.*, 408, 2985, 2010.
75. Cummings, A.M., Methoxychlor as a model for environmental estrogens. *Crit. Rev. Toxicol.*, 27, 367, 1997.
76. Lee, H.R., Hwang, K.A., Park, M.A., Yi, B.R., Jeung, E.B., Choi, K.C., Treatment with bisphenol A and methoxychlor results in the growth of human breast cancer cells and alteration of the expression of cell cycle-

related genes, cyclin D_1 and p_{21}, via an estrogen receptor-dependent signaling pathway. *Int. J. Mol. Med.*, 29, 883, 2012.

77. Akingbemi, B.T., Ge, R.S., Klinefelter, G.R., Gunsalus, G.L., Hardy, M.P., A metabolite of methoxychlor, 2,2-bis(p-hydroxyphenyl)-1,1, 1-trichloroethane, reduces testosterone biosynthesis in rat leydig cells through suppression of steady-state messenger ribonucleic acid levels of the cholesterol side-chain cleavage enzyme. *Biol. Reprod.*, 62, 571, 2000.
78. Zepp, R.G., Wolfe, N.L., Azarraga, L.V., Cox, R.H., Pape, C.W., Photochemical transformation of the DDT and methoxychlor degradation products, DDE and DMDE, by sunlight. *Arch. Environ. Toxicol.*, 6, 305, 1977.
79. Wolfe, N.L., Zepp, R.G., Paris, D.F., Baughman, G.L., Hollis, R.C., Methoxychlor and DDT degradation in water rates and products. *Environ. Sci. Technol.*, 11, 1077, 1977.
80. Baczynski, T.P., Pleissner, D., Grotenhuis, T., Anaerobic biodegradation of organochlorine pesticides in contaminated soile-significance of temperature and availability. *Chemosphere*, 78, 22, 2010.
81. McGuire, C.M., Hansen, A.M., Karty, J.A., Peters, D.G., Catalytic reduction of 4,4¢-(2,2,2-trichloroethane-1,1-diyl)bis (methoxybenzene) (methoxychlor) with nickel(I) salen electrogenerated at reticulated vitreous carbon cathodes. *J. Electroanal. Chem.*, 772, 66, 2016.
82. Zaleska, A., Hipka, J., Wiergowski, M., Biziuk, M., Photocatalytic degradation of lindane, p¢,p¢-DDT methoxychlor an aqueous environment. *J. Photochem. Photobiol. A Chem.*, 135, 213, 2000.
83. Bourguignon, N., Fuentes, M.S., Benimeli, C.S., Cuozzo, S.A., Amoroso, M.J., Aerobic removal of methoxychlor by a native streptomyces strain: Identification of intermediate metabolites. *Int. Biodeterior. Biodegrad.*, 96, 80, 2014.
84. Howard, P.H. (Ed.), *Handbook of Environmental Fate and Exposure Data: For Organic Chemicals*, Lewis Publishers, Ririe, 1991.
85. Welch, H.E., Muir, D.C., Billeck, B.N., Lockhart, W.L., Brunskill, G.J., Kling, H.J., Olson, M.P., Lemoine, R.M., Brown snow: A long-range transport event in the canadian arctic. *Environ. Sci. Technol.*, 25, 280, 1991.
86. Deziel, N.C., Friesen, M.C., Hoppin, J.A., Hines, C.J., Thomas, K., Freeman, L.E.B., A review of nonoccupational pathways for pesticide exposure in women living in agricultural areas. *Environ. Health Perspect.*, 123, 515, 2015.
87. Hoff, R.M., Muir, D.C.G., Grift, N.P., Annual cycle of polychlorinated biphenyls and organohalogen pesticides in air in Southern Ontario. 1. Air concentration data. *Environ. Sci. Technol.*, 26, 266, 1992.
88. Peng, Y., Cui, M., Zhang, Z., Shu, S., Shi, X., Brosnahan, J.T., Liu, C., Zhang, Y., Godbold, P., Zhang, X., Dong, F., Jiang, G., Zhang, S., Bimetallic composition-promoted electrocatalytic hydrodechlorination reaction on silver– palladium alloy nanoparticles. *ACS Catal.*, 9, 10803, 2019.
89. Fang, L., Xu, C., Zhang, W., Huang, L.-Z., The important role of polyvinylpyrrolidone and Cu on enhancing dechlorination of 2,4-dichlorophenol

by Cu/Fe nanoparticles: Performance and mechanism study. *Appl. Surf. Sci.*, 435, 55, 2018.

90. Wang, H., Yin, S., Li, Y., Yu, H., Li, C., Deng, K., Xu, Y., Li, X., Xue, H., Wang, L., One-step fabrication of tri-metallic PdCuAu nanothorn assemblies as an efficient catalyst for oxygen reduction reaction. *J. Mater. Chem.*, 6, 3642, 2018.
91. Yan, D., Li, Y., Huo, J., Chen, R., Dai, L., Wang, S., Defect chemistry of nonprecious-metal electrocatalysts for oxygen reactions. *Adv. Mater.*, 29, 1606459, 2017.
92. Arfin, T., Kate, P., Singh, A., Kumari, K., Alternatives to POPs for a healthy life, in: *Persistent Organic Pollutants: Gaps in Management and Associated Challenges*, K. Kumari (Ed.), pp. 201–226, CRC Press, USA, 2021.
93. Guo, Y.L., Hsu, P.C., Hsu, C.C., Lambert, G.H., Semen quality after prenatal exposure to polychlorinated biphenyls and dibenzofurans. *Lancet*, 35, 1240, 2000.
94. *Polybrominated dibenzo-p-dioxins and dibenzofurans. Environmental health criteria 205*, http://www.inchem.org/documents/ehc/ehc/ehc205. htm, 1998.
95. Huang, H. and Buekens, A., De novo synthesis of polychlorinated dibenzo-p-dioxins and dibenzofurans. Proposal of a mechanistic scheme. *Sci. Total Environ.*, 193, 121, 1996.
96. Tame, N.W., Dlugogorski, B.Z., Kennedy, E.M., Formation of dioxins and furans during combustin of treated wood. *Prog. Energy Combust. Sci.*, 33, 384, 2007.
97. Bogdanovsky, G.A., Vidovich, G.L., Kultin, D.Y., Lebedeva, O.K., Zakharov, A.N., Chemical model for dioxin destructive in aqueous medium: Electrocatalytic oxidation of dioxin-like substance simulating natural pollutant. *Appl. Cata. A*, 232, 137, 2002.
98. He, Z., Wang, X., Ling, Z., Zhao, J., Guo, H., Shao, M., Wang, Z., Contributions of different anthropogenic volatile organic compound sources to ozone formation at a receptor site in the pearl river delta region and its policy implications. *Atmos. Chem. Phys.*, 19, 8801, 2019.
99. Agents classified by the IARC monographs, https://monographs.iarc.fr/list-of-classifications, 2021.
100. Nielsen, C.C., Amrhein, C.G., Serrano-Lomelin, J.A., Osornio-Vargas, A.R., Geographical information assessment of maternal ambient health hazards and adverse birth outcomes in Canada. *Sci. Total Environ.*, 696, 134091, 2019.
101. Zhou, W., Cohan, D.S., Henderson, B.H., Slower ozone production in Houston, Texas following emission reductions: Evidence from texas air quality studies in 2000 and 2006. *Atmos. Chem. Phys.*, 14, 2777, 2014.
102. Indoor air quality (IAQ), https://www.epa.gov/indoor-air-quality-iaq/what-are-volatile-organic-compounds-vocs, 2022.

103. Stoll, T., Zafeiropoulos, G., Dogan, I., Genuit, H., Lavrijsen, R., Koopmans, B., Tsampas, M.N., Visible-light-promoted gas-phase water splitting using porous $WO_3/BiVO_4$ photoanodes. *Electrochem. Commun.*, 82, 47, 2017.
104. Zhang, B., Chen, M., Wang, L., Zhao, X., Hu, R., Chen, H., Xie, P., Zhang, C., He, H., Electrochemical oxidation of volatile organic compounds in all-solid cell at ambient temperature. *Chem. Eng. J.*, 354, 93, 2018.
105. Zhi, J.-F., Wang, H.-B., Nakashima, T., Rao, T.N., Fujishima, A., Electrochemical incineration of organic pollutants on boron-doped diamond electrode. Evidence for direct electrochemical oxidation pathway. *J. Phys. Chem. B*, 107, 13389, 2003.

12

Extraction and Purification of Cellulase Enzyme for Bioethanol Production and Its Usefulness as a Sustainable Biofuel

Ayush Madan[1*], Rakhi Dhiman[2], Rishabh Garg[3], Narotam Sharma[4] and Syed Mohsin Waheed[5]

[1]*Department of Biotechnology, School of Biological Engineering and Life Sciences, Shobhit Institute of Engineering and Technology (Deemed-to-be-University), Meerut, Uttar Pradesh, India*
[2]*School of Biological Engineering and Sciences, Shobhit University, Gangoh (Saharanpur), India*
[3]*Department of Bioengineering and Biotechnology, Birla Institute of Technology, Ranchi, Jharkhand, India*
[4]*DNA Labs – Centre for Applied Sciences (DLCAS), Dehradun, Uttarakhand, India*
[5]*Divacc Research Laboratories Pvt. Ltd. AIC-JNUFI, Jawaharlal Nehru University, New Delhi, India*

Abstract

Carboxymethylcellulose-modified Czapek agar medium was utilized for subculturing and propagating the isolated *Aspergillus niger* strain. Different tests clearly demonstrated that the strain produced cellulase enzyme. The clear zone around the fungal colony on the medium showed cellulase production and cellulose digestion. The isolated fungal strain was then subjected to fermentation for enzyme production by two methods. First, a CMC Czapek broth medium was utilized to perform surface fermentation in the presence of the isolated *A. niger* strain. The second method was solid-state fermentation, where moistened sawdust enriched with 0.25% glucose was used. This method is a cost-efficient process for the production of cellulase. The filtrate obtained from solid-state fermentation of sawdust was then processed by 80% ammonium sulfate solution for enzyme precipitation at 4°C overnight on a magnetic stirrer. The precipitate obtained was dissolved in a

**Corresponding author*: madanayushmadan@gmail.com

Sudheesh K. Shukla, Chaudhery Mustansar Hussain, Santanu Patra, Meenakshi Choudhary (eds.) Electrocatalytic Materials for Renewable Energy, (333–352) © 2024 Scrivener Publishing LLC

phosphate buffer solution to make a purified enzyme solution. The enzyme solution was tested for cellulase activity by cellulose filter paper digestion assay. The results thus obtained were impressive and showed a positive cellulose production, extraction, and purification. The crude or purified solution obtained in the above-mentioned process could be now used to convert cellulosic mass into fermentable sugars, which can be fermented into bioethanol or biofuel. The isolated fungal strain of *A. niger* was found to give a good result for cellulase production.

***Keywords*:** Carboxymethylcellulose, *Aspergillus niger*, fermentation, bioethanol

12.1 Introduction

Annually, green plants transform millions of tons of carbon in the form of CO_2 into organic matter photosynthetically. Approximately one-third of the organic matter produced is cellulose. Cellulose is an important structural polysaccharide that is primarily a component of plant cell walls. It makes the most of both cotton and wood. It is therefore one of the most abundant natural resources, especially as its supply is constantly being replenished. The conversion of biomass to liquid fuels, such as ethanol, has been the focus of much interest during the 20th century. The process of making alcohol from cellulose, in principle, is relatively simple: after hydrolysis of cellulose to glucose and subsequent fermentation, ethanol can be purified by distillation. The development of an economically feasible technology would reduce crude oil dependency. Ethanol was first made from cellulose in large-scale production during World War I and II using acid hydrolysis and was later refined during the 1940s.

However, due to low yields in the hydrolysis steps and corrosion problems caused by the acid catalyst used, the process was not competitive with petroleum-based processes [1]. In the wake of the oil crisis of the 1970s, alternative fuels came into the picture again and were considered as an economically viable option. When the supply and price of crude oil fell back to normal, the interest in biomass-derived ethanol diminished.

However, during the last decade, biomass-derived fuels have gradually gained increasing attention. The impact of the greenhouse effect on the Earth's climate has been recognized, and efforts have been made to build large-scale production units for the conversion of lignocellulosic materials into ethanol. Today, ethanol derived from lignocellulosic biomass (bioethanol) can produce a fuel which is cost-competitive with the market price of corn-starch ethanol. The economic competitiveness with gasoline remains an issue.

The attractiveness of bioethanol as a potential replacement for fossil fuels lies in its low CO_2 release compared to when fossil fuels are burned. No energy input from fossil fuels is required to convert lignocellulosic material into ethanol. Lignin, a by-product, can be burned to provide the energy needed. Carbon dioxide and bioethanol produced during fermentation can be converted back into biomass to serve as new raw materials for production in energy crop cultivation. As a result, the contribution to atmospheric greenhouse gases is negligible [2].

Cellulose abounds in nature as an abundant natural product in the biosphere. It is a major product of photosynthesis in terrestrial areas, as well as a renewable bioresource produced in the biosphere, i.e., 100 billion dry tons/year [3, 4]. Cellulose is a major component of the hard cell wall that surrounds plant cells and strengthens plant stems, leaves, and twigs. Cellulose performs many functions: connecting cells to form tissues, indicating cell growth and division, in addition to regulating plant cell size. The homopolymeric cellulose made of anhydro-D-glucose linked by β-1,4 bonds [5] constitutes a major part of plant biomass. Therefore, waste produced from forests contains large amounts of unused cellulose. These wastes usually accumulate in the environment and cause pollution problems [6]. At present, the previous negative policy of considering this waste as self-consciously useless has been largely replaced by a positive approach to waste disposal which is recognized as a major renewable natural resource [7] and converted into valuable products. Cellulose is a homologous polymer chain that contains D-glucose residues [8, 9] containing 10,000 glucose residues bound by a β-1,4-glycosidic bond. Since cellulose is an abundant and renewable feedstock for energy sources, it is important to use lignocellulosic biomass by converting alternative liquid fuels into biofuels, mainly bioethanol and methane, which are produced by biochemical methods [10].

Many biotransformation methods have been proposed and used, from direct chemical methods such as acid hydrolysis to biological methods such as the application of cellulase enzymes. Although acid hydrolysis of cellulosic materials may seem cheaper than cellulase hydrolysis, the former often requires higher temperatures and pressures; this leads to the accumulation of objectionable by-products that are highly corrosive and cause serious pollution. Enzymatic hydrolysis is an environmentally friendly approach. The conversion is carried out under mild conditions, thus greatly reducing the cost of hydrolysis equipment. Sugar decomposition is avoided, thus eliminating this cause for loss in yield and producing clear sugar streams for further processing. Costly neutralization and purification equipment are unnecessary, and disposal of waste resulting from acid neutralization is eliminated. Balancing these potential savings, extensive pretreatment to break lignin and increase cellulose accessibility is required to achieve good yields. The cost of high-activity cellulolytic enzyme solution is at present very high. This enzyme is produced industrially or by microorganisms. Although the industrial production of cellulose is very expensive [11], significant cost reduction is possible through the discovery of novel hypercellulolytic microorganisms. Cellulose is a major component of textiles made from cotton, linen, and other plant fibers. It is converted into rayon, known as "regenerative cellulose fiber"; it is similar to cellulose in chemical structure and is usually composed of pulp dissolved by viscose. The most recent environmentally friendly method for producing a type of rayon is the lyocell process.

Cellulose glycoside hydrolases (GHs) convert cellulose into hydrophilic, water-insoluble polymers, which are small-chain polysaccharides consisting of D-glucose bonds bound by β-1,4-glycosidic bonds [12]. Cellulase has an important role in generating sugar feedstock for the lignocellulosic-based biorefinery platform. Microbial cellulases have applications in many industries, mainly in the textile industry, paper and pulp industry, food industry, and detergent industry, as well as cellulosic waste management. In addition, their demand is steadily increasing in the textile, paper, feed, food, and laundry detergent industries. However, these industrial applications require thermostable, catalytically highly efficient cellulases to make the processes commercially viable [13]. Cellulase is produced from a variety of organisms, mainly plants, animals, fungi, bacteria, and protozoa, while *Aspergillus* sp. is one of the most important commercial cellulose manufacturers and is widely used in various industries. Due to their wide range of applications and ever-increasing demand, novel cellulases are being discovered or engineered with improved process compatibility and high specificity and stability from new lineages of cellulolytic organisms [14]. Cellulase is used in the conversion of biomass into biofuels [15] and for fiber modification; furthermore, they are even used for pharmaceutical applications.

Thus, the utilization of cellulosic biomass by saccharification to induce cellulase synthesis derived from cellulolytic organisms has been suggested as a feasible process [16]. One effective approach for reducing the cost of enzyme production is to replace pure cellulose with relatively cheaper substrates such as lignocellulose-based materials [17]. The development process of costly pretreatment technologies for lignocellulosic feedstock is one of the primary challenges related to commercialization [18]. Pretreatment is necessary to increase the accessibility of cellulose in lignocellulosic biomass to facilitate enzymatic hydrolysis because unlike traditional sources of fermentable sugars, such as starch and sucrose, the cellulose component of lignocelluloses is a structural polymer and is protected against enzymatic attack by the surrounding matrix of lignin and hemicellulose [17]. To fully degrade the abundant and available cellulosic material, it is imperative to purify and characterize the various components of the cellulase system [19]. Cellulase is used for commercial food processing in coffee. It performs hydrolysis of cellulose during the drying of beans. The optimal temperature for the activity of the cellulase system ranged from 50°C to 55°C, with an optimum for stability between 23°C and 37°C after a 72-h incubation period.

12.2 Ethanol as Fuel

Ethanol is one of the most appealing forms of fuel as it contributes less to greenhouse gases than the amount contributed by fossil fuels and also because of its flexibility and beneficial properties. It can be used in non-modified vehicles in low-grade alloys (E10) or in high-grade gasoline alloys that can be fired in modified vehicles (E85, E95). As a combination of high-quality fuel, this can be burned in vehicles converted as pure ethanol (E100). Since ethanol has higher octane than gasoline, there is no need for additional, usually very toxic, octane boosters. Furthermore, ethanol contains oxygen, which facilitates better combustion in the motor, reducing tail-pipe CO emission. Unfortunately, the vapor pressure of ethanol–gasoline mixtures increases with initial ethanol addition (non-ideal mixture of polar and non-polar components). The resulting increase in evaporative emission raises questions about interference with the ozone production cycle. Ethanol can also be mixed into gasoline in the form of ethyl tertiary butyl ether (ETBE), providing the same benefits as mixing with ethanol. In addition, it lowers the vapor pressure of the gasoline, thereby reducing the evaporation of smog-forming compounds [20].

At the start of the 20th century, the United States triggered the use of biofuels mainly bioethanol. For instance, ethanol fuel was used to power several Ford T vehicles. Then, until the first and second oil crises, the interest in biofuels began to wane from 1973 until 1979. Gasoline, a blend of petrol and bioethanol, is the main fuel the country is concentrating on. In 2005, the US surpassed all other countries as the world's top manufacturer of ethanol fuel. The United States produced increasingly up to 13.2 billion U.S. liquid gallons (49.2 billion liters) in 2010 [21].

Table 12.1 Production of cellulase enzyme from microorganisms using fermentation techniques and their purification methods.

Microorganism	Fermentation type	Production source used	Purification method	Reference
Aureobasidium pullulans, Neurospora crassa, and *Trichoderma reesei*	Submerged and solid-state fermentation	Raw and processed canola meal	Centrifugation, crude	[24]
Saccharomyces cerevisiae and *Pachysolen tannophilus*	Submerged fermentation	Spent mushroom substrate	Acetone precipitation	[25]
Geotrichum candidum	Submerged fermentation	Carboxymethylcellulose salt	Ammonium sulfate precipitation	[26]
Pleurotus florida	Solid-state fermentation	Cotton spinning wastes	Crude hydrolysis	[27]
Sporotrichum thermophile	Solid-state fermentation	Wheat straw and cotton oil cake	Ammonium sulfate precipitation, ion exchange, gel filtration	[28]
Rhizopus oryzae, Cladosporium cladosporioides, and *Emericella variecolor*	Submerged and solid-state fermentation	Rice straw (RS), rice husk (RH), sugarcane bagasse (SCB), wheat straw (WS), and wheat bran (WB)	Centrifugation, crude	[29]
Aspergillus niger	Surface and solid-state fermentation	CMC and sawdust	Ammonium sulfate precipitation	This work

By the start of 2022, 275 plants in the US will be able to produce 21 billion gallons of biofuel annually (gal/y). Fuel ethanol accounted for more than half of the nation's biofuel production capacity. Research studies estimate that the worldwide bioethanol market will be valued at around USD 124.5 billion by 2030, growing at a CAGR of 13.2% between 2022 and 2030 [22, 23]. The demand for bioethanol is increasing as the petroleum crisis and pollution are on the rise. In addition to lowering our reliance on foreign oil and maintaining American prosperity while preserving the environment, abundant and renewable bioenergy may help secure America's energy future. The Renewable Fuel Standard (RFS) requires the use of biofuels to increase the energy security of the country. Enzyme production and utilization at the commercial level using microbes has always been the major area of interest and has a superior market with obvious benefits of production. This study may include a pillar of enzymatic production from a microbe for biomass conversion to fermentable sugars which can be further used for bioethanol production. Table 12.1 shows a comparative study of the production of cellulose enzyme using microbes.

12.3 Materials and Methods

12.3.1 Biological Materials

The *A. niger* strain was procured from DNA Labs-A Centre for Applied Sciences (DLCAS), Main Kargi Road, Near Vidya Vihar Phase I, Pathribagh Chowk, Dehradun-248001, Uttarakhand.

12.3.2 Sample Procurement

The fungal culture was procured from DNA Labs-A Centre for Applied Sciences (DLCAS), Main Kargi Road, Near Vidya Vihar Phase I, Pathribagh Chowk, Dehradun-248001, Uttarakhand.

12.3.3 Media Preparation

Potato dextrose agar (PDA) was used, and pH was maintained at 5.6. The skin of potatoes was peeled off, chopped into small pieces, boiled (200 g) in 500 ml of water until they were easily penetrated by a glass rod, and filtered through a cheesecloth. Dextrose was added to the filtrate. Agar was dissolved in water, brought up to the required volume by the addition of water, and Autoclaved at 15 lb. pressure for 15 min. The media was poured into sterile petri dishes and allowed to solidify for about 2 h. A small tuft of healthy mature colony-bearing spores from a fungal colony was point-inoculated on the media surface with the help of a clean sterilized inoculation needle.

12.3.4 Glycerol Mounted Staining of Fungi

A clean glass slide was taken and a drop of glycerol was mounted over it. A small drop of fungus was transferred into the drop using a flame-sterilized loop. The contents were gently mixed using a loop or a sterile toothpick. A cover glass was placed carefully over the glycerol mount to avoid bubble formation and was examined under a microscope at ×15 and ×45 magnification.

12.3.5 Lactophenol Cotton Blue Mounted Staining

A drop of lactophenol cotton blue stain was placed over a clean glass slide, and a small drop of fungal culture was transferred to the drop using a flamed, sterilized loop. The material was gently mixed using a sterile toothpick. A cover slip was placed over the mount and was examined under a microscope at ×15 and ×45 magnification.

12.3.6 Screening of the *Aspergillus niger* for Cellulase Enzyme Production

This test was done on carboxymethylcellulase (modified CMC agar).

The fungus was inoculated by the point inoculation method on the media plate. The media was prepared as follows:

12.3.7 Modified Czapek Agar Medium (pH 6.4–7.0)

All the salts were dissolved in distilled water and CMC was dissolved separately. Then, they were mixed to make up the required volume. Autoclaving was carried out at 15 psi of pressure for 15 min.

12.3.8 Sawdust (Solid) Fermentation

Sawdust was taken into a wide mouth beaker of 500 ml volume and was sterilized by autoclaving at 15 psi for 15 min. An autoclaved water was added to the sawdust inside the laminar flow. About 20% of moisture was maintained in the flask containing sawdust. A freshly grown inoculum of *A. niger* was added to the sawdust containing the medium and was kept at 25°C for 7 days.

12.3.9 Czapek (Surface) Fermentation

Czapek broth was prepared in a 500-ml conical flask and autoclaved at 15 psi for 15 min. Then inside the laminar airflow, the *A. niger* culture from the previous step was used for inoculation of the Czapek broth under sterile conditions in a laminar flow hood and was subsequently incubated at 25°C for 7 days.

12.3.10 Extraction of Cellulase

The liquid medium incubated at 25°C for 7 days was used for the extraction of cellulase from the fungal culture grown over the surface of the inoculated broth. During this fermentation process, the fungus secreted the enzyme in the culture medium. The fermentation broth was filtered by using Whatman No. 1 filter paper. The fungal mycelia were separated from the culture medium containing the enzyme by the process of filtration. The filtrate was centrifuged at 8,000 rpm for 40 min to remove spores and to fully extract the enzymes into the supernatant. The pellet was discarded and the supernatant was collected. The supernatant containing the crude enzyme was obtained by the following steps: 80% of ammonium sulfate solution was prepared and added to the supernatant containing the crude enzyme, and the mixture was then mixed gently. The mixture was kept at 4°C overnight in a magnetic stirrer to obtain a clear precipitate. From the precipitated medium, the supernatant was collected and the mixture was again centrifuged at 10,000 rpm for about 10–12 min to get the pellet, then the pellet was dissolved in 50 mM of sodium phosphate buffer (pH 7.0) for further purification.

Sodium phosphate buffer (1,000 ml) was prepared in a suitable container. Sodium phosphate (20.21 g) was added to 800 ml of distilled water. Then, 3.39 g of potassium phosphate was added to the solution. The solution was adjusted to the final desired pH using HCl or NaOH. Finally, 200 ml of distilled water was added to make up the volume of 1 L. The partially purified enzyme was then collected.

12.3.11 Cellulase Activity Measurement

The cell-free supernatant was used to measure and analyze the cellulase activities using filter paper assay, and 1.5 ml, 0.1 M (pH 5.0) of sodium acetate buffer was used to suspend 50 mg of Whatman no. 1 filter paper and kept for incubation with the enzyme at 50°C for a few hours. Furthermore, the samples were centrifuged for 5 min at 10,000 rpm after incubation, and the supernatant (1 ml) was collected to calculate the cellulase activity [filter paper degradation (FPase) activity] [30, 31]. Next, 5.33 g of NaOH and 100 g of potassium sodium titrate (Rochelle salt) were added to stop the reaction [31] and kept in a water bath for boiling for 5 min and then cooled with cold water for 10 min. The apparent density of the reduced sugar is measured at 540 nm. The calibration curve for D-glucose was used to determine the generation of monomeric sugar by cellulase. The amount of enzyme that released 1 μmol of reducing sugars as glucose equivalent min^{-1} is termed as 1 unit of FPase enzymatic activity (U) [32, 33]. The formula by Thomas was used for the determination of enzymatic activity [34, 35]:

$$\begin{aligned} 1\ \text{U} &= 1\ \mu\text{mol/ min of glucose equivalent released} \\ &= 0.18\ \text{mg/min of glucose} \end{aligned}$$

The formula used for the calculation of the enzyme activity was as follows:

$$= \frac{amount\ of\ equivalent\ sugar\ generated}{0.18} \times 0.5(Volume\ of\ enzyme\ solution \times 120 min\left(reaction\ time\right)$$

12.4 Results

The following results clearly demonstrate that *A. niger* cultures produced and secreted cellulase in the medium from which the enzyme was extracted and characterized to determine its activity.

12.4.1 Propagation of Fungal Culture

The fungal culture was procured from DNA Labs-A Centre for Applied Sciences (DLCAS), Main Kargi Road, Near Vidya Vihar Phase-1, Pathribagh Chowk, Dehradun-248001, Uttarakhand.

12.4.2 Regrowth of the Fungal Strain

The procured strain of *A. niger* was point-inoculated on PDA and incubated at 25°C ± 2°C for 3 days. After incubation, the plate was found filled with the growth of the fungal strain as shown in Figures 12.1 and 12.2.

12.4.3 Colony Morphology

The selected strain was subcultured on a CMC agar medium. Initially, the colony of *A. niger* was found to have smooth and colorless conidiophores

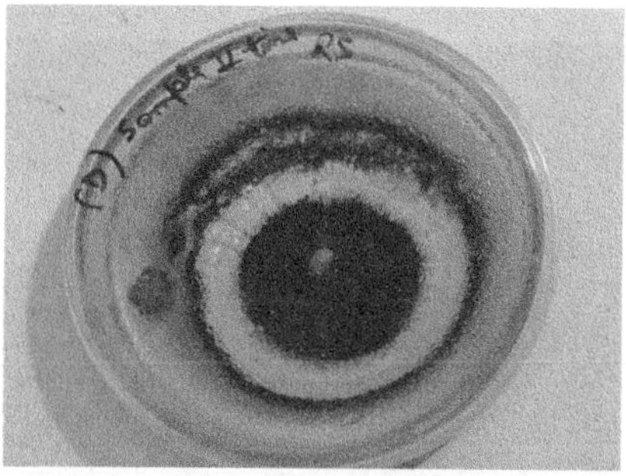

Figure 12.1 Procured Bengal Rose agar plate of *Aspergillus niger.*

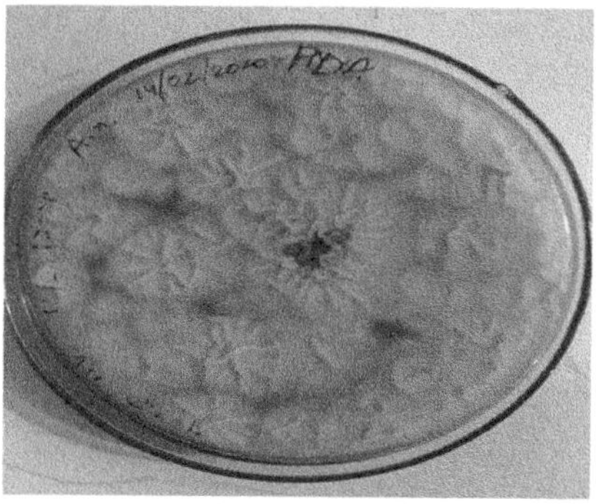

Figure 12.2 *Aspergillus niger* on PDA.

and spores, and the colony color changed into black as the culture matured. When immature, the colonies were covered with white fluffy aerial mycelia, while mature colonies showed a salt and pepper effect, being covered with black spores, and the reverse side of the colony was buff-colored (Figure 12.2).

12.4.4 Glycerol Mounted Staining

The slide was examined at ×15 and ×45 magnification. The mycelium was seen along with black round oval sporangia and a straight long stalk or sporangiophores (Figure 12.3).

12.4.5 Lactophenol Cotton Blue Mounted Strain

The slide was examined at ×15 and ×45 magnification, and with lactophenol cotton blue stain, the fungal mycelia appeared blue with the spore head having a dark purple-colored round structure (Figure 12.4).

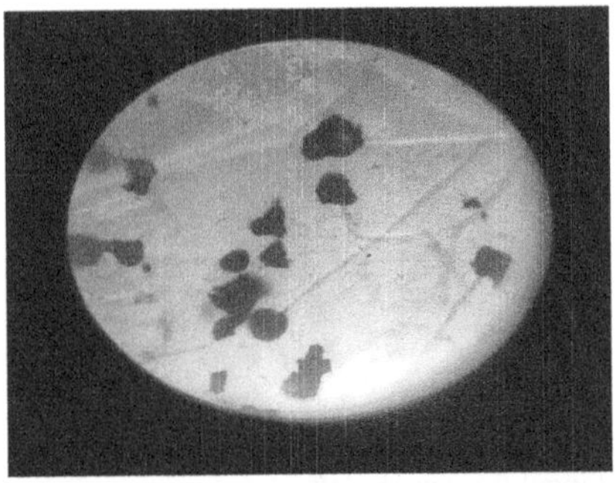

Figure 12.3 Glycerol mounting of the fungal strain (sporangium, sporangiophore, and mycelia).

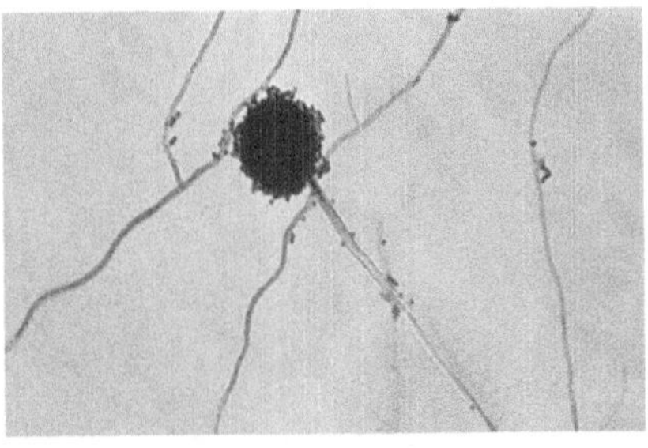

Figure 12.4 Lactophenol mount of the fungal strain (sporangium, sporangiophore, and mycelia).

12.4.6 Screening of the Procured Fungal Cellulase Enzyme Production

After incubation, the plates were examined. Clearance zones around the fungal colonies were observed indicating the degradation of cellulose with the cellulose being secreted by the fungus.

12.4.7 Sawdust (Solid) Fermentation

Solid-state fermentation was done on sawdust media, as shown in Figure 12.5.

12.4.8 Czapek (Surface) Fermentation

The fermentation process applied for growing *A. niger* cultures is called surface/Czapek fermentation. Within 2 to 3 days of inoculation, young

Figure 12.5 Sawdust solid-state fermentation.

Figure 12.6 Surface fermentation for cellulose production on CMC media.

colonies were seen floating over the surface, and within 10 days, the whole surface was covered with fully matured spore-bearing black colonies of *A. niger* (Figure 12.6).

12.4.9 Extraction and Isolation of the Enzyme

The cellulose enzyme was precipitated out by the ammonium sulfate method.

12.4.10 Enzyme Activity Assay

FPase activity from the fungal species was examined after cultivation in a 100-ml liquid medium provided with Whatman No. 1 filter paper. The media filter showed deterioration of the paper, while the control, i.e., the media without the fungal inoculum, exhibited a slight decline (Figure 12.7).

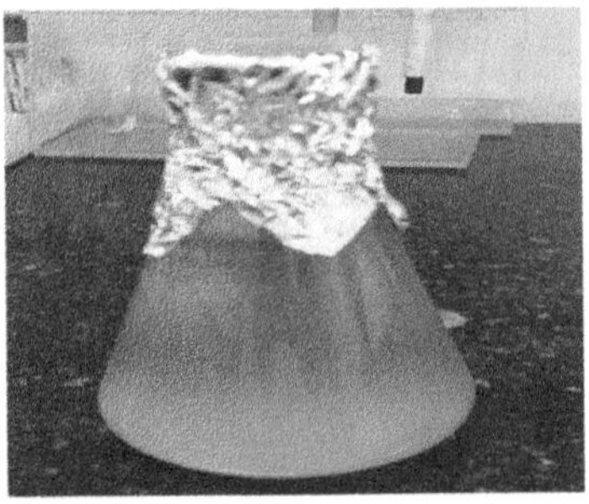

Figure 12.7 Enzyme supernatant after the removal of fungal mycelia.

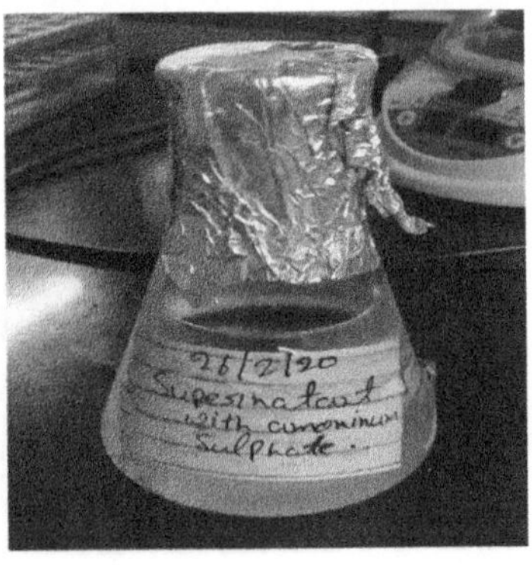

Figure 12.8 Crude enzyme supernatant under ammonium sulfate treatment.

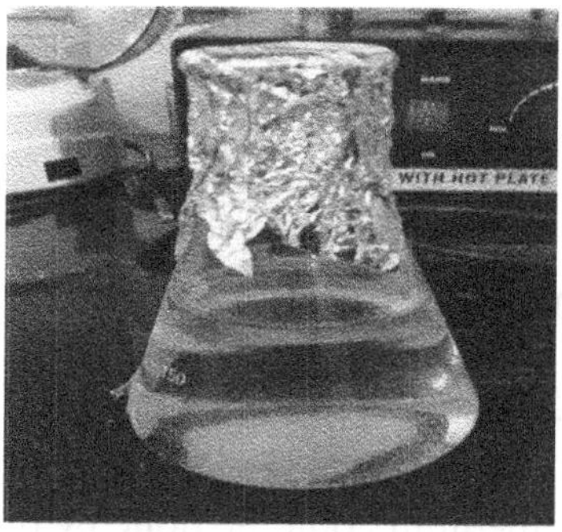

Figure 12.9 Enzyme crystals precipitating.

This suggests that the decline in FP was due to the activity of the acquired species of *A. niger*. Enzymatic analysis using two media supernatants showed that cellulose activity peaked after 5 days (Figures 12.8 and 12.9).

12.5 Discussion

The present study was dedicated to the production, extraction, and analysis of cellulase enzyme obtained from the fungal strain of *A. niger* for cellulose digestion in order to utilize the cellulosic digest thus obtained as a starting material for ethanol production. The procured fungal strain was grown on potato dextrose agar for fresh culture obtainment. The staining and partial identification of the matured colony was done, followed by the screening of cellulose production activity on carboxymethylcellulose-modified Czapek agar medium. The clear zone around the fungal colony on the medium showed cellulase production and cellulose digestion. The fungus was then subjected to fermentation for enzyme production by two methods: firstly, CMC Czapek broth medium was prepared and inoculated aseptically with the *A. niger* strain for surface fermentation, and the second method was solid-state fermentation in which sawdust was used which was moistened with sterilized water and enriched with 0.25% glucose, and this method was used as a cost-cutting strategy. After the incubation period, the solid-state medium was pressed to obtain the fermentation filtrate, and the liquid medium was filtered to obtain a filtrate devoid of fungal mycelia, cells, and spores. Both filtrates were then processed with 80% ammonium sulfate solution for enzyme precipitation. The mixture was kept at 4°C overnight using a magnetic stirrer. The precipitates were obtained and dissolved in phosphate buffer solution to make the purified enzyme solution.

The screening of this enzyme solution for cellulose presence was done by cellulose filter paper digestion assay. The results thus obtained were impressive and showed positive cellulose production, extraction, and purification.

12.6 Conclusion and Future Scope

The procured fungal strain of *A. niger* was found to give a good result for cellulase production. Fungal cellulase enzyme through surface fermentation on CMC Czapek medium could be done by the mentioned method, and the crude enzyme isolated could be used to digest pretreated cellulosic mass to convert the cellulose into fermentable sugars. The fermentation of sugar obtained by the digestion of cellulose achieved by the cellulase enzyme could be done by alcohol-fermenting yeasts or industrially important alcohol-producing bacteria like *Zymomonas mobilis* to obtain bioethanol in the future by the industries. The crucial part of using a common fungus for the production of an industrially important enzyme was the main motive of this study. The biochemistry of fungi for its cellulase activity has been done which showed a good result by cellulose filter paper enzyme assay. The extraction process done gave a high yield of the enzyme, and a cost-cutting strategy using sawdust as a solid-state fermentation medium could be developed in the future as mentioned in the study. As per the futuristic requirements, the cellulase enzyme produced is cost-effective and may be industrially produced in bulk for the conversion of lignocellulosic mass, biomass, and plant waste. The main hurdle in biomass-to-fuel conversion is limited toward its conversion to fermentable sugars, and cellulase is the rectification of this problem.

References

1. Chang, M.M., Chou, T.Y.C., Tsao, G.T., Structure, pretreatment and hydrolysis of cellulose, in: *Bioenergy*, pp. 15–42, Springer, Berlin Heidelberg, 1981.
2. Wyman, C.E., Biomass ethanol: Technical progress, opportunities, and commercial challenges. *Annu. Rev. Energy Env.*, 24, 1, 189–226, 1999.
3. Jarvis, M., Cellulose stacks up. *Nature*, 426, 6967, 611–612, 2003.
4. Zhang, Y.-H.P. and Lynd, L.R., Toward an aggregated understanding of enzymatic hydrolysis of cellulose: Noncomplexed cellulase systems. *Biotechnol. Bioeng.*, 88, 7, 797–824, 2004.
5. Kotchoni, S.O., Gachomo, E.W., Omafuvbe, B.O., Shonukan, O.O., Purification and biochemical characterization of carboxymethyl cellulase

(CMCase) from a catabolite repression insensitive mutant of bacillus pumilus. *Int. J. Agric. Biol.*, 8, 2, 286–92, 2006.
6. Abu, E.A., Onyenekwe, P.C., Ameh, D.A., Agbaji, A.S., Ado, S.A., Cellulase (EC 3.2. 1.3) production from sorghum bran by aspergillus niger SL1: An assessment of pretreatment methods, in: *Proceedings of the International Conference on Biotechnology: Commercialization and Food security*, Abuja, Nigeria, pp. 153–159, 2000.
7. Al-Ka'aby, T., Optimization for cellulase production by Aspergillus niger using saw dust as substrate. *J. Babylon Univ./Pure Appl. Sci.*, 20, 1300–10, 2012.
8. Fukuoka, A. and Dhepe, P.L., Catalytic conversion of cellulose into sugar alcohols. *Angew. Chem. Int. Ed.*, 45, 31, 5161–5163, 2006.
9. Teeri, T.T., Crystalline cellulose degradation: New insight into the function of cellobiohydrolases. *Trends Biotechnol.*, 15, 5, 160–167, 1997.
10. Horn, S.J., Vaaje-Kolstad, G., Westereng, B., Eijsink, V.G.H., Novel enzymes for the degradation of cellulose. *Biotechnol. Biofuels*, 5, 1, 1–13, 2012.
11. Solomon, B., Amigun, B., Betiku, E., Ojumu, T.V., Layokun, S.K., Optimization of cellulase production by aspergillus flavus linn isolate NSPR 101 grown on bagasse. *JNSChE J. Niger. Soc Chem. Eng.*, 16, 61–68, 1999.
12. Garvey, M., Klose, H., Fischer, R., Lambertz, C., Commandeur, U., Cellulases for biomass degradation: Comparing recombinant cellulase expression platforms. *Trends Biotechnol.*, 31, 10, 581–593, 2013.
13. Sani, R. K. and Krishnaraj, R. N. (eds.), *Extremophilic Enzymatic Processing of Lignocellulosic Feedstocks to Bioenergy*, Springer International Publishing, Gewerbestrasse, Cham, Switzerland, 2017.
14. Bajaj, B.K., Pangotra, H., Wani, M.A., Sharma, P., Sharma, A., Partial purification and characterization of a highly thermostable and pH stable endoglucanase from a newly isolated bacillus strain M-9. *Indian J. Chem. Technol. (IJCT)*, 16, 5, 382–387, 2009.
15. Cherry, J.R. and Fidantsef, A.L., Directed evolution of industrial enzymes: An update. *Curr. Opin. Biotechnol.*, 14, 4, 438–443, 2003.
16. Dale, B.E., Biobased industrial products: Bioprocess engineering when cost reality counts. *Biotechnol. Progr.*, 15, 5, 775–776, 1999.
17. Sridevi, A., Narasimha, G., Reddy, B.R., Production of cellulase by aspergillus niger on natural and pretreated lignocellulosic wastes. *Internet J. Microbiol.*, 7, 1, 580–592, 2009.
18. Wyman, C.E., Dale, B.E., Elander, R.T., Holtzapple, M., Ladisch, M.R., Lee, Y.Y., Coordinated development of leading biomass pretreatment technologies. *Bioresour. Technol.*, 96, 18, 1959–1966, 2005.
19. Huang, X.P. and Monk, C., Purification and characterization of a cellulase (CMCase) from a newly isolated thermophilic aerobic bacterium caldibacillus cellulovorans gen. nov., sp. nov. *World J. Microbiol. Biotechnol.*, 20, 85–92, 2004.

20. Wyman, C.E. and Hinman, N.D., Ethanol: Fundamentals of production from renewable feedstocks and use as a transportation fuel. *Appl. Biochem. Biotechnol.*, 24, 735–7535, 1990.
21. Hoang, T.-D. and Nghiem, N., Recent developments and current status of commercial production of fuel ethanol. *Fermentation*, 7, 4, 314, 2021.
22. Therasme, O., Volk, T.A., Fortier, M.-O., Kim, Y., Wood, C.D., Ha, H.S., Ali, A., Brown, T., Malmsheimer, R., Carbon footprint of biofuels production from forest biomass using hot water extraction and biochemical conversion in the Northeast United States. *Energy*, 241, 1228535, 2022.
23. Madan, A., Kumar, S., Waheed, S.M., Isolation & characterization of hydrocarbon degrading bacteria: A bio-remedial approach to cleanup oil spills. *Adv. Mater. Lett.*, 14, 4, 2304–1736, 2023.
24. Alhomodi, A.F., Gibbons, W.R., Karki, B., Estimation of cellulase production by aureobasidium pullulans, neurospora crassa, and trichoderma reesei during solid and submerged state fermentation of raw and processed canola meal. *Bioresour. Technol. Rep.*, 18, 101063, 2022.
25. Devi, R., Kapoor, S., Thakur, R., Sharma, E., Tiwari, R.K., Joshi, S.J., Lignocellulolytic enzymes and bioethanol production from spent biomass of edible mushrooms using saccharomyces cerevisiae and pachysolen tannophilus. *Biomass Convers. Biorefin.*, 12, 2, 1–15, 2022.
26. Gad, A.M., Suleiman, W.B., El-Sheikh, H.H., Elmezayen, H.A., Beltagy, E.A., Characterization of cellulase from geotrichum candidum strain gad1 approaching bioethanol production. *Arabian J. Sci. Eng.*, 47, 6, 6837–6850, 2022.
27. Ranjithkumar, M., Uthandi, S., Kumar, P.S., Muniraj, I., Thanabal, V., Rajarathinam, R., Highly crystalline cotton spinning wastes utilization: Pretreatment, optimized hydrolysis and fermentation using pleurotus florida for bioethanol production. *Fuel*, 308, 1220525, 2022.
28. Singh, B., Bala, A., Anu, A., Kumar, V., Singh, D., Biochemical properties of cellulolytic and xylanolytic enzymes from sporotrichum thermophile and their utility in bioethanol production using rice straw. *Prep. Biochem. Biotechnol.*, 52, 2, 197–209, 2022.
29. Srivastava, N., Singh, R., Srivastava, M., Syed, A., Pal, D.B., Bahkali, A.H., Mishra, P.K., Gupta, V.K., Impact of mixed lignocellulosic substrate and fungal consortia to enhance cellulase production and its application in NiFe2O4 nanoparticles mediated enzymatic hydrolysis of wheat straw. *Bioresour. Technol.*, 345, 126560, 2022.
30. Gopinath, S. M., Shareef, I., Ashalatha, Ranjit, S., Isolation, screening and purification of cellulase from cellulase producing klebsiella variicola RBEB3 (KF036184.1). *Int. J. Sci. Res.*, 3, 6, 1398–1403, 2014.
31. Shareef, I., Satheesh, M., Christopher, S. X., Isolation and identification of cellulose degrading microbes. *Int. J. Innov. Res. Sci. Eng. Technol.*, 8, 4, 6788–6793, 2015.

32. Gupta, P., Samant, K., Sahu, A., Isolation of cellulose-degrading bacteria and determination of their cellulolytic potential. *Int. J. Microbiol.*, 2012, 1–5, 2012.
33. Kim, Y.-K., Lee, S.-C., Cho, Y.-Y., Oh, H.-J., Ko, Y.H., Isolation of cellulolytic Bacillus subtilis strains from agri- cultural environments. *Int. Sch. Res. Notices*, 2012, 1–9, 2012.
34. Wood, T.M. and Bhat, K.M., Methods for measuring cellulase activities. *Methods Enzymol.*, 160, 87–112, 1988, Academic Press.
35. Madan, A. and Prakash, D., Optimization of production parameters for endoglucanase enzyme by isolated fungal strain/s and its application in saccharification of cellulosic materials, in: *Advances in Mechanical Engineering: Select Proceedings of CAMSE 2020*, pp. 777–786, 2021.

13

A Sustainable Catalytic Approach for Wastewater Bodies: An Innovation and Technological Point of View

Anupama Rajput[1], Sudheesh K. Shukla[2], Ravi Kumar[3], Gaurav Jha[4], Vikas Kalia[5] and Bindu Mangla[3]*

[1]*Department of Chemistry, FET, Manav Rachna International Institute of Research and Studies, Faridabad, Haryana, India*
[2]*School of Chemical Engineering and Physical Sciences, Lovely Professional University, Jalandhar, Punjab, India*
[3]*Department of Chemistry, J.C. Bose University of Science and Technology, YMCA, Faridabad, Haryana, India*
[4]*Chemical Information Science Laboratory, Department of Chemistry, Pondicherry University, Puducherry, India*
[5]*Department of Chemistry, MD University, Rohtak, Haryana, India*

Abstract
The utilization of living organisms or simply bioremediation (microbial consortium) in the detoxification process and in the removal of toxic contaminants from oil spills and contaminated groundwater as well as pollutants and toxins from both biotic and abiotic environments is the most popular and sustainable approach. Biocatalysis using a wide range of microorganisms such as molds, bacteria, yeast, fungi, and algae is highly beneficial for wastewater treatment. Textile wastewater includes a broad range of dyes and compounds that can cause a variety of environmental problems. This article will tackle the role of different microorganisms, various strategies, and factors affecting the rates of bioremediation and detoxification and will also discuss the *ex-situ* and *in-situ* bioremediation processes.

***Keywords*:** Bioremediation, organism, pollutants, textile wastewater, detoxification, microbes, treatment process

**Corresponding author*: bindumangla@gmail.com

Sudheesh K. Shukla, Chaudhery Mustansar Hussain, Santanu Patra, Meenakshi Choudhary (eds.) Electrocatalytic Materials for Renewable Energy, (353–376) © 2024 Scrivener Publishing LLC

13.1 Introduction

Bioremediation is a specific branch of biotechnology that incorporates the use of microorganisms in various processes like the degradation of contaminants that pose both environmental and human risks. The degradation process is initiated with the microbial consortium belonging to different groups acting in a parallel manner to finally complete the process. The science of bioremediation is categorized into *in-situ* (in place) and *ex-situ* (removal and treatment in another place) strategies. The ease of degradation of a wide range of pollutants generally makes bioremediation a flexible technology that can be applied in terms of various soil applications and conditions.

A specific approach to bioremediation includes exploiting and stimulating the existing microbial colonies by providing them with specific vitamins and nutrients and specific requirements. Furthermore, the biostimulation process includes various changes in terms of pH scale, water activity, oxygen supply, and incorporation of negatron donors, acceptors, and nutrients.

Another biological approach related to bioremediation is termed "bioaugmentation," where organisms with degradation abilities are chosen to inoculate the contaminated site. These two approaches do not seem to be reciprocally exclusive—they will be used at the same time. Bioremediation with the help of microbial consortium has sprouted out as a potential technology to treat and manage the acerbating textile effluents [1–3].

The microalgal species of algae such as *Chlorella vulgaris* and *Chlorella pyrenoidosa* and the bacteria *Oscillatoria tenuis* degrade dyes with toxic chemical groups into fewer toxic forms of aromatic amines. Today, textile wastewater presents a typical case of water contamination sources, where different technologies have been implemented for the treatment of malignant-related neoplastic diseases.

Among the various developed technologies, the sorption process is one of the most promising and supportive carbon-based assisted cost-effective and energy-efficient sorption technology which has been used for the treatment of wastewater. The incorporation of biopolymers like chitosan is one of the advanced sorption technologies being used for textile effluents (dyes and metal ions). As a result, different studies have recently been in deep dilemma regarding the utilization of chitosan-based materials using the chemisorptive approach [4].

The textile business generally utilizes a large volume of water and proportionately produces a Brobdingnagian volume of wastewater [5].

Textile wastewater contains forms of toxic dyes and chemicals that pose a serious threat to the environment [6–8], increasing issues related to energy demand and non-tariff bindings concerning environmental conservation. Furthermore, the local textile industry generally contributes 7%–9% to the global textile process [9] and more than 50%–60% to the total country exports [10].

The sector of textile attained $12–13.7 billion in 2013 to 2014 which changed to $12–12.5 billion in 2015 to 2016 [11] chiefly because of energy issues and liabilities. Textile wastewater treatment may be a huge challenge as textile wastewater contains toxic dyes and huge amounts of phosphates and nitrates, resulting in multiscale injury to the atmosphere and the environment [12, 13].

Textile wastewater is differentiated in terms of many important parameters like color, temperature, pH, biological oxygen demand, chemical oxygen demand, total dissolved solids, total gas, total phosphorus, and additional non-biodegradable elements (Table 13.1).

Bioremediation of textile wastewater (TWW) is based on some specific compositions related to the presence of salts, dyes, salts, significant metals, reagents, etc. [15–18]. The different upcoming sorption processes like the utilization of activated charcoal, natural action, specificity in activity, filtration procedures, ozonation, electrochemical-assisted destruction, and irradiation [19–22] do not promise to treat TWW alone in an effective manner due to their expensive nature, energy liabilities, and poor economics and utility [23, 24]. Fungi stand as the potential candidate in facilitating the

Table 13.1 Features of textile wastewater [14].

Metals	Amount	Quantity
Pb	0.5	mg/l
Cd	0.3	mg/l
Zn	1.0	mg/l
Mn	2.0	mg/l
Cr	1.5	mg/l
Ni	1.0	mg/l
Cu	0.2	mg/l

process of bioremediation by secreting various extracellular enzymes and acting as powerful natural decomposers of waste material.

"White-rot fungi," a group with potential from the fungal kingdom, have the capability to be utilized in the bioremediation process by degrading insoluble chemicals. They have a variety of benefits in terms of the degradation of insoluble chemicals and harmful environmental pollutants that may be exploited in bioremediation systems [25]. The employed methodologies used to treat textile wastes such as sorption specificity (Figure 13.1); membrane filtration processes like reverse osmosis (RO), nanofiltration (NF), and ultrafiltration; ozonation; evaporation; chemistry chemical reaction; natural action; phytoremediation; and crystallization have been used for the last decades [26].

The utilization of microalgae in bioremediation (Figure 13.1) has proven to be an excellent source for biodiesel production. The integration of biodiesel with bioremediation has been observed with fruitful results in terms of biodiesel efficiency and wastewater management. In addition, studies should be carried out to screen and optimize conditions like the selection of algal species, harvesting strategies, cultivation strategies, etc. [27]. TWW conjointly includes significant trace metals such as chromium (Cr), arsenic (Ar), copper (Cu), and Zn (Zn), and the concentrations of nutrients in textile wastewaters are known to be supply-dependent. In this context, the American state accounts for approximately 20–21 to 56–57 mg L^{-1} in textile waste matter [28–30]. The Islamic Republic of Pakistan emphasizes textile waste matter treatment and management according to domestic and industrial water necessities [31–33].

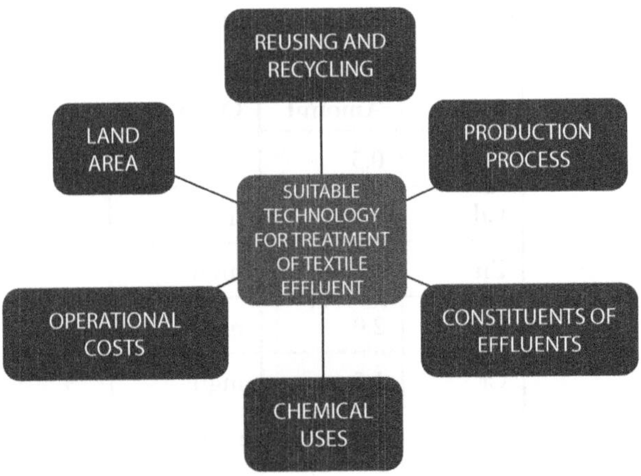

Figure 13.1 Suitable technology for the treatment of textile effluent.

13.1.1 Organisms Involved in Bioremediation

Bioremediation utilizes living organisms or a microbial consortium to break down toxic pollutants into harmless natural compounds. The choice of "bioremediators" (the microorganisms involved in bioremediation are illustrated in Figure 13.2) is often based on the rapid growth rate, specific metabolic needs, and variability in genetic modification, and the prime choice involves bacteria and fungi. They can utilize environmental pollutants as their choice of food and can easily metabolize them to less toxic remains, thus ensuring environmental safety and protection.

13.1.2 Bacteria Species Involved in Bioremediation

(a) Pseudomonas putida

Pseudomonas putida is a group of gram-negative soil bacteria incorporated in processes leading to the bioremediation of components involved in paint thinner known as "toluene." In addition, it also results in the degradation of various products like naphthalene obtained through processes like petroleum refining.

(b) Dechloromonas aromatica

Dechloromonas aromatica, a rod-shaped bacterium, is very well known to harbor various aromatic oxidizing processes like toluene, benzoate, and chlorobenzoate. It holds an efficient potential to oxidize benzene through

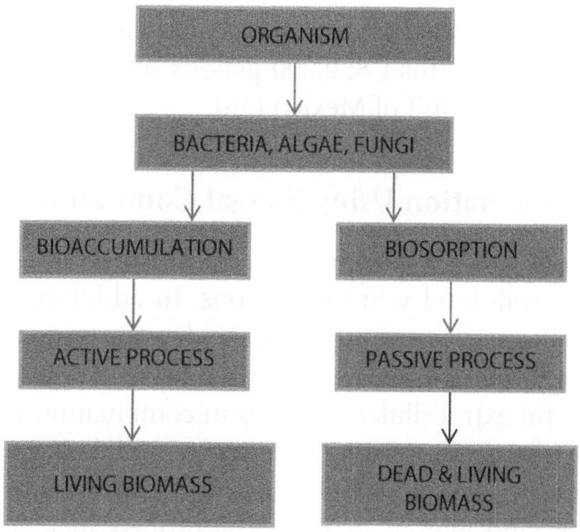

Figure 13.2 Organisms involved in different processes.

an anaerobic-assisted pathway. Due to the various advantages offered by this microbe, it is potentially being incorporated into upcoming *in-situ* strategies of bioremediation especially in treating wastewater effluents.

(c) Nitrifiers and Denitrifiers

The removal of excess nitrogen includes a tagged two-way process, namely, nitrification and denitrification. Ammonium is oxidized to nitrite through nitrification with the help of *Nitrosomonas europaea*, and finally, nitrite is further oxidized to nitrate by *Nitrobacter hamburgensis*. Through this process, the pollutants responsible for eutrophication in natural waters can be remediated positively.

(d) Deinococcus radiodurans

Deinococcus radiodurans is a genetically modified microbe for the bioremediation process of solvents and heavy metals. It is a radiation-resistant extremophile bacterium, which is known to degrade a radioactive mixture of waste environmental effluents targeting mercury and toluene [34].

(e) Methylibium petroleiphilum

Methylibium petroleiphilum, a PM1 strain, possesses the ability to degrade methyl tert-butyl ether (MTBE), by utilizing carbon as a sole energy source [35].

(f) Alcanivorax borkumensis

Alcanivorax borkumensis, a rod-shaped marine bacterium that acts on hydrocarbons, is found in fuel stocks and gives off carbon dioxide. It grows profusely in damaged environments marked with oil fuel stocks and is known to clean more than 850,000 gallons of oil from the Deepwater Horizon oil spill in the Gulf of Mexico [36].

13.1.3 Bioremediation Using Fungal Communities

Fungi project out as an important link connecting the essential components related to soil–food web interactions. In addition, it is featured as the potential influential agent inducing the decomposition of waste material [37]. The key role of fungi is to break down leaf litter by excretion of the most powerful extracellular enzymes in combination with the specific acids targeting the decomposition process of cellulose and lignin (powerful plant sources). Fungi play an integral role in various processes like biodeterioration and bioremediation. The process includes selecting the appropriate fungal specie targeting a particular pollutant.

Many fungi have been found significant in bioremediation with an effective remediation process. The group includes species like *Aspergillus niger*, *Mucor thermohyalospora*, *Fusarium ventricosum*, and *Trichoderma harzianum*, which have an important ability to degrade endosulfan [38]. Some fungal groups do have the capability to perform various other processes like dehydrogenation, deoxygenation, hydroxylation, and esterification during the conversion of 3-phenoxybenzoic acid into 3-hydroxy-5-phenoxybenzoic acid which is finally transformed into products like gallic acid and phenol in terms of pesticide transformation procedures [39].

Similarly, a fungal strain "*Rhizopus oryzae*" has been reported to have a biosorption potential for pentachlorophenol incorporating processes like dichlorination and methylation [40]. Furthermore, fungi belonging to Zygomycetes and especially the "white-rot fungi" have been known to detoxify and decolorize textile waste. In addition, fungi like *Penicillium digitatum* and *Fusarium solani* are also observed with exceptional capabilities for the degradation of polychlorinated biphenyls (PCBs) with the help of non-lignolytic enzymes.

13.2 Microbial Processes

13.2.1 Biological Factors

The competition between various microbial communities for the utilization of various sources like carbon sources and various types of antagonistic strategies has a direct influence on the process of degradation [41–43]. However, the degradation speed is directly proportional to the concentration of contamination and the quality of the catalyst involved. The quality of catalyst {the quantity|the quantity|the number} represents the quantity of microorganisms ready to metabolize and utilize the source of contamination. In addition, the specificity and expression of the special enzyme presented by the cell will influence the speed of degradation. The act of "bacterial taxis" is an important feature for assisting the degradation of recalcitrant organic compounds [44].

13.3 Factors Affecting the Rates of Bioremediation

TWW comprises an advanced composition containing incorporated salts, dyes, integral metals, reagents, etc. Many upcoming procedures like sorption technologies related to flocculation, filtration procedures, irradiation and ozonation [45], and ozonation and irradiation [46–48] have, however, not proven excellent in terms of cost efficiency, energy issues, effective sludge formation, etc. [49–51]. Bioremediation has been emerging as a promising technology to treat textile effluents [52–54] with the incorporation of different microorganisms like bacteria, fungi, yeast, and algae. The utilization of microalgae is a promising alternative strategy for friendly microalgal bioremediation process [56–60].

13.3.1 The Process of Utilization of the Primary Substrate

Primary substrate utilization is a process in which the effective transformation of a specified substrate has been witnessed for utilizing the energy and carbon supply effectively. This kind of biodegradation pattern is commonly used for the breakdown of oil components and pesticidal mixtures, where an electron acceptor (anaerobic/aerobic) is needed for the processing and transformation procedures.

13.3.2 The Act of Co-Metabolism: Utilization of the Secondary Substrate

The process of co-metabolism includes the progressive transformation of a substrate at its primary energy and carbon supply criteria [61]. Throughout the particular process of reaction, which includes degrading the substance, it is observed that the concerned organism has no integral

Table 13.2 Microorganisms important for bioremediation.

Microorganism	Compound	References
Micrococcus luteus, *Listeria denitrificans*, and *Nocardia atlantica*	Textile dyes, azo dyes	[63]
Bacillus spp. ETL-2012, *Pseudomonas aeruginosa*, *Bacillus pumilus* HKG212	Textile dye, azo dye, RNB dye	[64–66]
Exiguobacterium indicum, *Exiguobacterium aurantiacums*, *Bacillus cereus*, and *Acinetobacter baumanii*	Azo dye effluents	[67]
Bacillus firmus, *Bacillus macerans*, *Staphylococcus aureus*, and *Klebsiella oxytoca*	Vat dyes, textile effluents	[68]

carbon or energy gain [62]. Microorganisms have a very significant role in bioremediation (shown in Table 13.2).

13.3.3 Detoxification

The persistent nature of organic pollutants is related to their high chemical stability and the difficulties encountered in terms of transformations including biotic and abiotic phases. They have a tendency to accumulate in large amounts in plants as well as animals and finally are interlinked in the food chain posing the greatest threat in terms of health and safety. In order to cope with the new burning ecological criteria like unwanted and excessive emission of pollutants in the environment, the plant and microbial world has started exhibiting special characteristics related to the adsorption and degradation of toxic pollutants.

These special features hold an important pathway for exploiting these phytoremedial technologies to degrade a wide range of chemical pollutants (organochlorine pesticides) through enzymatic transformation. The remedial solutions and approaches through biological processes are cost-effective, eco-friendly, and profitable, but they do offer some drawbacks like issues related to low productivity in some bacterial strains. As a result, only some of the bacterial strains are found to have potential in terms of carrying out the mineralization process. The incorporation of plant–microbial consortiums is helpful in terms of degradation processes associated with organochlorine pesticides [69]. To improve the various processes related to the detoxification process, research needs to focus on specific strains with selectively high degradation and detoxification abilities, completely eradicating the pollutant from the environment. The research should be technology-oriented in framing concepts and pathways directed toward specificity in the selection of strain, secretion of oxidative enzymes, and finally degradation of the toxic carbon skeleton into stable and non-toxic cellular compounds.

13.4 Bioremediation Treatment Processes

The bioremediation process is influenced by abiotic and organic issues that are directly related to the degradation abilities of the microbial communities that are monitored by specific tests related to toxicity, targeting a single specie generally to confirm that there are no possible changes that result in residual toxicity. However, working on the community response rather than considering a single species can be a more justified finding regarding toxicity analysis. The techniques and procedures provided by human interventions are categorized into *in situ* (removing contamination problems at the concerned source) and *ex situ* (by removing the contaminated soil or water and treating it outside the supply).

There are some techniques that aim at handling the correct mobilization of pollutants. In addition, microbial communities play a major role as biocatalysts, facilitating the process of degradation of the specified waste source. There are certain important factors that potentially influence these bioremediation processes like the nature, kind, and growth of the population of microorganisms involved; the effective target screening of contaminants for specific microbes; type of soil; temperature; pH; and presence and absence of negatron receptors and nutrients. The typical organic phenomenon and abiotic factors also affect the nature and growth of microorganisms, thus influencing the bioremediation strategies [70]. The textile industry has a negative influence on the environment due to the continuous discharge of toxic colored dyes and contaminated effluents which are discharged into surface water creating threats to living forms. Textile wastewater includes detergents, solvents, biological oxygen demand (BOD), chemical oxygen demand (COD), salts, surfactants, toxic dyes, chlorinated compounds, heavy metals (Zn, Cu, Al, and Cr), and sometimes sulfide and formaldehyde [71]. Furthermore, the textile industry adds about tons of toxic waste worldwide, thus adding a broad ratio of toxic dyes, where 90% of dyes are reported by the ETAD (Ecological and Toxicological Association of the Dyestuffs Manufacturing Industry) as lethal. These dyes due to higher stability and longevity expand their periods and do not get extracted easily. To solve these issues related to contamination of wastewater or to clean the water flow from the textile industry, many methods like physical methods have been utilized, but due to the various glitches in terms of environment and cost-effectiveness, a move has been made to adopt methods which are effective, cost-effective, and most importantly, eco-friendly.

For example, the microbial method like incorporating laccase enzyme in *Streptomyces* can be effective in detoxifying wastewater from toxic dyes [72]. The eco-friendly biological remedies for targeting dye degradation at a large scale include the oxidation of a large group of phenolic compounds by using the association of enzymes with specific mediators like acetosyringone and methyl syringate, 1-hydroxybenzotriazole (HBT), and oxime violuric acid (VLA) for enhancing the oxidization process of refractile aromatic compounds. For example, the SilA process has a very low ability to decolorize dyes like Acid Black 48, Acid Orange 63, and Reactive Black 5, but the addition of mediators with high redox potential like acetosyringone (AS) and methyl syringate (MeS) enhances the capabilities and activity related to the degradation process. It has also been observed that the dye has the potential to increase the toxicity ratio even after the degradation procedure, but due to the attachment to a mediator, the activity and efficiency of degradation associated with a complex like the SilA–acetosyringone system, increases so the dye is not left with any type of toxicity residues [73].

Another procedure including enzymatic processes using laccases and laccase–mediator systems in the removal of toxic dyes presents a viable and commercially effective and environmentally friendly alternative to conventional mechanical and physicochemical methods, which require a high load of energy and investment issues. There are various kinds of dyes used in the textile industry, but the most harmful is the azo dye because its metabolic intermediate product is toxic and carcinogenic. The main drawback of azo dye is that it has been reported as the most potential cancer-suspecting agent, and other azo dyes like p-phenylenediamine (p-PDA) and triphenylmethane are also reported to leave a negative impact. Treatment with the microbial consortium for degradation conjointly and during bioremediation of textile wastewater consortia provides a powerful metabolic network to decompose the dyes, and single strains attack dye molecules from different sides in the microbial consortium. During this process, microbes get into toxic waste, and furthermore, naturally new resistant strains are generated which are used to convert toxic chemicals into a less toxic nature. In mixed culture, the remaining toxicity is also treated and completely degrades the organism by synergistic action, e.g., azo dye. Direct and indirect are the mechanisms through which the decolorization of dyes takes place, but in both cases, enzymes are used for the degradation. Laccase and azoreductase enzymes are mostly used for industrial dyes.

The bacterial consortium can be used in aerobic or anaerobic conditions for the decolorization of azo dyes with variable pH and temperature. The bacterial consortium is more effective than pure culture because of quorum sensing which according to bacteria population density will regulate gene expression. Biodegradation of dyes in textile wastewater by using bacterial consortium prevails due to optimization in abiotic situations. Yeast and fungal consortia are also a variety of bacterial consortium used for many kinds of macromolecule organics for degradation. UV spectra, thin layer chromatography, high-performance thin layer chromatography, and Fourier transform infrared spectroscopy (FTIR) techniques can used in the detection of intermediates, metabolites, and azo group attached with azo dye, and nuclear magnetic resonance (NMR) is used for a defined quantitative study. In this way, the microbial consortium proved to be a more beneficial and appropriate technique than pure culture because of microbial functionality. The textile industry is responsible for the hazardous effect on the environment because of toxic discharged material which contains high loads of non-biodegradable organic matter, heavy metals, and pathogenic microorganisms. The effluents of the textile industry are a major issue because they contaminate water to a high degree. The dyes used in the industry have higher solubility in water, which makes them difficult to remove, resulting in polluted water. The dyes used in textiles are Azure B dye, Disperse Red 1 dye, Disperse Orange 1 dye, dye Basic Red 46, Poly R-478, Methyl Orange, Methylene Blue, and many more. During the dyeing process, azo dye mainly does not adhere to the fabric completely and is released in water. Azo dye has an adverse impact on plant growth and microbial communities. The dyes have heavy metals when they are discharged, and the positive charge of metals gets attracted toward negative charge species in the water and gets accumulated through the food chain gaining access to the human body where they cause serious diseases.

The reason for the damage to water bodies is the color used in textile dyes which leads to the reduction in photosynthesis rate because it stops the penetration of light through water, increasing the BOD and COD levels, mutagenicity, toxicity, and carcinogenicity in water because of the effluents of dyes. Central nervous system disorder and deactivation of enzyme activity in the body are the diseases caused by textile dyes. Azure B dye has the tendency to interact with the helical structure of DNA and duplex RNA [74] and also segmented the lipid membrane of the cell. Disperse Red 1 dye and Disperse Orange 1 dye show mutagenic potential and are responsible for DNA adduct and damage to the DNA which leads to cancer [75]. Basic Red 9 dye breaks down into aromatic amines which exhibit carcinogenic properties and high toxicity, resulting in skin irritation, mutation, and cancer [76]. Bioremediation through microorganisms and enzymes is a sustainable solution for this problem. To remove these hazardous dye effects, one of the methods commonly used is photoremediation. It is a natural technology used to degrade and detoxify dyes using plants and their enzymes. Photoremediation includes the combination of various techniques like rhizodegradation and filtration and phytoaccumulation, phytodegradation, and rhizodegradation. *Nasturtium officinale* and *Pennisetum purpureum* biodegrade the dyes Basic Red 46 and Poly R-478, respectively [77, 78].

Bioremediation plays a role in the reduction of damage caused by hazardous chemicals using microorganisms, i.e., algae, filamentous fungi, and yeast in soil and wastewater. Biostimulation or bioaugmentation morphology is involved in microorganism bioremediation. Fungal biosorption is an effective technique for the decoloration of dyes. The phosphate group from fungi and the amino group from chitosan (organic matter) interact due to the opposite charge [79], and it is a fast process. For example, Methyl Orange dye is decolorized by *Aspergillus flavus* and Methylene Blue by *Aspergillus fumigates*. Enzymes used for the degradation of textile dyes are azoreductase, lignin peroxidase, Mn peroxidase, reductase and tyrosinase, phenoloxidase, etc. However, microbiology, bioremediation, enzymatic action, and microorganism bioremediation are the processes used to acerbate the degradation process and, finally, decolorize the toxic textile dyes. Phytoremediation and the fungal treatment process proved to be economically effective alternatives for decolorization involving processes such as biosorption, bioaccumulation, and biodegradation. Textile industry effluents contain various compositions of toxic elements and compounds and create a risk for the environment. The solubility of the dye, the high organic content of the complex structure, and the heavy metals found in dyes generate contamination in wastewater by resisting biodegradation. Azo dyes are the most commonly used dyes and have the highest toxicity in nature due to an amine group present. Many enzymes have been used for the bioremediation process of azo dyes, but until now, some other enzymes may also be approached with a redox mediator to deal with target contaminants like Amaranth and Acid Black 1 types of azo dye. Vasileva and Y. Topalova [80] showed that to minimize the contaminants in wastewater, algorithms are used with anthraquinone-2-sulfonate (AQS) as a redox mediator to enhance the activity in the azo bond reduction process by maintaining denitrification and decolorization. Using this mediator, only 13 to 20 days are required for effective results to be seen. The decolorization of Amaranth dye was performed using a cell-free extract and an isolated bacterial strain. The ability to remove color decides which type of microbial strain is used from activated sludge. Two types of mechanisms were applied for azo reduction: one is working around the cytoplasmic zone, which is the cytoplasmic-mediated enzyme system, and the second is working around the membranes, which is the membrane-mediated enzyme system, and both processes have good efficiency for decolorization using the AZO29 culture. Yeast extract and peptone were added as bioreactors during biostimulation of microbial contamination. Bioremediation through enzymatic action shows a good effect in preventing the harmful effects of effluents in water [81].

13.4.1 *Ex Situ* (Composting)

Various eco-friendly methods have been evaluated for the detoxification of textile wastewater. Anaerobic and aerobic microorganisms were identified as an appropriate and effective treatment for detoxification including algae, fungi, and bacteria with enzyme activity. Aerobic microorganisms have enzymes that are oxidative in nature and that can destroy the aromatic amines and release them during the removal process. Bioremediation treatment is always preferred for detoxification [82]. In the *ex-situ* process, polluting contaminants are taken to a location other than the contamination site where bioremediation is carried out with the aid of microorganisms. The use of various bioreactors is a popular technique in this area. Different kinds of recalcitrant organic pollutants can be degraded anaerobically by using white-rot fungi. Hossain *et al.* [83] focused on the use of membrane bioreactors using the white-rot fungus for enhancing the detoxification and bioremediation process for treating textile wastewater and effluents. White-rot fungi possess lignin as an energy source and are utilized for the degradation of pollutants. Stable discoloration activity (approx. 90%–98%) and removal of biological oxygen demand (35%–50%), chemical oxygen demand (45%–67%), and total organic carbon (TOC) (90%–95%) were achieved using white-rot fungi-based membrane bioreactors. White-rot fungi used in membrane bioreactors showed absorption phenomenon and detoxified wastewater as a consequence.

13.4.2 *In Situ* (Biostimulation)

In-situ bioremediation can be understood as an advanced process of on-site bioremediation where the decomposition of contaminants is carried out by implanting or cultivating a suitable type of microorganism with the help of a favorable level of nutrients and fertilizers. The textile industry utilizes a huge amount of water during dyeing, bleaching, and printing processes, and large amounts of synthetic dyes are released in the water during discharge and create contamination in surface water. Even a very low concentration of colored liquid water can cause harmful effects to humans, agriculture, and the environment. Heavy metals, salts, acids, alkalis, and chlorine exist in effluents of the textile industry. High amounts of chlorine can lead to the lowering of dissolved oxygen amount in water bodies. A variety of methods are already in use for the purification of wastewater which is contaminated due to textile effluents. An eco-friendly technique is always preferred, and the potential of *Asparagus densiflorus* to treat dyes in Ribin GFL vertical subsurface flow (VSbF) phytoreactor for decolorization received attention. However, an eco-friendly, reliable, and approachable technology is required to deal with such issues. Phytoremediation is the first choice for this type of case for significant effectiveness.

As per Watharkar *et al.*, this method not only just decolorizes the Rubin GFL dye but also minimizes the other hazardous organic contents from wastewater. The vertical subsurface flow (VSbF) phytoreactor technology was introduced to reduce BOD, COD, total dissolved solids (TDSs), and total suspended solids (TSSs) of real textile effluents by 60%, 65%, 66%, 45%, and 60%, respectively, within 6 days. The tuberous structured root system of *Asparagus densiflorus* is the main reason for its use in phytoremediation. *Asparagus densiflorus* is used in the photovolatization process to remove benzene, toluene, and trichloroethylene from the atmosphere. The leaf area of any plant indicates its photovolatization ability. *Asparagus densiflorus* has various enzymatic and anatomical properties for the treatment of RGFL, and studying its behavior before or after the treatment helps us to know about its functionality. FTIR spectra evaluated the samples of *A. densiflorus* roots with RGFL before and after the process, and the changes in the peaks at various frequencies indicated that RGFL was degraded by *A. densiflorus. Asparagus densiflorus* shows a high ability to degrade the RGFL dye up to 90%–95%, and the VBSF phytoreactor has a great tendency to reduce other toxic contents (60%–70%) from textile effluents. Heavy metals like Hg, Cr, and Cd were also minimized to standard value through phytoremediation [84].

However, the treatment of textile effluents using *A. densiflorus* and the VBSF phytoreactor is a beneficial methodology to reduce and eliminate the harmful impacts on the environmental system. Many kinds of phytoreactors are used to treat textile effluents such as *Portulaca grandiflora* (Mexican rose) for the degradation of Navy Blue HE2R dye, *Alternanthera philoxeroides* for sulfonated dye Remazol Red, and *Ipomoea aquatica* for the Brown 5R dye.

13.5 Bioremediation

Polluted contaminated effluents released into the oceans affect plants and wildlife such as fish, mammals, and birds. Bioremediation technology preferably plays a great role in these conditions. Basically, bioremediation is the process of adding a material into a contaminated system to enhance the activity of biodegradation. Bioremediation mainly involves two strategies for treatment: *ex situ* and *in situ.* The materials used in bioremediation are considered microbiological cultures or enzyme additives by the U.S. EPA. In *in-situ* bioremediation, the aqueous solution is distributed to the contaminated environment to excite indigenous bacteria for degradation. The growth of microorganisms speeds up by the increase of physicochemical conditions by adding oxygen and nutrients. On the other hand, the *ex-situ* process needs a hollow surface for treating contaminated water, and this involves land farming, composting, and biopiles for a solid-phase system as well as bioventing, biosparging, bioreactors, and bioaugmentation for slurry-phase treatment.

Various kinds of microbes like bacteria, fungi, and yeast are known for contaminant degradation, whereas for marine pollution degradation, microbes such as *Mycobacterium, Pseudomonas, Rhodococcus, Aeromonas, Corynebacterium, Acinetobacter*, and cyanobacteria are used. For the degradation of pesticides and organic hydrocarbons like alkane and cyclic compounds, microbes like *Sphingomonas, Rhodococcus*, and *Mycobacterium* are used. pH values of 6–9 are reported as the optimum range for nutrient and oxygen concentration. The white-rot fungus, *Phanaerochaete chrysosporium*, is a category of fungi, which has the potential to degrade toxic pollutants. *Prototheca zopfi* is an alga with the ability to degrade alkanes and aromatic hydrocarbons. *Bacillus subtilis, B. licheniformes, B. cereus, B. coagulans*, and *Phenibacillus polymyxa* are bacteria appropriate for the bioremediation of organic detritus and the removal of water turbidity. For aquaculture systems, the removal of ammonia is done using bacteriological nitrification methods.

Some factors, which can affect bioremediation technology, are weather, temperature, oxygen content, etc. Weather impacts the concentration of hydrocarbons by reducing them through evaporation. A decrease in temperature will result in a decrease in the rate of biodegradation, and 15°C–20°C is the optimum temperature for a marine environment. Oxygen reduction also leads to a reduction in biodegrading activity. Many techniques and various methods were developed for the detoxification of textile wastewater and the degradation of dyes such as the use of bacteria and microorganisms such as white-rot fungi, laccase, lignin peroxidase (LiP), and manganese peroxidase (MnP). To detoxify wastewater using fungal biomass such as *Trametes pubescens* MUT fungus is also an effective technology. The degradation of dye through fungi and sludge treatment reduced COD and toxicity [85].

Decolorization is analyzed through UV absorption phenomenon. The toxicity of the samples was analyzed using the green unicellular alga *Pseudokirchneriella subcapitata* (Korshikov-Hindak) (method UNI EN ISO 8692:2005) with some more fungal species. *Trametes pubescens, Lepidium sativum, T. versicolor, B. adusta* MUT 2295, *P. spadiceum, Pseudokirchneriella subcapitata, gallica, S. ochraceum*, and *T. hirsuta* are the types of fungi and alga species which result in colorless wastewater through the absorption phenomenon. The COD level was also reduced when evaluated with different fungi, but activated sludge was a more appropriate method for the reduction of COD and for biostimulation [86]. In strain removal and for detoxification of recalcitrant compounds, *T. pubescens* is the most effective among all nine fungi. Various kinds of bioreactors were applied during bioremediation processes. In the biological process, two methods were applied to examine textile wastewater, in which fungal treatment was used for color and strain removal and activated sludge was used for reduction in toxicity and COD.

13.5.1 Advantages and Disadvantages

Bioremediation is advantageous over conventional technologies, as it is cost-effective, less intrusive to the contaminated site, and more environmentally friendly and has less hazardous end-product formation. Biological methods are better than chemical and physical methods because they are more economical and efficient.

13.6 Conclusion

The release of toxic effluents through contaminated textile wastewater sources can be attributed to both natural and anthropogenic activities, which pose a great risk of contamination to other environmental sources. Looking for potential, cost-effective, and eco-friendly solutions has been successfully adopted to reduce the hazards of wastewater contamination and toxic residues. Addressing the constraints of the current traditional combined chemical technology for wastewater management, biological processes provide insight through the efficient utilization of microbial consortiums that are cost-effective, selective, specific in action, and most importantly, eco-friendly. The potential of microorganisms or microbial consortiums to remove and extract metals from solutions via multiple mechanisms, comprising active and passive modes of action, has been highlighted to be one of the viable techniques to effectively eliminate hazardous metal effluent. The application and implementation of bioremediation processes on a large scale or toward a commercialization process, however, remains a questionable criterion due to the lack of knowledge regarding characterization, identification, and genetic engineering. In order to more effectively address the various environmental threats and risks of contamination, a better understanding of specific eukaryotes and prokaryotes is required. It will also allow for the identification of potential target detoxification pathways and the fulfillment of cost, feasibility, specificity, selectivity, and ease of cultivation specifications for the straightforward application of bioremediation processes in wastewater management.

References

1. Taştan, B.E., Ertuğrul, S., Dönmez, G., Effective bioremoval of reactive dye and heavy metals by aspergillusversicolor. *Bioresoure Technol.*, 101, 3, 870–876, 2010.
2. Dellamatrice, P.M., Silva-Stenico, M.E., Moraes, L.A.B., Fiore, M.F., Monteiro, R.T.R., Degradation of textile dyes by cyanobacteria. *Braz. J. Microbiol.*, 48, 25–31, 2017, http://dx.doi.org/10.1016/j.bjm.2016.09.012.
3. Ponraj, M., Gokila, K., Zambare, V., Bacterial decolorization of textile dye-orange R abstract. *Int. J. Adv. Biotechnol. Res.*, 2, 168–77, 2011.
4. Lim, S. L., Chu, W. L., Phang, S. M., Use of chlorella vulgaris for bioremediation of textile wastewater. *Bioresource Technol.*, 101, 19, 7314–7322, 2010.
5. Khatri, A., Peerzada, M.H., Mohsin, M., White, M., A review on developments in dyeing cotton fabrics with reactive dyes for reducing effluent pollution. *J. Clean Prod.*, 87, 50–7, 2015, doi.org/10.1016/j.jclepro.2014.09.017.
6. Firmino, P., da Silva, M., Cervantes, F.J., dos Santos, A.B., Colour removal of dyes from synthetic and real textile wastewaters in one- and two-stage anaerobic systems. *Bioresour. Technol.*, 101, 7773–9, 2010, http://dx.doi.org/10.1016/j.biortech. 2010.05.050.
7. Bhuiyan, M.A.R., Rahman, M.M., Shaid, A., Bashar, M.M., Khan, M.A., Scope of reusing and recycling the textile wastewater after treatment with gamma radiation. *J. Clean Prod.*, 112, 3063–71, 2016, http:// dx.doi.org/10.1016/j.jclepro. 2016.10.029.
8. Maqbool, Z., Hussain, S., Ahmad, T., Nadeem, H., Imran, M., Khalid, A., Use of RSM modeling for optimizing decolorization of simulated textile wastewater by pseudomonas aeruginosa strain ZM130 capable of simultaneous removal of reactive dyes and hexavalent chromium. *Environ. Sci. Pollut. Res.*, 23, 11224–39, 2016, http://dx.doi.org/10.1007/s11356-016-6275-3.
9. Taneja, N., Ray, S., Pande, D., *India – Pakistan Trade: Textiles and Clothing*, Indian Council for Research on International Economic Relations, 2016, Available at: http://hdl.handle.net/11540/6700 [Accessed on April 2020]
10. Afzal, H.M.Y., Impact of electricity crisis and interest rate on textile industry of Pakistan. *Acad. Contemp Res. J.*, 1, 36–41, 2012.
11. Khan, M. J., Agriculture and textile sector: Challenges and way forward, 2017, https ://peri.punja b.gov.pk/.
12. Lim, S.L., Chu, W.L., Phang, S.M., Use of Chlorella vulgaris for bioremediation of textile wastewater. *Bioresour. Technol.*, 101, 7314–22, 2010, http:// dx.doi.org/10.1016/j. biortech.2010.04.092.

13. Pathak, V.V., Singh, D.P., Kothari, R., Chopra, A.K., Phycoremediation of textile wastewater by unicellular microalga chlorella pyrenoidosa. *Cell Mol. Biol.*, 60, 35–40, 2014, http://dx.doi.org/10.14715/cmb/2014.60.5.7.
14. Rajasimman, M., Babu, S. V., Rajamohan, N., Biodegradation of textile dyeing industry wastewater using modified anaerobic sequential batch reactor–Start-up, parameter optimization and performance analysis. *J. Taiwan Inst. Chem. Eng.*, 72, 171–181, 2017.
15. Jayakumar, S., Yusoff, M.M., Rahim, M.H.A., Maniam, G.P., Govindan, N., The prospect of microalgal biodiesel using agro-industrial and industrial wastes in Malaysia. *Renew. Sustain Energy Rev.*, 72, 33–47, 2017, http:// dx.doi.org/10.1016/j.rser.01.002.
16. Padmanaban, V.C., Prakash, S.S., Sherildar, P., Jacob, J.P., Nelliparambil, K., Biodegradation of antraquinone based compounds: Review. *Int. J. Adv. Res. Eng. Technol.*, 4, 74–83, 2013.
17. Srinivasan, A. and Viraraghavan, T., Decolorization of dye wastewaters by biosorbents- A review. *J. Environ. Manage.*, 91, 1915– 29, 2010, http://dx.doi.org/10.1016/j. jenvman. 2010.05.003.
18. Logroño, W., Pérez, M., Urquizo, G., Kadier, A., Echeverría, M., Recalde, C., Single chamber microbial fuel cell (SCMFC) with a cathodic microalgal biofilm: A preliminary assessment of the generation of bioelectricity and biodegradation of real dye textile wastewater. *Chemosphere*, 176, 378– 88, 2017, http://dx.doi.org/10. 1016/j.chemosphere.2017.02.099.
19. Tehrani-Bagha, A.R., Mahmoodi, N.M., Menger, F.M., Degradation of a persistent organic dye from colored textile wastewater by ozonation. *Desalination*, 260, 34–8, 2010, http://dx.doi.org/10.1016/j. desal.2010.05.004.
20. Gupta, V.K., Agarwal, S., Saleh, T.A., Synthesis and characterization of alumina-coated carbon nanotubes and their application for lead removal. *J. Hazard Mater.*, 185, 17–23, 2011, http://dx.doi.org/10.1016/j. jhazmat.2010.08.053.
21. Kamaroddin, M.F., Hanotu, J., Gilmour, D.J., Zimmerman, W.B., *In-situ* disinfection and a new downstream processing scheme from algal harvesting to lipid extraction using ozone-rich microbubbles for biofuel production. *Algal Res.*, 17, 217–26, 2016, http://dx.doi.org/10.1016/j. algal.2016.05.006.
22. Devi, S., Murugappan, A., Kannan, R.R., Textile dye wastewater treatment using freshwater algae in packed-bed reactor: Modeling. *Desalin. Water Treat.*, 57, 38, 17995–18002, 2016.
23. Sartape, A.S., Mandhare, A.M., Jadhav, V.V., Raut, P.D., Anuse, M.A., Kolekar, S.S., Removal of malachite green dye from aqueous solution with adsorption technique using limonia acidissima (wood apple) shell as low cost adsorbent. *Arabian J. Chem.*, 10, 3229– 3238, 2017, https://doi.org/10.1016/j. arabjc.2013.12.019.
24. Khandare, R.V. and Govindwar, S.P., Phytoremediation of textile dyes and effluents: Current scenario and future prospects. *Biotechnol. Adv.*, 33, 1697–714, 2015, http://dx.doi.org/10.1016/j.biotechadv.2015.09.003.

25. Gupta, S., Wali, A., Gupta, M., Annepu, S. K., Fungi: An effective tool for bioremediation. *Plant-Microbe Interactions in Agro-Ecological Perspectives: Volume 2: Microbial Interactions and Agro-Ecological Impacts*, pp. 593–606, Springer, 2017.
26. Kasiri, M. B., Application of chitosan derivatives as promising adsorbents for treatment of textile wastewater. *The impact and prospects of green chemistry for textile technology*, pp. 417–469, Elsevier, 2019.
27. Fazal, T., Mushtaq, A., Rehman, F., Khan, A. U., Rashid, N., Farooq, W., ... Xu, J., Bioremediation of textile wastewater and successive biodiesel production using microalgae. *Renewable Sustainable Energy Rev.*, 82, 3107–3126, 2018.
28. Im, S.L., Chu, W.L., Phang, S.M., Use of chlorella vulgaris for bioremediation of textile wastewater. *Bioresour. Technol.*, 101, 7314–22, 2010, http://dx.doi.org/10.1016/j. biortech.2010.04.092.
29. Ahmad, S. *et al.*, Optimization of nutrient stress using *C. pyrenoidosa* for lipid and biodiesel production in integration with remediation in dairy industry wastewater using response surface methodology. *3 Biotech.*, 8, 326, 2018, doi: 10.1007/s13205-018-1342-8.
30. Kahsay, A., Gebremedhin, M.S., Ayele, B., Effects of Bahir Dar textile factory effluents on the water quality of the head waters of blue nile river, Ethiopia. *Int. J. Analytical Chem.*, 5, 1–7, 2015.
31. Murtaza, G. and Zia, M. H., Wastewater production, treatment and use in Pakistan. In *Second Regional Workshop of the Project 'Safe Use of Wastewater in Agriculture*, pp. 16–18, University of Agriculture, Faisalabad, Pakistan, 2012, May.
32. Khan, A. A. and Khan, M., Pakistan textile industry facing new challenges. *Res. J. Int. Stud.*, 14, 14, 21–29, 2010.
33. Walayat, S., Warraich Usman, A., Kabeer, K., Challenges faced by textile industry of Pakistan: Suggested solutions. *J. Kasbit Bus.*, 39, 33–9, 2012.
34. Biello, D., Slick solution: How microbes will clean up the deepwater horizon oil spill. *Sci. Am.*, 25, 1–4, 2010.
35. Bellou, S., Baeshen, M.N., Elazzazy, A.M., Aggeli, D., Sayegh, F., Aggelis, G., Microalgal lipids biochemistry and biotechnological perspectives. *Biotechnol. Adv.*, 32, 1476–93, 2014, http://dx.doi.org/10.1016/j. biotechadv.2014.10.003.
36. Olguín, E.J., Dual purpose microalgae-bacteria-based systems that treat wastewater and produce biodiesel and chemical products within a Biorefinery. *Biotechnol. Adv.*, 30, 1031–46, 2012, http://dx.doi. org/10.1016/j.biotechadv. 2012.05.001.
37. Rhodes, C.J., Applications of bioremediation and phytoremediation. *Sci. Prog.*, 96, 4, 417–427, 2013.
38. Bhalerao, T.S., Bioremediation of endosulfan-contaminated soil by using bioaugmentation treatment of fungal inoculant aspergillus niger. *Turk J. Biol.*, 36, 561–567, 2012.
39. Pinto, A.P., Serrano, C., Pires, T., Mestrinho, E., Dias, L., Teixeira, D.M., Caldeira, A.T., Degradation of terbuthylazine difenoconazole and

pendimethalin pesticides by selected fungi cultures. *Sci. Total Environ.*, 435, 402–410, 2012.
40. Deng, W., Lin, D., Yao, K., Yuan, H., Li, J.W., Zou, J.L., Han, X., Zhou, K., He, L., Hu, X., Liu, S., Characterization of a novel β-cypermethrin-degrading aspergillus niger YAT strain and the biochemical degradation pathway of β-cypermethrin. *Appl. Microbiol Biotechnol.*, 99, 19, 8187–8198, 2015.
41. Neethu, C.S., Rahiman, K.M.M., Saramma, A.V., Hatha, A.A.M., Heavy-metal resistance in gram-negative bacteria isolated from Kongsfjord, Arctic. *J. Microbiol.*, 61, 6, 429–35, Jun2015, doi: 10.1139/cjm-2014-0803. Epub 2015 Apr 2.
42. Madhavi, G.N. and Mohini, D.D., Review paper on – Parameters affecting bioremediation. *Int. J. Life Sci. Pharma. Res.*, 2, 77–80, 2012, https://goo.gl/tBP2C6.
43. Adams, G.O., Fufeyin, P.T., Okoro, S.E., Ehinomen, I., Bioremediation, biostimulation and bioaugmention-A review. *Int. J. Environ. Bioremediat. Biodegrad*, 3, 28–39, 2015.
44. Han, X. and Gu, J., Sorption and transformation of toxic metals by microorganisms, in: *Environmental Microbiology*, R. Mitchell and J. Gu (Eds.), Wiley Blackwell Pub, Hoboken, 2010.
45. Tharmer, M., Al-Kubaisi, A.R., Ahraw, Z., Abdullah, H.A., Hindy, I., Khadium, A.A., Biodegradation of Kirkuk light crude oil by *Bacillus thuringiensis*, northern of Iraq. *Nat. Sci.*, 5, 7, 865–873, 2013.
46. Tascione, V., Raffaele, M., Raggi, A., Optimizing the environmental performance of integrated waste management scenarios by means of linear programming: A case study. *J. Cleaner Prod.*, 112, 4, 3086–3096, 2016.
47. Bapat, S. A. and Jaspal, D. K., Parthenium hysterophorus: Novel adsorbent for the removal of heavy metals and dyes. *Glob. J. Environ. Sci. Manag.*, 2, 2, 135–144, 2016.
48. Monari, C., Righi, S., Olsen, S.I., Greenhouse gas emissions and energy balance of biodiesel production from microalgae cultivated in photobioreactors in denmarka life-cycle modeling. *J. Clean Prod.*, 112, 4084–92, 2016, http://dx.doi.org/10. 1016/j.jclepro.2015.08.112.
49. Mahmoud, M.S., Decolorization of certain reactive dye from aqueous solution using baker's yeast (saccharomyces cerevisiae) strain. *HBRC J.*, 12, 1, 88, 2016.
50. Alalewi, A. and Jiang, C., Bacterial influence on textile wastewater decolorization. *J. Environ. Protect.*, 3, 889–903, 2012.
51. Srivastava, A., Rani, R.M., Patle, D.S., Kumar, S., Emerging bioremediation technologies for the treatment of textile wastewater containing synthetic dyes: A comprehensive review. *J. Chem. Biotechnol.*, 9, 26–41, 2021, https://doi.org/10.1002/jctb.6891.
52. Taştan, B.E., Ertuğrul, S., Dönmez, G., Effective bioremoval of reactive dye and heavy metals by aspergillus versicolor. *Bioresour. Technol.*, 101, 870–6, 2010, http://dx. doi.org/10.1016/j.biortech.2009.08.099.

53. Piaskowski, K., Świderska-Dąbrowska, R., Zarzycki, P.K., Dye removal from water and wastewater using various physical, chemical and biological processes. *J. AOAC Int.*, 101, 1371–1384, 2018, doi: 10.5740/jaoacint.18-0051.
54. Saratale, R.G., Saratale, G.D., Chang, J.S., Govindwar, S.P., Bacterial decoorization and degradation of azo dyes: A review. *J. Taiwan Inst. Chem. Eng.*, 42, 138–57, 2011.
55. Wu, L.F., Chen, P.C., Huang, A.P., Lee, C.M., The feasibility of biodiesel production by microalgae using industrial wastewater. *Bioresource Technol.*, 113, 14–8, 2012, http:// dx.doi.org/10.1016/j.biortech.2011.12.128.
56. Roberts, G. W., Fortier, M. O. P., Sturm, B. S., Stagg-Williams, S. M., Promising pathway for algal biofuels through wastewater cultivation and hydrothermal conversion. *Energy Fuels*, 27, 2, 857–867, 2013.
57. Chisti, Y., Constraints to commercialization of algal fuels. *J. Biotechnol.*, 167, 201–14, 2013, http://dx.doi.org/10.1016/j.jbiotec.2013.07.020.
58. Lam, M.K. and Lee, K.T., Microalgae biofuels: A critical review of issues, problems and the way forward. *Biotechnol. Adv.*, 30, 673–90, 2012, http:// dx.doi.org/10.1016/j. biotechadv.2011.11.008.
59. Seo, Y.H., Park, D., Oh, Y.K., Yoon, S., Han, J.I., Harvesting of microalgae cell using oxidized dye wastewater. *Bioresour. Technol.*, 19, 2, 802, 2015, http:// dx.doi.org/10. 1016/j.biortech.2015.05.074.
60. Janarthanan, M. and Senthil Kumar, M., The properties of bioactive substances obtained from seaweeds and their applications in textile industries. *J. Ind. Text.*, 48, 1, 361–401, 2018.
61. Djokic, L.N., Kenny, S.T., Connor, K.E., Rodulovic, V., Nikodinovic-Runic, J., Vasilijevic, B., Metabolic versatility of gram positive microbial isolates from contaminated river sediments. *J. Hazard Mater.*, 15, 215–216, 243–251, 2012.
62. Skariyachan, S., Megha, M., Kini, M. N., Mukund, K. M., Rizvi, A., Vasist, K., Selection and screening of microbial consortia for efficient and ecofriendly degradation of plastic garbage collected from urban and rural areas of Bangalore. *India. Environ. Monit. Assess.*, 187, 1–14, 2015.
63. Hassan, M.M., Alam, M.Z., Anwar, M.N., Biodegradation of textile azo dyes by bacteria isolated from dyeing industry effluent. *Int. Res. J. Biol. Sci.*, 2, 8, 27–31, 2013.
64. Shah, M.P., Patel, K.A., Nair, S.S., Microbial degradation of textile dye (remazol black b) by bacillus spp ETL-2012. *J. Bioremed Biodeg.*, 4, 2, 1–5, 2013.
65. Yogesh, P. and Akshaya, G., Evaluation of bioremediation potential of isolated bacterial culture YPAG-9 (pseudomonas aeruginosa) for decolorization of sulfonated di-azodye reactive red HE8B under optimized culture conditions. *Int. J. Curr. Microbiol. App Sci.*, 5, 8, 258–72, 2016.
66. Das, A., Mishra, S., Verma, V.K., Enhanced biodecolorization of textile dye remazol navy blue using an isolated bacterial strain Bacillus pumilus HKG212 under improved culture conditions. *J. Biochem. Tech.*, 6, 3, 962–9, 2015.

67. Kumar, S., Chaurasia, P., Kumar, A., Isolation and characterization of microbial strains from textile industry effluents of Bhilwara, India analysis with bioremediation. *J. Chem. Pharm. Res.*, 8, 4, 143–50, 2016.
68. Adebajo, S.O., Balogun, S.A., Akintokun, A.K., Decolourization of vat dyes by bacterial isolates recovered from local textile mills in southwest. *Microbiol. Res. J. Int.*, 18, 1, 1–8, 2017.
69. Kurashvili, M.V., Adamia, G.S., Ananiashvili, T.I., Varazi, T.G., Pruidze, M.V., Gordeziani, M.S., Khatisashvili, G.A., Plants astools for control and remediation of the environment pollutedby organochlorine toxicants. *Ann. Agrar. Sci.*, 12, 3, 84–87, 2014.
70. Blánquez, A., Rodríguez, J., Brissos, V., Mendes, S., Martins, L.O., Andrew, S.B., María, E.A., Hernández, M., Decolorization and detoxification of textile dyes using a versatile streptomyces laccase-natural mediator system. *Saudi J. Biol. Sci.*, 26, 913–920, 2019.
71. Ali, H., Biodegradation of synthetic dyes—A review. *Water Air Soil Pollut.*, 213, 251–273, 2010, https://doi.org/10.1007/s11270-010-0382-4.
72. Holkar, C.R., Jadhav, A.J., Pinjari, D.V., Mahamuni, N.M., Pandit, A.B., A critical review on textile wastewater treatments: Possible approaches. *J. Environ. Manage.*, 182, 351–366, 2016, doi: 10.1016/j.jenvman.2016.07.090.
73. Chung, K.T., Azo dyes and human health: A review. *J. Environ. Sci. Health C Environ. Carcinog Ecotoxicol Rev.*, 34, 4, 233–261, 2016, doi: 10.1080/10590501.2016.1236602, [published correction appears in J Environ Sci Health C Environ Carcinog Ecotoxicol Rev. 2017 Jan 2;35(1):67].
74. Sghaier, I., Guembri, M., Chouchane, H. *et al.*, Recent advances in textile wastewater treatment using microbial consortia. *J. Textile Eng Fashion Technol.*, 5, 3, 134–146, 2019, DOI: 10.15406/jteft.2019.05.00194.
75. Haq, A.R., Biodegradation of azure-B dye by serratia liquefaciens and its validation by phytotoxicity, genotoxicity and cytotoxicity studies. *Chemosphere.*, 196, 58–68, 2018.
76. Khan, A.Y. and Kumar, G.S., Spectroscopic studies on the binding interaction of phenothiazinium dyes, azure A and azure B to double stranded RNA polynucleotides. *Spectrochim. Acta, Part A Mol. Biomolecular Spectrosc.*, 152, 417–425, 2016.
77. Chequer, F.M.D., Lizier, T.M., deFelício, R., Zanoni, M.V.B., Debonsi, H.M., Lopes, N.P. *et al.*, Analyses of the genotoxic and mutagenic potential of the products formed after the biotransformation of the azo dye disperse red *1. Toxicol. Vitro*, 25, 8, 2054–2063, 2011.
78. Lacasse, K. and Baumann, W., *Textile chemicals: Environmental data and facts*, Springer, Dortmund, 2012.
79. Torbati, S., Khataee, A. R., Movafeghi, A., Application of watercress (Nasturtium officinale R. Br.) for biotreatment of a textile dye: Investigation of some physiological responses and effects of operational parameters. *Chem. Eng. Res. Des.*, 92, 10, 1934–1941, 2014.

80. Naja, G. and Volesky, B., The mechanism of metal cation and anion biosorption, in: *Microbial Biosorption of Metals*, P. Kotrba, M. Mackova, T. Macek (Eds.), pp. 19–58, Springer, Heidelberg, 2011.
81. Lellis, B., Fávaro-Polonio, C.Z., Pamphile, J.A., Polonio, J.C., Effects of textile dyes on health and the environment and bioremediation potential of living organisms. *Biotechnol. Res. Innov.*, 3, 275–290, 2019.
82. Watharkar, A.D., Kadam, S.K., Khandare, R.V., Kolekar, P.D., Jeon, B.H., Jadhav, J.P., Govindwar, S.P., Asparagus densiflorus in a vertical subsurface flow phytoreactor for treatment of real textile effluent: A lab to land approach for *in situ* soil remediation. *Ecotoxicol. Environ. Saf.*, 161, 70–77, 2018.
83. Khandare, R.V., Kabra, A.N., Kadam, A.A., Govindwar, S.P., Treatment of dye containing wastewaters by a developed lab scale phytoreactor and enhancement of its efficacy by bacterial augmentation. *Int. Biodeterior Biodegrad.*, 78, 89–97, 2013.
84. Rane, N.R., Chandanshive, V.V., Khandare, A.D., Patil, T.S., Pawar, P.K., Govindwar, S.P., Phytoremediation of sulfonated remazol red dye and textile effluents by alternanthera philoxeroides: An anatomical, enzymatic and pilot scale study. *Water Res.*, 83, 271–281, 2015.
85. Rane, N.R., Patil, S.M., Chandanshive, V.V., Kadam, S.K., Khandare, R.V., Jadhav, J.P., Govindwar, S.P., Ipomoea hederifolia rooted soil bed and ipomoea aquatic rhizofiltration coupled phytoreactors for efficient treatment of textile wastewater. *Water Res.*, 96, 1–11, 2016.
86. Hossain, K., Quaik, S., Ismail, N., Rafatullah, M., Maruthi, A., Rameeja, S., Bioremediation of textile wastewater with membrane bioreactor using the white-rot fungus and reuse of wastewater. *Iran J. Biotechnol.*, 14, 3, 124–16, 2016, https://doi.org/10.15171/ijb.1216.

14

Electrocatalytic Materials for Renewable Energy: Perspectives and Initiatives

Trupti R. Das[1]*, Rashmiprava Sahoo[2], Meenakshi Choudhary[3], Santanu Patra[4] and Sudheesh K. Shukla[3]

[1]*CIPET: Institute of Petrochemicals Technology (IPT) - Bhubaneswar, Bhubaneswar, India*
[2]*CSIR-Central Institute of Mining and Fuel Research, Dhanbad, Jharkhand, India*
[3]*School of Chemical Engineering and Physical Sciences, Lovely Professional University, Jalandhar, Punjab, India*
[4]*IDUN, Department of Health Technology, Technical University of Denmark, Kgs. Lyngby, Denmark*

Abstract

The shift to renewable energy sources is one of the greatest challenges to addressing global energy demand and climate change. Electrocatalytic materials are crucial to the success of many renewable energy technologies because of their capacity to boost their potential, decrease their prices, and improve their efficiency. An overview of electrocatalytic materials and their applications in renewable energy is provided in this chapter. The classification, characteristics, and uses of electrocatalytic materials are discussed, along with a variety of renewable energy technologies such as fuel cells, electrolyzers, carbon capture, batteries, and solar cells. Many organizations and countries have made steps to encourage the use of electrocatalytic materials in the renewable energy industry, and this chapter discusses such efforts as well as the role electrocatalytic materials play in reaching climate neutrality and net-zero targets. The chapter concludes that electrocatalytic materials are essential to achieving a sustainable and clean energy future and that more research and development of these materials will be required to meet the global energy demand and address climate change.

**Corresponding author*: trdas1893@gmail.com

Sudheesh K. Shukla, Chaudhery Mustansar Hussain, Santanu Patra, Meenakshi Choudhary (eds.) Electrocatalytic Materials for Renewable Energy, (377–396) © 2024 Scrivener Publishing LLC

***Keywords*:** Energy demand, climate change, electrocatalytic materials, net-zero goals, renewable energy sector

14.1 What is the Importance of Renewable Energy in the Current Context?

Renewable energy is defined as that which is derived from natural sources that either are not depletable or regenerate quickly [1]. Energy from renewable sources like the sun, wind, and water is never depleted and may be replenished indefinitely. Solar, wind, hydro, geothermal, biomass, and tidal power are just a few examples of widely known renewable energy sources [2]. These renewable energy sources have less of a negative impact on the environment than traditional, non-renewable energy sources like fossil fuels [3]. Technologies based on renewable energy harvest energy from the environment and convert it into usable forms like electricity, heat, or mechanical power [4, 5]. It is essential to transition to renewable energy sources in order to mitigate climate change and cut back on greenhouse gases. Therefore, the development and implementation of renewable energy technology are crucial to the transition to a more sustainable

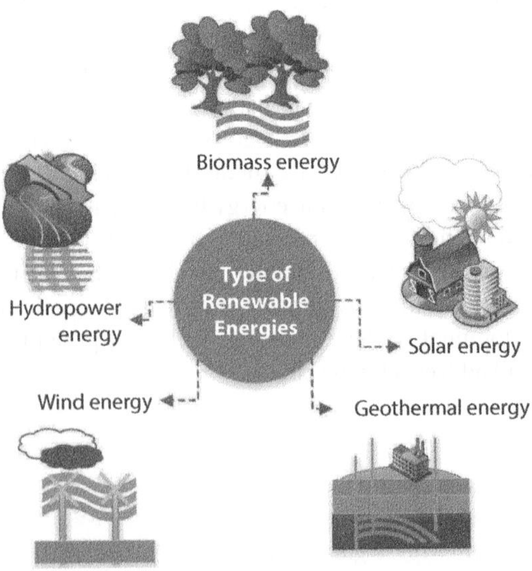

Figure 14.1 Various types of renewable energy sources (reproduced with permission from ref. [6]).

and low-carbon energy system [6]. Figure 14.1 emphasizes the most well-known renewable energy options currently available.

14.2 Renewable Energy Perspective: Connecting Net-Zero and Climate Neutrality Agendas

To achieve climate neutrality, a system or society must ensure that the amount of greenhouse gases emitted into the atmosphere is equal to the amount of greenhouse gases removed from the atmosphere, hence the terms "net zero" and "climate neutral" [7–9]. Getting to net zero or climate neutrality is a crucial aspect of promoting renewable energy in order to cut down on greenhouse gas emissions and switch to a more sustainable and low-carbon energy system (Figure 14.2) [10, 11]. In order to

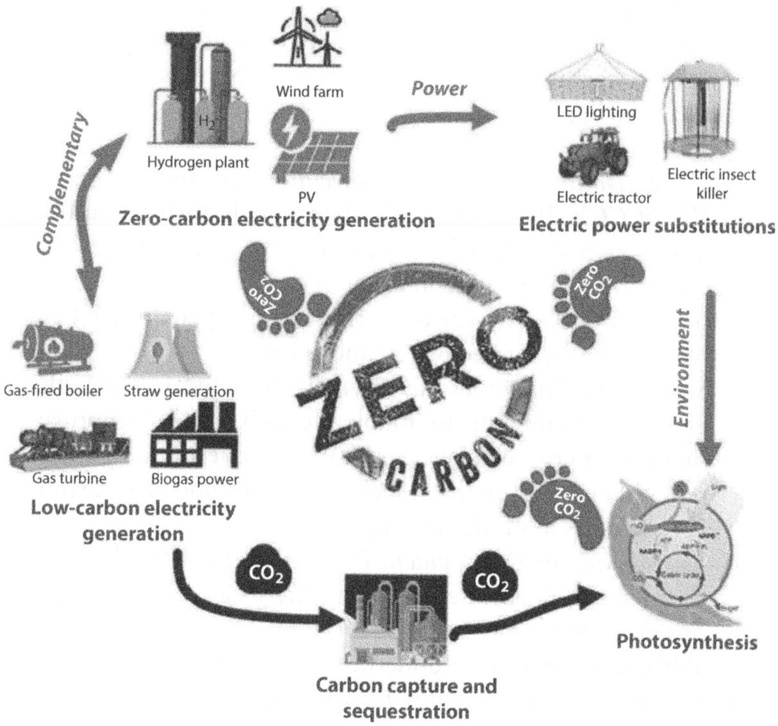

Figure 14.2 A prosperous net-zero-emissions energy cycle (reproduced with permission from ref. [13]).

reduce greenhouse gas emissions and transition to a more sustainable and low-carbon energy system, reaching net zero or climate neutrality is crucial [12, 13].

This can be achieved through both reducing emissions (for instance, by promoting the use of renewable energy sources) and doing away with emissions entirely (which can be done, for example, by using carbon capture and storage technologies). Climate neutrality is a related concept, but it typically refers to the push for worldwide net-zero emissions. This means that there will be no overall rise in greenhouse gas concentrations in the atmosphere, as the whole amount of world greenhouse gas emissions created is equal to the total amount of global emission removal.

14.3 Efforts of the United Nations to Promote Renewable Energy

The United Nations (UN) has launched several programs to encourage the global use of renewable energy sources [14]. These initiatives aim to reduce greenhouse gas emissions and support sustainable growth to mitigate the effects of global warming. The United Nations (UN) most crucial projects include the following [15]:

- *United Nations Framework Convention on Climate Change (UNFCCC)*: The UNFCCC's stated goals include cutting down on emissions of greenhouse gases and pushing for a transition to a low-carbon economic system. The agreement has been ratified by 197 countries, including major emitters like China, the United States, and India.
- *Sustainable Energy for All (SE4All)*: Initiated in 2011, the SE4All program aims to double the global rate of improvement in energy efficiency by 2030, double the amount of renewable energy in the global energy mix, and provide universal access to modern energy services.
- *United Nations Development Programme (UNDP)*: In order to aid in sustainable development, energy access, and reducing greenhouse gas emissions, the UNDP finances renewable energy projects in underdeveloped countries.
- *International Renewable Energy Agency (IRENA)*: The International Renewable Energy Agency (IRENA) is a group of countries that collaborate to increase renewable energy usage.

The organization monitors the growth of the renewable energy industry and reports its findings and recommendations to policymakers and other stakeholders.

The goals of the Paris Agreement and the Sustainable Development Goals (SDGs) to increase access to and usage of renewable energy sources depend on these kinds of initiatives [16]. The United Nations continues to be a driving factor behind efforts to build out a global network of low-carbon, ecologically friendly power plants.

14.4 Goals for Promoting Renewable Energy in the Sustainable Development Agenda

To further the cause of sustainable development on a global scale, the United Nations General Assembly approved a set of 17 goals in 2015, termed as Sustainable Development Goals (SDGs) (Figure 14.3) [17]. Most of these goals emphasize the relevance of renewable energy in creating a low-carbon and sustainable energy system [18]. Here are some examples.

1. *Goal 7*: The goal of this initiative is to ensure that all people may obtain the power they require in a way that is equitable, sustainable, low cost, and technologically advanced. The goal, more specifically, is to increase the share of renewable

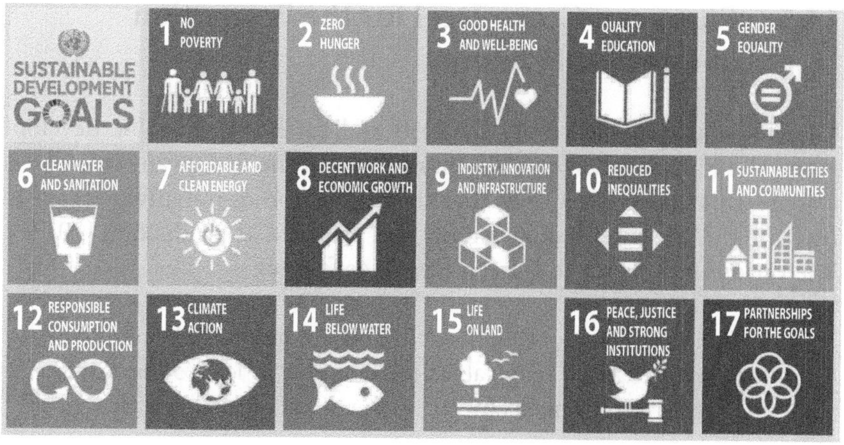

Figure 14.3 The United Nations' Sustainable Development Goals (source: United Nations webpage).

energy in global energy production while simultaneously boosting efficiency.
2. *Goal 9*: To accomplish this, we must promote ecologically responsible industry and work toward constructing more sturdy infrastructure. More frequent usage of renewable energy in infrastructure and industry is necessary to reach this objective.
3. *Goal 13*: This goal has as its overarching purpose the prompt implementation of measures to lessen global warming's devastating effects. Explicitly mobilizing financial resources to promote the use of renewable energy sources is essential to achieve this objective.
4. *Goal 17*: This goal recognizes the importance of teamwork in achieving sustainable development. Encouragement of technology transfer and capacity building to help developing countries adopt renewable energy sources is a key part of achieving this goal.

To establish a sustainable and low-carbon energy system, these goals highlight the need of encouraging the worldwide development and use of renewable energy sources. Countries and organizations that work toward these objectives will help the world adopt an energy system that is better for people and the earth, as well as being more sustainable, affordable, and easy to implement.

14.5 European Green Deal for the Promotion of Renewable Energy

To achieve carbon neutrality by 2050 and further sustainable development, the European Union (EU) has proposed the European Green Deal (Figure 14.4) [19]. The Green Deal is a collection of initiatives meant to increase the use of renewable energy and promote its production [20, 21]. The various agenda for the Green Deal is as follows:

- *Renewable Energy Directive (RED II)*: In line with the binding goal of RED II, renewables must supply at least 32% of the world's energy needs by 2030. The directive also encourages

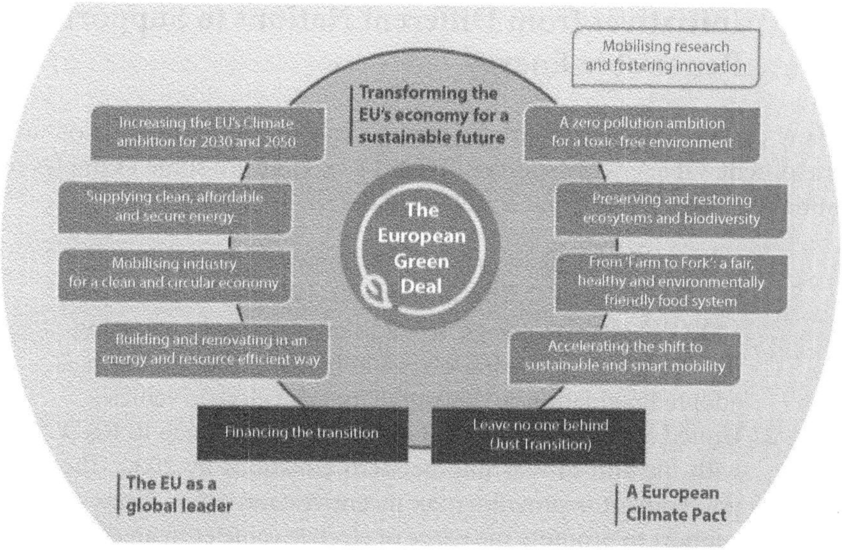

Figure 14.4 The European Green Deal (source: European Union).

the use of renewable energy in HVAC systems and the development of local energy projects.

- *European Climate Law*: The goal of reaching climate neutrality by the year 2050 is articulated in the European Climate Law, which also provides a framework for getting there. This includes expanding the use of renewable energy sources.
- *Clean Hydrogen Alliance*: The Clean Hydrogen Alliance's end goal is to promote the development of a hydrogen economy across Europe, with an emphasis on the use of renewable energy in hydrogen electrolysis.
- *Offshore Renewable Energy Strategy*: The EU Offshore Renewable Energy Strategy aims to increase the use of offshore wind, wave, and tidal energy and has set a target of at least 60 GW of offshore wind capacity by 2030 and up to 300 GW by 2050.

The EU's transition to a low-carbon, more sustainable energy system relies heavily on the success of these programs, which promote the development and use of renewable energy sources [21].

14.6 Initiatives from Different Nations to Support Renewable Energy

Many countries have started programs to encourage the production and use of renewable energy sources (Table 14.1). Here are some examples of initiatives taken by different countries [22–25]:

1. India: India plans to increase its renewable energy capacity to 175 GW by 2022. This includes 60 GW of wind power and 100 GW of solar power. In addition to actively encouraging the growth of hydropower and biofuels, the country has adopted regulations to boost the usage of electric vehicles.
2. United States: The United States has invested much in R&D with the aim of reaching net-zero emissions by 2050. The government is providing tax incentives and other forms of support to promote the usage of electric vehicles in an effort to increase the use of renewable energy sources.
3. China: China has ambitious plans to increase its usage of green power. The country has made significant investments in wind and solar energy with the aim of meeting 35% of its energy needs from non-fossil sources by 2030.

Table 14.1 Top 10 countries that invest in renewable sources (reproduced with permission from ref. [14]).

General information			Technology-specific scores						
Rank	Country	RECAI score	Onshore wind	Offshore wind	Solar PV	Solar CSP	Biomass	Geothermal	Hydro
1	USA	70.2	58.6	59.4	58.1	45.8	44.5	48.9	48.0
2	China	67.8	54.5	55.0	60.0	53.8	52.1	30.3	54.7
3	Australia	66.4	54.5	30.6	57.0	36.0	39.6	18.0	42.6
4	India	65.2	54.1	27.5	62.7	48.9	47.1	26.9	47.8
5	UK	64.2	57.3	60.5	47.1	11.2	55.9	30.1	44.2
6	Germany	63.7	49.9	51.0	52.1	11.4	50.4	34.5	44.3
7	France	63.5	55.0	52.2	53.5	22.2	46.8	32.6	51.9
8	Japan	61.3	50.3	51.1	48.0	19.2	55.8	50.8	50.1
9	Netherlands	58.8	49.1	45.6	47.1	10.7	45.1	18.6	32.3

4. Japan: Japan's goal is to achieve a carbon-neutral society by the year 2050. In order to achieve this goal, it promotes the development of alternative energy sources such as solar, wind, and biomass. Hydrogen is widely supported because of its potential as a clean energy source, particularly in the transportation sector.
5. Brazil: Brazil has been an early adopter of bioenergy, particularly from sugarcane, and has committed to meeting 45% of its energy demands from renewable sources by 2030. Regulations have been put in place to promote the usage of electric vehicles and the expansion of renewable energy sources like wind and solar power in the country.

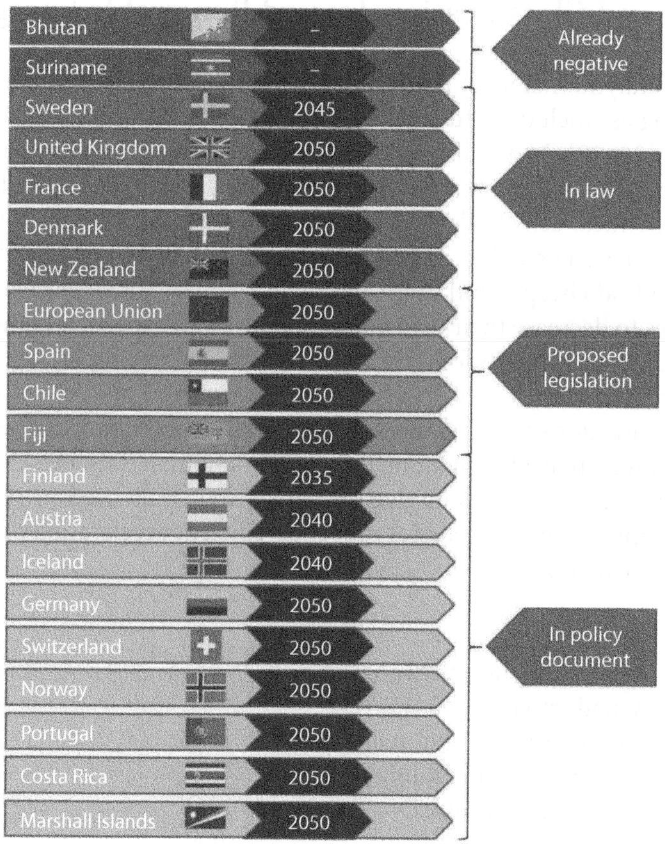

Figure 14.5 Net-zero race of various countries (reproduced with permission from ref. [27]).

6. Germany: Germany is a pioneer in the movement to adopt alternative energy sources like wind and solar. The country has created a number of incentives and support mechanisms to encourage the growth of renewable energy with the goal of having it account for 65% of energy consumption by 2030.

These are just a few of the various ways in which different countries are working to promote the study and use of renewable energy [14]. Other countries are also taking steps to wean themselves off of fossil fuels and shift to a low-carbon energy system [26]. The current status of the countries striving toward a net-zero future is shown in Figure 14.5.

14.7 Electrocatalytic Materials: Properties and Classification Toward Renewable Energy

Electrocatalytic materials play crucial roles in various renewable energy technologies, including fuel cells, electrolyzers, and photoelectrochemical cells [28, 29]. Materials with electrocatalytic capabilities are crucial to renewable energy systems including fuel cells, electrolyzers, and photoelectrochemical cells [30]. There is always new research being conducted and new materials being developed and evaluated [31]. Researchers are trying to find cheaper and more abundant catalysts for renewable energy processes to decrease their dependency on rare and expensive components [32, 33]. Electrocatalysts for sensors, separation, and energy storage devices are only some of the many possible uses for nanosized materials. Common examples include carbonaceous nanomaterials, metal nanoparticles, metal oxide, sulfide, phosphite nanomaterials, 2D nanomaterials, etc. [34–38].

Certain characteristics of electrocatalytic materials used in renewable energy applications are necessary for their efficient acceleration of electrochemical reactions:

a. Selectivity
b. High catalytic activity
c. Conductivity
d. Durability and high stability
e. Electrolyte compatibility
f. Cost-effectiveness

Researchers are still digging into the specifics of these characteristics in an effort to improve the efficacy and efficiency of electrocatalytic materials used in renewable energy systems [39]. There are a number of ways to categorize electrocatalytic materials. Some examples of categories are as follows (Figure 14.6):

a. *Material type*: Metals, metal oxides, metal alloys, organic compounds, and other substances can all be used as electrocatalytic materials.
b. *Active site*: It is possible to classify electrocatalytic materials based on their active sites, which catalyze the electrochemical process, and where they are located (either on the surface, at the edge, or at defect sites).
c. *Synthesis method and application*: It is possible to categorize electrocatalytic compounds based on their synthesis strategy and their applications. Batteries, fuel cells, electrolyzers, and sensors are just some of the reported applications of different synthesis techniques such sol–gel, electrodeposition, chemical vapor deposition, and hydrothermal synthesis.

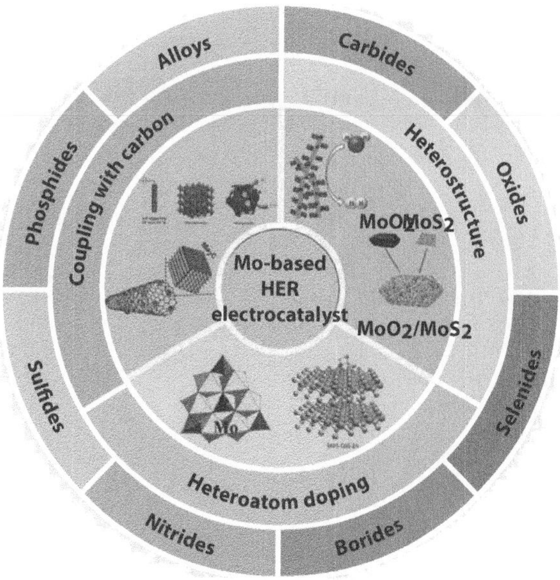

Figure 14.6 An example of the classification of electrocatalytic materials (reproduced with permission from ref. [40]).

d. *Reaction type*: Electrochemical processes can also be used to classify electrocatalytic materials, with hydrogen/oxygen evolution, oxygen reduction, carbon dioxide reduction, and electrochemical water splitting being some of the most common.

14.8 Electrocatalytic Materials: Various Applications in Renewable Energy

Given the importance of electrochemical processes in many renewable energy systems, electrocatalytic materials have several possible uses in this field [41]. Here are some specific applications of electrocatalytic materials in the field of renewable energy (Figures 14.7 and 14.8):

i. *Solar cells*: Electrocatalytic materials are used as catalysts in dye-sensitized solar cells, which use a photoactive dye to turn sunlight into electricity through an electrochemical reaction. Many dye-sensitized solar cells use

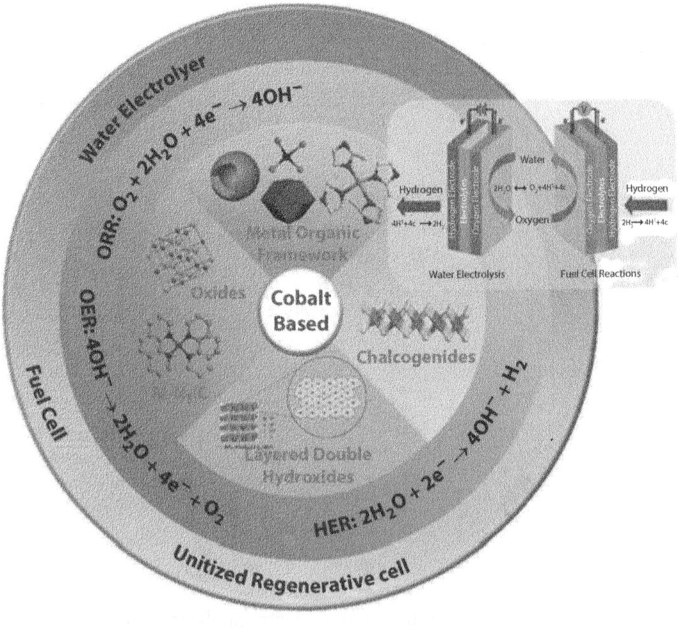

Figure 14.7 The concept of fuel cell, electrolyzer, and metal–air batteries (reproduced with permission from ref. [41]).

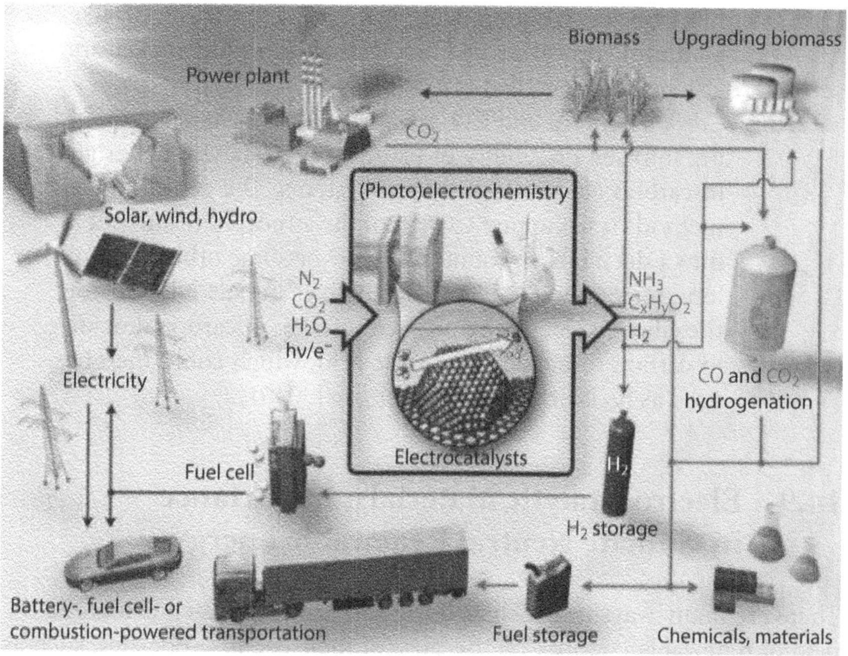

Figure 14.8 Real-time applications of electrocatalytic materials (reproduced with permission from ref. [56]).

electrocatalytic materials like ruthenium-based dyes although cobalt-based dyes are also being investigated as a potential replacement [42, 43].

ii. *Fuel cells*: Hydrogen or other fuels are used in an electrochemical reaction catalyzed by electrocatalytic materials in fuel cells. Currently, platinum is the most popular choice for electrocatalytic materials in fuel cells, but palladium, nickel, and cobalt are also being investigated as potential alternatives [44, 45].

iii. *Batteries*: Electrocatalytic materials are employed as catalysts in certain batteries, such as metal–air batteries, which generate energy through an electrochemical interaction between metal and oxygen. Most metal–air battery electrocatalysts are carbon-based, but transition metal oxides and others are also under investigation [46, 47].

iv. *Electrolyzers*: Electrocatalytic materials are used as catalysts in water electrolyzers, which use an electric current to break water molecules into hydrogen and oxygen.

Electrolyzers commonly use electrocatalytic materials like platinum; however, iridium, ruthenium, and nickel are also being researched [48, 49].

v. *Carbon capture and CO_2 reduction*: Researchers are looking into the possibility of using electrocatalytic materials in carbon capture and storage systems. The goal of these methods is to capture CO_2 from the atmosphere and put it into a deep underground reservoir or some other form of long-term storage. Electrocatalytic materials allowed for the efficient electrochemical conversion of carbon dioxide to carbon monoxide or formic acid, which could then be used as fuels or stored for later use [50, 51].

14.9 Electrocatalytic Materials: Importance in Climate Neutral Renewable Energy

There are many ways in which electrocatalytic materials contribute to the field of renewable energy [52]. Because of their useful properties, such as their capacity to speed up electrochemical reactions, these materials are integral to many forms of renewable energy production [53]. Some specific benefits of electrocatalytic materials in the field of renewable energy are outlined below [54, 55].

I. *Improved efficiency*: Electrocatalytic materials can improve the performance of electrochemical reactions in many types of renewable energy systems by speeding up these processes, leading to improved efficiency.

II. *Lower costs*: Carbon-based compounds and transition-metal oxides are two examples of electrocatalytic materials that are both abundant and inexpensive in comparison to platinum and other valuable metals.

III. *Sustainable and environmentally friendly*: The conventional catalysts may contain harmful compounds; however, many of the elements used in electrocatalysis are renewable and environmentally friendly. For instance, carbon-based compounds can be made from biomass or waste, making them a promising alternative to traditional catalysts.

IV. *New possibilities*: Modern renewable energy technologies, such as highly efficient perovskite solar cells, would not be possible without the development of electrocatalytic materials. The versatility of electrocatalytic materials is expanding the range of renewable energy technologies that can be used to convert sun and wind into usable fuels and chemicals.

The pressing nature of these objectives has increased the demand for tools that can provide renewable energy while cutting emissions. Electrocatalytic materials play a crucial role in certain of these technologies, and their widespread adoption can be sped up with their use [57]. Research into these materials and the application of related technologies are essential in order to have a sustainable and clean energy future (Figure 14.8) [58].

14.10 Conclusion

In conclusion, electrocatalytic materials have the potential to have a major impact on the renewable energy sector. Because of their ability to boost efficiency, lower costs, and expand the scope of renewable energy technologies, these materials are crucial to achieving global climate targets of climate neutrality and net-zero emissions. However, more research and development of these materials is required to improve their efficacy and decrease their cost. Electrocatalytic materials have shown promise in the renewable energy market, thanks to the efforts of many countries and organizations, but more study is needed to reach their full potential. In the future, sustainable and clean energy may be provided through fresh ways that may be sparked by the ongoing investigation of electrocatalytic materials.

References

1. Wang, S., Sun, L., Iqbal, S., Green financing role on renewable energy dependence and energy transition in E7 economies. *Renew. Energy*, 200, 1561–1572, 2022.
2. Nchofoung, T.N., Fotio, H.K., Miamo, C.W., Green taxation and renewable energy technologies adoption: A global evidence. *Renew. Energy Focus*, 44, 334–343, 2023.

3. Ji, M. *et al.*, Optimisation of multi-period renewable energy systems with hydrogen and battery energy storage: A P-graph approach. *Energy Convers. Manage.*, 281, 116826, 2023.
4. Dincer, I. and Bicer, Y., Chapter 6 - Integration of renewable energy systems for multigeneration, in: *Integrated Energy Systems for Multigeneration*, I. Dincer and Y. Bicer (Eds.), pp. 287–402, Elsevier, Qatar, 2020.
5. Strielkowski, W. *et al.*, Renewable energy in the sustainable development of electrical power sector: A review. *Energies*, 14, 24, 8240, 2021, https://doi.org/10.3390/en14248240.
6. Avtar, R. *et al.*, Exploring renewable energy resources using remote sensing and GIS—A review. *Resources*, 8, 3, 149, 2019, https://doi.org/10.3390/resources8030149.
7. Satola, D. *et al.*, Comparative review of international approaches to net-zero buildings: Knowledge-sharing initiative to develop design strategies for greenhouse gas emissions reduction. *Energy Sustain. Dev.*, 71, 291–306, 2022.
8. Christiansen, K.L. *et al.*, "Our burgers eat carbon": Investigating the discourses of corporate net-zero commitments. *Environ. Sci. Policy*, 142, 79–88, 2023.
9. Mishra, A. *et al.*, Progress in paper-based analytical devices for climate neutral biosensing. *Biosens. Bioelectron.: X*, 11, 100166, 2022.
10. Chen, L. *et al.*, Strategies to achieve a carbon neutral society: A review. *Environ. Chem. Lett.*, 20, 4, 2277–2310, 2022.
11. Rabbi, M.F. *et al.*, Energy security and energy transition to achieve carbon neutrality. *Energies*, 15, 21, 8126, 2022, https://doi.org/10.3390/en15218126.
12. Liu, Z. *et al.*, Challenges and opportunities for carbon neutrality in China. *Nat. Rev. Earth Environ.*, 3, 2, 141–155, 2022.
13. Fu, X., Viewpoints on net-zero emissions of agricultural energy internet. *Front. Energy Res.*, 10, 919001, 2022, doi: 10.3389/fenrg.2022.919001.
14. Bhuiyan, M.A. *et al.*, Renewable energy consumption and economic growth nexus—A systematic literature review. *Front. Environ. Sci.*, 10, 2022.
15. Nations, U., Key takeaways from COP 27, [cited 2023 02-03-2023], Available from: https://unfccc.int/process-and-meetings/conferences/sharm-el-sheikh-climate-change-conference-november-2022/five-key-takeaways-from-cop27?gad_source=1&gclid=CjwKCAiA_tuuBhAUEiwAvxkgT-vHr9q2EqtLfRWtUjkKO68SJgwbzjElDP5AvY87OvzQAj_c81-yYaxo-CU9sQAvD_BwE.
16. Akrofi, M.M., Okitasari, M., Kandpal, R., Recent trends on the linkages between energy, SDGs and the Paris agreement: A review of policy-based studies. *Discover Sustainability*, 3, 1, 32, 2022.
17. Gielen, D. *et al.*, The role of renewable energy in the global energy transformation. *Energy Strategy Rev.*, 24, 38–50, 2019.
18. Chen, J. *et al.*, Does renewable energy matter to achieve sustainable development goals? The impact of renewable energy strategies on

sustainable economic growth. *Front. Energy Res.*, 10, 829252, 2022, doi: 10.3389/fenrg.2022.829252.

19. Union, E., European green deal, [cited 2023 0.03.2023], Available from: https://commission.europa.eu/strategy-and-policy/priorities-2019-2024/european-green-deal_en.
20. Rybski, R., Energy in the european green deal: Impacts and recommendations for MENA countries. *J. World Energy Law Bus.*, 16, 2, jwac033, 2022.
21. Ringel, M., Bruch, N., Knodt, M., Is clean energy contested? Exploring which issues matter to stakeholders in the european green deal. *Energy Res. Soc. Sci.*, 77, 102083, 2021.
22. Tryndina, N. *et al.*, Renewable energy incentives on the road to sustainable development during climate change: A review. *Front. Environ. Sci.*, 10, 1016803, 2022, doi: 10.3389/fenvs.2022.1016803.
23. Choukri, K., Naddami, A., Hayani, S., Renewable energy in emergent countries: Lessons from energy transition in Morocco. *Energy Sustainability Soc.*, 7, 1, 25, 2017.
24. Grafström, J. *et al.*, Government support to renewable energy R&D: Drivers and strategic interactions among EU member states. *Econ. Innov. New Technol.*, 32, 1, 1–24, 2023.
25. Ntanos, S. *et al.*, Renewable energy and economic growth: Evidence from european countries. *Sustainability,* 10, 8, 2626, 2018, https://doi.org/10.3390/su10082626.
26. Goggins, G. *et al.*, The role of culture in advancing sustainable energy policy and practice. *Energy Policy*, 167, 113055, 2022.
27. Lu, Y. *et al.*, A critical review of sustainable energy policies for the promotion of renewable energy sources. *Sustainability,* 12, 12, 5078, 2020, https://doi.org/10.3390/su12125078.
28. Vazhayil, A. *et al.*, A comprehensive review on the recent developments in transition metal-based electrocatalysts for oxygen evolution reaction. *Appl. Surf. Sci. Adv.*, 6, 100184, 2021.
29. Mistry, H. *et al.*, Nanostructured electrocatalysts with tunable activity and selectivity. *Nat. Rev. Mater.*, 1, 4, 16009, 2016.
30. Jung, H. *et al.*, Graphene-based materials as electrocatalysts for the oxygen evolution reaction: A review. *Sustain. Energy Fuels*, 6, 640–663, 2022.
31. Debata, S. *et al.*, Controlled hydrothermal synthesis of graphene supported NiCo2O4 coral-like nanostructures: An efficient electrocatalyst for overall water splitting. *Appl. Surf. Sci.*, 449, 203–212, 2018.
32. Ma, R. *et al.*, A review of oxygen reduction mechanisms for metal-free carbon-based electrocatalysts. *NPJ Comput. Mater.*, 5, 1, 78, 2019.
33. Yang, M. *et al.*, Nanostructured carbon electrocatalysts for clean energy conversion and storage: A mini review on the structural impact. *Front. Mater.*, 9, 1090412, 2022, doi: 10.3389/fmats.2022.1090412.

34. Patra, S. *et al.*, Heteroatom-doped graphene 'Idli': A green and foody approach towards development of metal free bifunctional catalyst for rechargeable zinc-air battery. *Nano Energy*, 30, 118–129, 2016.
35. Roy, E. *et al.*, Shape effect on the fabrication of imprinted nanoparticles: Comparison between spherical-, rod-, hexagonal-, and flower-shaped nanoparticles. *Chem. Eng. J.*, 321, 195–206, 2017.
36. Patra, S. *et al.*, 2-dimensional graphene as a route for emergence of additional dimension nanomaterials. *Biosens. Bioelectron.*, 89, 8–27, 2017.
37. Roy, E. *et al.*, Europium doped magnetic graphene oxide-MWCNT nanohybrid for estimation and removal of arsenate and arsenite from real water samples. *Chem. Eng. J.*, 299, 244–254, 2016.
38. Gonçalves, J.M. *et al.*, Sensing performances of spinel ferrites MFe2O4 (M = Mg, Ni, Co, Mn, Cu and Zn) based electrochemical sensors: A review. *Anal. Chim. Acta*, 1233, 340362, 2022.
39. Radwan, A. *et al.*, Design engineering, synthesis protocols, and energy applications of MOF-derived electrocatalysts. *Nano Micro Lett.*, 13, 1, 132, 2021.
40. Gao, X. *et al.*, Smart designs of Mo based electrocatalysts for hydrogen evolution reaction. *Catalysts*, 12, 1, 2, 2022, https://doi.org/10.3390/catal12010002.
41. Zhong, H. *et al.*, Recent advances of cobalt-based electrocatalysts for oxygen electrode reactions and hydrogen evolution reaction. *Catalysts*, 8, 11, 559, 2018, https://doi.org/10.3390/catal8110559.
42. Kwon, W., Kim, J.-M., Rhee, S.-W., Electrocatalytic carbonaceous materials for counter electrodes in dye-sensitized solar cells. *J. Mater. Chem. A*, 1, 10, 3202–3215, 2013.
43. Wang, L. *et al.*, The search for efficient electrocatalysts as counter electrode materials for dye-sensitized solar cells: mechanistic study, material screening and experimental validation. *NPG Asia Mater.*, 7, 11, e226–e226, 2015.
44. Acres, G.J.K. *et al.*, Electrocatalysts for fuel cells. *Catal. Today*, 38, 4, 393–400, 1997.
45. Kong, Z. *et al.*, Advanced cathode electrocatalysts for fuel cells: Understanding, construction, and application of carbon-based and platinum-based nanomaterials. *ACS Mater. Lett.*, 3, 12, 1610–1634, 2021.
46. Wang, P. *et al.*, Recent progress of carbon-based electrocatalytic materials in lithium-based batteries. *Sustain. Mater.Technol.*, 32, e00384, 2022.
47. Zahoor, A. *et al.*, Electrocatalysts for lithium–Air batteries: Current status and challenges. *ACS Sustain. Chem. Eng.*, 7, 17, 14288–14320, 2019.
48. Yang, H., Driess, M., Menezes, P.W., Self-supported electrocatalysts for practical water electrolysis. *Adv. Energy Mater.*, 11, 39, 2102074, 2021.
49. Sapountzi, F.M. *et al.*, Electrocatalysts for the generation of hydrogen, oxygen and synthesis gas. *Prog. Energy Combust. Sci.*, 58, 1–35, 2017.
50. Wang, J. *et al.*, Electrocatalytic reduction of carbon dioxide to high-value multicarbon products with metal–Organic frameworks and their derived materials. *ACS Mater. Lett.*, 4, 11, 2058–2079, 2022.

51. Liu, A. *et al.*, Current progress in electrocatalytic carbon dioxide reduction to fuels on heterogeneous catalysts. *J. Mater. Chem. A*, 8, 7, 3541–3562, 2020.
52. da Silva Freitas, W., D'Epifanio, A., Mecheri, B., Electrocatalytic CO2 reduction on nanostructured metal-based materials: Challenges and constraints for a sustainable pathway to decarbonization. *J. CO2 Util.*, 50, 101579, 2021.
53. Electrocatalysis for the generation and consumption of fuels. *Nat. Rev. Chem.*, 2, 4, 0125, 2018.
54. Jiang, Q. *et al.*, Editorial: Electrocatalysis towards carbon-neutral future. *Front. Chem.*, 11, 1159716, 2023, doi: 10.3389/fchem.2023.1159716.
55. Masa, J., Andronescu, C., Schuhmann, W., Electrocatalysis as the nexus for sustainable renewable energy: The gordian knot of activity, stability, and selectivity. *Angew. Chem. Int. Ed.*, 59, 36, 15298–15312, 2020.
56. Yu, F. *et al.*, A review on the promising plasma-assisted preparation of electrocatalysts. *Nanomaterials*, 9, 10, 1436, 2019, https://doi.org/10.3390/nano9101436.
57. O'Mullane, A.P. *et al.*, The role of electrocatalysis in a sustainable future: From renewable energy conversion and storage to emerging reactions. *ChemPhysChem*, 20, 22, 2900–2903, 2019.
58. Kaartemo, V. and Gonzalez-Perez, M.A., Renewable energy in international business. *Crit. Perspect. Int. Bus.*, 16, 4, 325–336, 2020.

Index

Printed and bound by CPI Group (UK) Ltd, Croydon, CR0 4YY
17/09/2026
14964275-0003